AF540285

BIOCHEMISTRY OF NUTRITION

BIOCHEMISTRY
OF
NUTRITION

By
Dr. Lata Bhattacharya
Professor
School of Studies in Zoology & Biotechnology
Vikram University
Ujjain (M.P.)
(India)

DISCOVERY PUBLISHING HOUSE PVT. LTD.
NEW DELHI-110 002

Published by:
DISCOVERY PUBLISHING HOUSE PVT. LTD.
4383/4B, Ansari Road, Darya Ganj
New Delhi-110 002 (India)
Phone: +91-11-23279245, 43596064-65
Fax: +91-11-23253475
E-mail: discoverypublishinghouse@gmail.com
sales@discoverypublishinggroup.com
web: www.discoverypublishinggroup.com

First Edition: 2010

Repriented: 2014

ISBN: 978-81-8356-616-2

Biochemistry of Nutrition

Printed at:
Infinity Imaging Systems
Delhi

Preface

Biochemistry today has made spectacular progress in unraveling the mysteries of animate nature. This progress has allowed us to gain deeper insight into the principles of vital activity and has to a very significant extent stimulated the development of applied disciplines, especially medicine. The present title *"Biochemistry of Nutrition"* is intended for those who wish to understand living organisms, especially man. Biochemistry is essential for this purpose, but it would be almost impossible for a student to survey on his own the massive body of existing knowledge, constantly augmented by a remarkable torrent of brilliat discoveries. The purpose of the book, then is to organize our knowledge into something that can be comprehended in a relatively short time and still convey a reasonable complete picture of the chemical structure and function of man. Readability without sacrifice of coverage has been a prime goal. An important device in gaining that goal is to keep attention constantly focused on function, with repeated use of rationalization to show that the chemical facts are not isolated, but part of a whole.

Biochemistry has two major goals as a fundamental science, it treats the vital functions from the standpoint of physical chemistry, and as an applied discipline, it points out practical applications for the wealth of scientific knowledge it has acquired. The dual purpose has been, as far as possible, take into account in the writing of this book. We have also tried to summarize our pedagogical experience in teaching biochemistry to students specializing in medicine and pharmacy. The material of this book has been organized according to the principle of functionally in order to trace the close relationship between the functions of the living organism and the structures of its constituents molecules as well as the chemical and physico-chemical process in which they are involved.

The aim of this book is to present a core of biochemical knowledge that is desirable for undergraduate and postgraduate students and also those involved in the field of medical, microbiology, biotechnology and pharmaceutical. Every attempt has been made to keep abrest of the advances in the subject and at the same time to include the fundamentals.

To make the work more comprehensive and informative, the author has consulted many authoritative books, research journals, abstracts, monographs etc. He is grateful to all those great scholars whose work are cited or substantially reproduced.

There can be no claim to originality except in the manner of treatment and much of the information has been obtained from the books and scientific journals available in the different libraries.

The author expresses his thanks to his friends and colleagues whose continue inspirations have initiated him to bring out this book.

The author expresses his gratitude to Mr. Wasan and Staff of M/s Discovery Publishing House Pvt. Ltd., for their whole hearted co-operation in the publication of this book.

In the mean time, the author will remain sincerely responsible for any shortcomings of the book and be grateful to the readers for their suggestions and constructive criticism for the continuous betterment of the book. He takes this opportunity to appeal to the readers to send their suggestions straightway to his publisher.

Author

CONTENTS

CHAPTER 1

Introduction

As nutrition books go, this one is distinctly unusual in that it was written in one millenium and published in another. Perhaps, therefore, this is an excellent juncture to reflect on a century of tremendous advance and peer forward to a century of improved nutrition and health opportunities. Much can be gained by looking back and learning from the past as to what types of nutrition investigations have been most fruitful. In our opinion, three types of study stand out: epidemiology, animal studies, and human intervention trials. Epidemiology includes both population comparisons as well as studies on individuals (prospective, case-control, and cross-sectional).

Animal studies, and here we mean testing whether an aspect of diet or lifestyle directly affects risk of disease, are most often used to test hypothesis generated by epidemiology. Intervention trials typically follow from the two previous types of study and are rightly referred to as the "gold standard." It cannot be stressed too strongly that conclusions, whether tentative or (more or less) definitive, must reflect the totality of the evidence. We can define a triad of evidence where ideally a conclusion is supported by evidence from all three types of study. Table 1 illustrates examples of these three types of study and their relationship to specific diseases. It is important to bear in mind the value and limitations of epidemiology. This is well illustrated by the controversy in recent years concerning the efficacy of β-carotene as a preventive agent against cancer.

The foundation of this hypothesis was epidemiological evidence, namely the negative association between nutrient intake and the risk of various types of cancer. However, intervention trials failed to demonstrate any reduction in cancer risk after supplementation with β-carotene. In actuality, what the epidemiological evidence showed was association, not causation. A better interpretation of the evidence is that fruit and vegetables prevent cancer and that any substance commonly present in these foods will also manifest such an association. The apparently spurious association between β-carotene and cancer is a common problem in epidemiology.

Table 1.1. Triad of Evidence for Diet-Disease Relationships

Relationship	*Epidemiological*	*Animal studies*	*Intervention studies*	*Overall evidence*
Sodium-blood pressure	XX	XX	XX	XX
Saturated fat-hypercholesterolemia	XX	XX	XXX	XXX
β-carotene-cancer	XX	X	O	X
Fat-breast cancer	X	XX	—	X
Selenium-cancer	XX	XXX	XX	XX
Fat-obesity	XX	XXX	XX	XX

Single X indicates weak evidence; XX indicates fairly strong evidence; XXX indicates convincing evidence. An O indicates evidence against. A dash indicates absence of evidence (either for or against). Evidence refers to whether a causal association exists, not to the strength of the association.

OBSERVING DISEASE MECHANISMS

Although the foregoing types of study have contributed the bulk of our practical knowledge, most resources, in fact, do not go to that type of research. The lions' share goes instead to studies of disease mechanisms. This unbalanced distribution of resources has delayed vital discoveries in the area of how nutrition can prevent disease. Different types of research can be divided into two groups: "simple research," which includes the three types outlined previously; and "complex research," which includes most studies on disease mechanisms.

Complex research most often leads not to light at the end of the tunnel but to more tunnel where the light should be. The words "simple" and "complex" refer to the degree of complexity in translating observations into practical knowledge that can be applied to problems of human health. We can illustrate this principle with the following examples. There has been much debate in the last few years regarding the role of salt in hypertension and the efficacy of a low-salt diet as a treatment for the condition. The critical evidence that has helped resolve this question is direct studies of the relationship between salt intake and blood pressure, both those done epidemiologically and those done by intervention studies.

By contrast, the great number of studies of disease mechanisms, such as those attempting to comprehend the interaction between salt, hormone levels, and kidney function, have contributed little to the debate. Returning to the question of β-carotene and the prevention of cancer, it is difficult to find any evidence of where research into disease mechanisms, such as studies of carcinogen metabolism and of oncogenes, have illuminated the problem and thereby paved the way for effective changes in nutrition habits. The same may be said of the relationship between selenium and cancer risk. Although complex research has told us little of value, international correlation studies, animal experiments, and an intervention study have indicated that the nutrient is a potent anticarcinogen. Another example that illustrates how simple research has given information of superior value is provided by studies of n-3 fats.

Evidence from epidemiological and intervention studies indicate that these lipids may protect against common heart disease (CHD). By contrast, the extensive body of evidence from complex research that attempts to elucidate how these lipids affect metabolism, although expensive, has done little to answer the question as to whether an increased intake of n-3 fats prevents CHD. The central aim of research into disease mechanisms is to understand the detailed functioning of the body.

This is sometimes disparagingly referred to as "reductionism" as it assumes that the whole is no more than the sum of the parts. But it may well be that nutrients and phytochemicals often induce their health-protecting effects by extremely complex interactions. For this reason, it is likely that studying substance-disease interactions one at the time may fail to show the whole picture. Perhaps instead of asking: "Do supplements of β-carotene prevent cancer and how do they work?" we should be asking: "Do vegetables prevent cancer?"

THE GREAT LEAP FORWARD

In nutrition and the medical sciences, as much as anywhere else in science, we occasionally see a great leap forward when an investigator with especially well developed powers of insight grapples with a problem and sees things that were under everyone's nose but were otherwise out of sight. An excellent example of such a person is the late Denis Burkitt, who one of us (NT) had the great fortune to know and collaborate with. Burkitt carried out an epidemiological study in Africa into the geographical distribution of a type of cancer common in children. This cancer is now known as Burkitt's lymphoma. Although the cancer was there for anyone to see, it was Burkitt who put one and one together.

Quite remarkably, he did this using the resources available in Africa in the 1950s on a grant worth under $ 1000 in today's money. This work led to the identification of the Epstein-Barr virus. In collaboration with Hugh Trowell and others, Burkitt later went on to establish the vital importance to health of dietary fiber and, hand in hand with this, developed the concept of Western disease. Again, this was done using little more than great perceptive powers and a shoestring budget. Is the age of the great leap forward behind us? We doubt that very much. Buried in the next section, perhaps, are the seeds of one or two great leaps forward.

CHALLENGES AHEAD

Rather than making any bold predictions it may be useful to give the following quote:

So much precise research has been done in the laboratory and so many precise surveys have been made that we know all we need to know about the food requirements of the people...The position is perfectly clear-cut [with respect to Britain].

These prophetic words were penned by Drummond and Wilbraham in *The Englishman's Food,* which was published in 1939. Jack Drummond was a major nutrition authority in the 1920s and 1930s and coined the term "vitamin," It would be foolhardy to believe that we can be any more accurate today in our predictions.

Nutrients and Disease Prevention

As discussed earlier, consumption of fruits and vegetables has a strong protective association with various types of cancer. We still do not know which substances in these foods are responsible for the anticarcinogenic action. Although most attention has hitherto been directed to β-carotene, other substances likely play more important roles, especially phytochemicals. One class of possibly anticarcinogenic phytochemicals is soy isoflavones. The phytoestrogenic action of these substances and how they may be protective against breast cancer (and CHD) is explored by Wilson and Murphy. Similarities between these phytoestrogens and pharmacological estrogen agonists continue to blur the differences between foods and drugs.

With a better understanding of the anticarcinogenic potential of such substances as phytochemicals, folate, and selenium, it should be possible to produce a supplement that provides an effective, safe, and cheap way to prevent a great many cases of cancer. Vitamin C has a negative association with various conditions, including cataracts, asthma, and loss of pulmonary function. As in the case of β-carotene and cancer, it is critically important to bear in mind that what these data really reveal is not that vitamin C protects against these conditions, but rather that ascorbater-ich fruit and vegetables do.

As with cancer, therefore, once the responsible substances have been identified, we can better understand how to enrich our dietary intake, perhaps using supplements, so as to prevent these conditions. A nutrient that merits further investigation is vitamin E. There is evidence that it may have some prophylactic effectiveness against cancer and CHD. These benefits are generally seen only at intakes several times higher than the RDA. These levels cannot be achieved through food consumption alone. This suggests that vitamin E is acting as a nutraceutical rather than as a vitamin.

We have growing evidence that optimal dietary intake (especially vitamin E) may enhance immune function and thereby reduce the burden of infectious disease. Clarification of these relationships may be of enormous value, especially to the elderly and malnourished and to populations at high risk of infectious disease.

Optimal Intake of Nutrients

Information that has emerged in recent years challenges our concepts of nutrient requirements. Although the details are still hazy, it is becoming increasingly clear that many substances have disease-preventing actions in ways that do not fit into the classic model of nutrient/deficiency disease. In some cases, these substances are nutrients, notably vitamin E and selenium, but are apparently most effective at intakes far above what is required for preventing deficiency diseases. In other cases, the phytochemicals, there is no requirement, as the substances are not nutrients in the classic sense; possible examples of this are particular carotenoids and flavonoids by Weinberges and Heaney. Indicate that this concept applies to potassium and calcium by virtue of their hypotensive action. In a previous work, one of us (NT) argued that it is short sighted to define recommended nutrient intakes narrowly in terms of preventing deficiency diseases. For various nutrients, we can more accurately speak of three levels of intake.

A low intake is clearly deficient by any definition and will produce clinical symptoms or, at least, subclinical deficiency. A somewhat higher intake will prevent deficiency symptoms but will not fully protect health; it is best described as a suboptimal intake. A yet greater intake is required for maximal health preservation: an optimal intake. In the case of the phytochemicals, of course, we cannot talk of deficiency, although there does appear to be an optimal intake. Taking vitamin E as an example, a daily intake of under 5 mg is deficient, 10 mg (the US RDA) is suboptimal, whereas an optimal intake (assuming the previously cited evidence is correct) is between 50 and 500 mg. The reason for these different levels of requirement stems from the relationship between the nutrient or phytochemical and the cause of the disease. With classic deficiency diseases, such as scurvy, the disease is solely due to lack of the nutrient.

As a result, one or more specific biochemical pathways fail to operate normally. But with diseases prevented by relatively high doses of nutrients and phytochemicals, the true cause is a factor, such as tobacco, which the protective substances can counter to some extent. However, they typically do this in ways where the required dose is much higher than that pertaining to the prevention of a deficiency disease. Just as the discovery of the relationship between nutrients and deficiency diseases necessitated the establishment of the concept of recommended nutrient requirement, so the foregoing arguments suggest the necessity of a new concept based on optimal intake. In this regard, it matters little whether the substance in question is a nutrient or not. Gey's proposed term, recommended optimum intake, closely corresponds to the aforementioned concept (16).

Nutrition and Aging

Many features associated with aging—for instance, atherosclerosis, impaired immune function, declining respiratory function, decreasing lens opacity (leading to cataract), and increasing blood pressure—are all associated with diet, and therefore preventable to a greater or lesser extent. Recent evidence suggests that age-related cognitive decline and memory loss may also come into this category. Indeed, optimal nutrition may be acting not so much by a different mechanism in each of these conditions but by a common pathway. Oxidative stress is the mechanism that first springs to mind, but the supporting evidence is far from overwhelming.

Diet and Fetal Programming

This has been a contentious issue for decades. Determining its importance with respect to the development of obesity, CHD, diabetes, cancer, and other diseases in adults remains a challenge. More work is clearly needed to clarify the ideal fetal and infant diet for the prevention of these diseases in adulthood.

Diet and CHD

The relationship between diet and CHD cannot help but appear as a confusing picture. There are several areas of potential importance which require further investigation before we can formulate an overall strategy for the prevention of this ongoing pandemic. These areas include homocysteine, high doses of vitamin E (discussed previously), n-3 fatty acids, and determining the maximum safe intake of saturated fat and *trans* fatty acids.

Diet and Obesity

The chapter by Richards and colleagues authoritatively reviews our present position with regard to obesity. Four centuries ago, William Shakespeare wrote: "They are as sick that surfeit with too much as they that starve with nothing" *(Merchant of Venice).* Despite decades of intensive research, effective dietary therapy for the condition remains as out of reach as it was two-thirds of a century ago when Somerset Maugham wrote what is arguably the finest picture painted of the pain and frustration of weight loss in his story *The Three Fat Women of Antibes.* Here, in more ways than one, "An ounce of prevention is worth a pound of cure." The area of prevention and treatment of obesity will hopefully see ongoing progress.

The Ideal Diet

Central to any discussion on the subjects of CHD is the debate regarding the ideal level of fat, saturated fat, and dietary fiber. This also has major implications in regard to obesity, diabetes, and colon cancer. It has often been suggested that the ideal diet is one based largely on low-fat plant foods with little or no meat. Research may well prove that this is the best means to incorporate into the diet the optimal quantities of fruit, vegetables, and perhaps soy foods. We should also keep our minds open to the possibility that there is no ideal diet for a population, but rather unique nutritional needs of the individual.

Herbal Medicine and Bioengineering

As discussed by Craig, traditional medicine continues to predominate in less developed regions of the world. In the next millenium, we can expect to see continued developments in these low-technology supplements and this will lead to promotion of health in the developed regions of the world. Our scientific understanding of herbal medicine and nutraceuticals remains in its infancy; clearly some provide reproducible health benefits and others are useless. Their nutritional significance remains hazy at present. Herbals that can be incorporated into our diet and do prevent diseases will become important given their generally low cost. Nutritional impacts of high-technology changes have been taking place since the days of Norman Borlaug and his "Green Revolution." The impact of biotechnology was discussed by Beitz. Genetically modified crops and livestock have become fairly common, although they are as yet poorly accepted elsewhere. The risks and benefits of the use of these technologies to increase yields comes with risks that are still poorly understood and risks that may yet to be appreciated. Until there is a much better understanding of the nutritional benefits/risks, the use of this technology is likely to remain controversial.

Dietary Assessment

An ongoing problem in nutritional epidemiology is the quality of dietary assessment. So many of the controversies in the area of nutrition and disease emanate from inaccuracies in measuring people's dietary intake. Well-known examples include sodium/blood pressure, energy intake/obesity, and fat intake/breast cancer.

Health Promotion and Government Policy

All the foregoing research is worth little if people cannot be persuaded to actually follow dietary advice. One approach is that discussed by Jacobson, *i.e.,* how a private organization can be highly effective at promoting healthier diets. Temple and Nestle discusses a crucial area for investigation, *i.e.,* the limitations of health promotion interventions and why government policy may prove to be a more effective vehicle of change. Clearly, this whole area offers great challenges and cries out for radical new ideas.

The Internet and Nutrition Information

We are now entering the information age. Given the widespread access to Internet web sites, as discussed by Halman, our access to nutrition information will undoubtedly grow rapidly. In the 21st century, the long time gaps between data collection, writing, publishing, and the reading of articles will undoubtedly shrink. Answers in an instant will be the norm in the future.

CHAPTER

2 Nutrition on the Internet

The Internet is the fastest growing communication medium in history. It has the potential to profoundly impact our commercial, personal, and professional lives. The health arena is no exception, and in this chapter, we examine some of the current and potential applications of the Internet in the world of nutrition.

The Internet is a network that connects computers. There is nothing remarkable in this—most organizations with more than a handful of computers have a network. What makes the Internet unique are two things:

1. Size: Nobody knows for sure exactly how many computers are permanently connected to the Internet nor how many people have access to the Internet through those computers. Current industry estimates are that the total number of Internet users as of the end of 1999 is around 260 million worldwide, and that this number is doubling every 24 mo *(la)*. Although the Internet has spread to cover many countries, access is still very much lopsided toward the developed world, with 80% of users being based in just 15 countries. In the United States, which is one of the world's most Internet-connected countries, it is estimated that over 40% of the population now have Internet access.
2. Decentralized organization: Because of its origins as a communication medium for institutions engaged in military research, the Internet was deliberately organized to be free of central administration as much as possible (by diffusing control, it is more protected from deliberate attack or malfunction). Thus, there is no central controlling body to set standards of content or acceptable use.

IMPLICATIONS OF THE INTERNET

Stemming directly from these two characteristics, the Internet provides in the broadest sense a brand new medium with three striking features: (1) distance-independent communication, (2) universal availability of information, and (3) immediate interactivity.

Distance-Independent Communication

The Internet is the first communications medium in history where the cost and ease of use is essentially independent of distance. It is now possible for a person to send an e-mail (electronic mail) to another person almost instantaneously without any regard for (or indeed knowledge of) where or at what distance the person lives. Thus, for example, if I have read an interesting research paper in which the author's e-mail address is given (increasingly the case), I can contact the author with a question or comment and not unreasonably expect to hear back from him/her the next day.

E-mail also differs from the telephone/fax in two other ways. First, it does not matter whether the other person is at home or not (one can pick up one's e-mail from any Internet-connected computer anywhere in the world). This has helped to considerably speed up communication exchanges between itinerant academics! E-mail technology is such that it is a simple matter to send the same message to large numbers of people at once. This has engendered a new form of communication called a *mailing list* (sometimes known as a discussion list) in which large numbers of people can discuss topics of mutual interest by sharing e-mails amongst the whole group.

A closely related resource is the *newsgroup* in which messages are posted on a common "bulletin board," which must be specifically sought out when one wishes to read the messages.

"Universal" Availability of Information

Although "universal" in this context means "universal amongst those with Internet access," the ability of the Internet to make information available instantaneously across the globe is quite unparalleled in history. The Internet has always allowed documents to be shared between computers on its network, and if I chose to update the content of my document, the new content will be instantly available to any reader who chooses to download it, wherever they may live. But it is the development of *web sites* with their multimedia and interactive capacity which has moved the Internet to becoming a part of everyday life in countries with high levels of Internet use. A web site is a collection of web pages [illegible] documents, each of which is written in a standard language (*html* and its more recent developments such as *vrml*), which allow rich multimedia and interactive capabilities, such as animated graphics, sound, and interactive input and responses, etc.

Hyperlinking is a feature by which clicking on a highlighted word or image automatically takes the viewer to another section of the text, or another resource altogether. In this way, it is possible to weave an almost endless web of interconnected material from many different sites. Furthermore, the presentation of this rich environment takes place on the user's screen (and computer speakers in the case of audio files!) largely independent of what type of computer they have (*e.g.*, PC, Macintosh, Unix). A web site may contain anything from transcripts of the latest decisions of the US Supreme Court, to downloadable tracks from a hot band's new rock video and album, to an individual's philosophical musings. Although nobody knows how many web sites and pages are on the Internet, it is believed to be at least several hundred million and growing at a rapid rate. Writing a web site is now a relatively trivial technical exercise. As a result, the web has effectively democratized the flow of information.

Anyone can put whatever they wish on the web. In essence, it is a gigantic library, with little or any organization. Even so, by using an appropriate *search engine* it is possible to read about almost anything. The only problem being, as mentioned before, is that there is nobody to vouch for the accuracy of the material. Just three relatively trivial uses for such an instantly editable, universally viewable library will suffice to illustrate the potential of the web. The author is a fan of the great English game of cricket. No matter where he travels, he is now able to tune into instant score updates complete with live commentary and video highlights of an important game as it takes place. Before setting out on his travels, he can obtain a complete weather report for any possible location of interest, as well as a check on that day's exchange rate between any two currencies.

Immediate Interactivity

The Internet's true potential for interactivity is really only beginning to unfold. Two obvious examples: online share trading and interactive auction sites where a bidder and a seller may interact in real time from anywhere in the world.

NUTRITION APPLICATIONS

There are nutrition applications paralleling all of the foregoing features of the Internet. Broadly these can be divided into resources for health professionals and resources for the lay public. This chapter focuses mainly on the former. At the same time, it is important to note that one of the more interesting aspects of the Internet is its democratization of information, which allows lay users to have much the same access to health professional resources as do health professionals.

E-Mail-Based Nutrition Applications

Private Communication

Many biomedical journals now publish the e-mail address of the lead or contact author of their articles, thus facilitating private communication between readers and authors. This includes such prominent journals as the *British Medical Journal*. Unfortunately, this practice is not particularly common among nutrition journals as yet. For example, the *American Journal of Clinical Nutrition* used to publish author e-mail addresses, but currently does not do so.

Mailing Lists and Newsgroups

There are a considerable number of nutrition-related mailing lists and newsgroups, allowing discussion among small to large numbers of like-minded individuals. These include lay or patient mailing lists for people suffering from or interested in such nutrition issues as weight problems, eating disorders, diabetes, vegetarianism, and so forth. Topics of discussion can vary from patient support to the latest news on treatment, to alternative therapies, to philosophical debates. The easiest way to find such resources is through a web site dedicated to cataloging mailing lists.

Mailing lists for nutrition professionals are also in plentiful supply. There is a certain level of turnover with new lists starting up, whereas other lists die a natural death through lack of interest; Table 1 highlights some of the most important and enduring of these.

Newsletters

There are some useful, free, one-way nutrition e-mail communications that mirror the traditional function of the printed newsletter. For the nonprofessional reader, *Nutrition News Focus* is a daily newsletter providing lay interpretation on current nutrition news as provided by a well-known US-based professor of nutrition. For the nutrition or other health professional, the *Arbor Clinical Nutrition Update* is the Internet's most widely read nutrition communication. It contains weekly summaries of the latest clinical nutrition research, together with information on the best nutrition resources available on the Internet. Readership at the end of 1999 totaled more than 15,000 health professionals based in 150 countries.

Web-Based Nutrition Applications

There are an enormous number of web sites concerned with nutrition. Entering "nutrition" as the search term into the general search engine Alta Vista returns over 1.3 million listings (using the term "food" adds another 8.5 million!). It is impossible to say how many of these web sites are commercial (although it is clear from casual inspection that many are), how many represent the private musings of nonprofessional individuals, how many are targeted at a noncommercial lay audience, and how many are intended for nutrition professionals. In the following sections, we examine some of the more useful categories of nutrition web sites from the health professional perspective. They can conveniently be broken down into portal sites, guides and search engines, institutional home pages, nutrition science (including research), dietetics, healthy diet, clinical nutrition, and food science.

Portal Site, Guides, and Search Engines

One of the greatest problems in approaching the Internet is knowing where to find things. For this reason, some of the most popular web sites are those that tell the user what is available and present in such an organized manner that the desired information source can be found quickly. Such sites may operate as directories, search engines, or both. They are often the first port of call en route to finding the resource one actually wants, and are therefore often referred to as *portal sites*. Provided they are updated regularly, portal sites can be an invaluable resource for the novice and experienced user alike. Because of the rapidly changing nature of the subject, no printed publication will ever replace portal sites in providing current information on what is available on the Internet.

The reader is advised to consult such portal sites for current and more detailed information to accompany the content of this chapter. The *Arbor Nutrition Guide* is the largest portal site specifically focused on resources for health professionals. It has over 3000 listings and covers the broad range of categories referred to before. It includes descriptions of the web sites listed and a search engine, and is updated regularly. Perhaps the best of all the portal sites aimed at the lay public is the *Nutrition Navigator* site. Produced by staff from the nutrition department at Tufts University, it has fewer listings than the Arbor site but provides an indopendent rating score for each of them, based on measures including how often the listings are updated.

It also has a search engine, and has found widespread acceptance as a portal site. Other portal sites have a more food and food science approach. For the lay audience, this includes *the Martindale "Virtual" Nutrition Center,* whereas for nutrition professionals there are two useful web sites from Australia and the US National Agricultural Library.

Institutional Home Pages

A *home page* refers to a web site focused on providing information about its owner. A large number of institutions relevant to nutrition professionals now have home pages on the Internet. This includes government organizations, universities, NGOs, book publishers, and many commercial bodies. The usefulness of these institutional home pages varies enormously. In some cases, they have little more than descriptions of what the organization does and the names of personnel.

In other cases, they carry information of more general nutrition interest, which can range from Extension (educational) newsletters to the actual content of courses as well as reports and lay education resources. A few of the more interesting such institutional web sites with rich nutrition content include US Department of Agriculture (USDA), Health Canada, United Nations Committee on Nutrition, International Life Sciences Institute (ILSI), and US National Academy of Science. There are several web sites that provide reasonably comprehensive lists of the home pages of university nutrition and food science departments, both internationally and within North America. These inform the user of those universities that have accredited nutrition science graduate or dietetic training programs.

Nutrition Science (Including Research)

The Internet has significantly enhanced access for those who wish to keep up with the latest in nutrition knowledge, including published research. Improvements have come about in both the ability to access this kind of information much more quickly after it is published, and the ability to filter and sort the vast amount of such information for what is relevant to the individual. In this respect, nutrition science has benefited from developments in biomedical science generally. Various journal abstract services are now available on-line, many without charge. These include *Cancerlit,* a collection from the National Cancer Institute, *andMedline,* the journal abstract service of the National Library of Medicine, which provides references and abstracts from 4300 biomedical journals.

A wide variety of third-party providers has made Medline access freely available to Internet users in many guises, and it is possible to tailor the service to specific needs. For example, *Infotrieve* allows the user to register a specific Medline search and receive weekly e-mail updates of all new Medline entries matching that search. *Pubmed,* which is the National Library of Medicine's own Internet Medline service, allows full Boolean queries of the complete range of Medline fields and can produce very large results going back many years.

A simplified version from the same source is *Grateful Med,* which also incorporates pre-Medline for rapid access to literature not yet incorporated into the main Medline index. *Biomednet* has a literature search service that provides "intelligent Medline," combining the results of a normal Medline search with other resources some of which have been evaluated by independent assessors. "On-searching" is provided, which is the ability to search for articles similar to a selected article (based on MESH headings) or for articles by the same author. In some cases, these sites allow the user to order full-text articles, as well as viewing the abstracts, but this is normally a commercially charged add-on feature.

At the same time, individual journals have been making use of the Internet to expand the reach and scope of the printed journals. One of the best examples of this is the *British Medical Journal* with its *eBMJ* web site. As each new issue of the journal is posted to eBMJ,

Table 2.1. Some Mailing Lists for Nutrition Professionals[a]

Mailing list	*Contact address*	*Purpose / subscriber base*	*Started*	*Web page*	*Subscribers (@ 12/99)*	*Digest available*	*Level of[b] activity*
NGONUT	Owner-ngonut	Those who work in @abdn.ac.uk in developing countries	1997 nutrition programs	www.univ-lille 1.fr/pfeda	525	No	Moderate
FNSPEC	Eversb@purdue.edu	Extension, nutrition professional providing information of the public	1993	hermes.ecn.purdue.edu/ Links/fnspec_mg	800	No	Heavy
NUTNET	Majordomo @dietitians.net	Australian nutritionists	1995	www.daa.net/cgi-bin/lwgate/NUTNET-LIST%40dietitians.NET/ archives/	460	No	Heavy
PHNUTR-L	Larsson@u. washington.edu	Public health nutrition edu: 70/11public/phnutr-1	1996	gopher://lists.u. washington.	600	Yes	Heavy
NUTR-MED	Nutrmed@ ozemail.com.au	Medical and nutrition professionals interested in primary care nutrition	1996	—	250	Yes	Moderate
NUTSCI	Owner-nutsic@hc-sc.gc.ca	Candain food/ nutrition regulatory issues	1997	www.hc-sc.gc.ca/food -aliment/	425	No	Light
MEALTALK	Owner-mealtalk @nal.usda.gov	USDA School Meals Initiative for Healthy Children		www.nal.usda.gov:8001/ Discussion/index.html	900+	Yes	Heavy
FOODSAFE	Croberts	Professionals interested		www.nal.usda.gov/fnic/	1200+	Yes	—

Mailing list	Contact address	Purpose / subscriber base	Started	Web page	Subscribers (@ 12/99)	Digest available	Level of activity[b]
	@nal.usda.gov	in food safety issues		cgi-bin/mealtalk.pl			
PEDI-RD	Susan-carlson @uiowa.edu	Neonatal and pediatric nutrition		www.uihc.uiowa.edu/ pubinfo/pedi-rd.htm		—	—
FEEDING-CHILDREN	Eanderson@ uidaho.edu	Information related to FeedingChildren	1996	—	80	—	—
DIET-EPI	Gblock@uclink2. berkeley.edu	Nutritional epidemiology research	1996	—	220	—	—
ASPENE	www.clinnutr.org/list serv/aspenet.html	Nutrition and metabolic support	1999	www.clinnutr.org/ listserv/aspenet.html	Yes	Heavy	—

[a]Dash means information or feature not available.

[b]Light = < 1 messages/d; Moderate = 1-4 messages/d; Heavy = >4 messages/d.

Table 2.2. Some Starting Points for Web Browsing

Category	*Site*	*Web address*	*Comments*
Portals, guides	Arbor Nutrition Guide	arborcom.com	For health professionals
	Tufts Nutrition Navigator	navigator.tufts.edu	For lay public
Home pages	USDA	www.usda.gov	Professional and lay resources
	Food and Nutrition Information Center	www.nalusda.gov:80/fnic	Professional and lay resources
	Health Canada	www.hc-sc.gc.ca/ main/hppb/nutrition/	Especially food for lay resources
	ILSIro fessional and lay resources	www.ilsi.org	Food safety and general nutrition
Nutrition science	Am J Clin Nutr	www.ajcn.org	Nutrition specific
	British Medical Journal	www.bmj.com	General but with nutrition "collection"
Dietetics	American Dietetics Association	www.eatright.org	Professional and lay resources
	Dietitians of Canada	www.dietitians.ca	Professional and lay resources
	Dietetics Online	www.dietetics.com/	Dietetic professional issues
Clinical nutrition	ASPEN	www.clinnutr.org	Nutrition support
	Family physician nutrition articles	arborcom.com/ frame/clinical.htm	Various topics
	HINS	www.hins.org	Pediatric
Food science	USDA food composition tables	www.nal.usda.gov /fnic/cgi-bin/nutsearch.pl	Individual food values
	Drive Thru Diet	www.bgsm.edu/ nutrition/FFMainF.htm	Fast food chain foods
	You Are What You Eat	library.advanced.org/11163 /gather/cgi-bing/ wookie.cgi/?id=5CMO	Dietary intake analysis

at the same time if not a little before the printed version is available, on-line visitors have immediate access to the latest content. A number of articles are available to all visitors in full text, whereas most of the rest appears in abstract form with full text for subscribers. References cited within an eBMJ article are linked to the relevant original abstract within Medline. Many biomedical journals now allow e-mailed letters to the editor, but in eBMJ you can read all the letters pertaining to a particular topic and participate in a forum on the topic. The site also has a *Customised @ lerts* feature, which sends e-mailed tables of contents on request for each new issue. This feature can be further customized by selecting from one or more of the 100 plus clinical categories into which journal content has been arranged.

A search engine allows ready access to archived material, which is also grouped within "clinical collections" corresponding to the same clinical categories as the e-mail alerts. As *Nutrition* is one of those collections, it is a simple matter to track nutrition-related content on what is one of the world's leading medical journals. Although the *British Medical Journal* is perhaps the best implementation of an electronic biomedical journal site, a substantial number of nutrition journals have provided on-line sites that provide some combination of these features. More than 50 nutrition and 30 food science journals currently offer at least free on-line tables of contents. A comprehensive list of these is maintained on the *Arbor Nutrition Guide.* The *American Journal of Clinical Nutrition* web site is a well-implemented example, providing abstracts of the current issues with archives back to 1996, as well as a search engine and e-mail table of contents. The *Journal of Nutrition* offers an interesting twist to e-mail updates, offering alerts whenever another journal article cites the paper in question. Exciting as these developments are, they are only the beginning. A recent trend is the prepublishing of research articles on the Internet prior to acceptance by the printed journal. This not only cuts down enormously on the time lag between submission of a paper and its availability to the health professional community, but it also allows the democratization of the peer review process. The *Medical Journal of Australia* was the first major medical journal to experiment with this concept, allowing any qualified viewer to submit peer review comments on articles that have been accepted in principle.

The comments and the authors' responses to those comments are all publicly posted. In this way, it is hoped that the peer review process will become a more open and transparent one. Two web sites initiated toward the end of 1999, *Clinical Medicine Net Prints* and *Biomedcentral,* offer anyone who wishes the ability to put up a paper, whether or not it has been accepted by any particular journal. Both have excellent academic credentials, one being sponsored by the British Medical Journal and Stanford University Libraries. Like eBMJ, it allows browsing by medical specialty, of which nutrition is one area. The ultimate conclusion of Internet publishing may be journals that exist only on the Internet.

However, although there are a scattering of such e-journals already, current trends suggest that Internet publishing is going to enhance rather than take over from the traditional hard copy journals. From the lay perspective, there are many ways to keep up with recent developments in nutrition science in a less formal way. Commercial news agencies such as *Reuters* and CNN have health news web sites, both of which allow the user to specify a food and nutrition focus. For health professionals, several of the web sites established specifically for physicians run regular nutrition articles and case reviews, for example *Doctors Guide.*

Dietetics, Healthy Diet

There are a very large number of web sites containing information on healthy diet, dietary guidelines, recommended dietary intakes, and the like. The *Food and Nutrition Information Center* of the US Department of Agriculture has one of the better collections on its web site, including, for example, both the text and graphics of the US dietary guidelines, along with the report of the committee that produced them, and the Recommended Dietary Allowances as produced by the National Academy of Sciences. For lay information sheets, a good starting point is the *American Dietetic Association* web site, which has a large range of "Nutrition Fact Sheets".

The International Food Information Council has a good collection of lay information, with material on food safety and on the general healthy diet. Dietitians have not been slow to take advantage of the Internet, both to promote their services commercially and to communicate among themselves. The *American Dietetic Association* web site has a good deal of material about the dietetic profession, position statements, and a database of dietitians, which can be searched by location and specialty interest. *Dietitians of Canada* also has a well thought-out web site with professional and lay resources. It is host to the home page of the *International Committee of Dietetic Associations,* within which can be found a list of dietitians' associations around the world. Surprisingly few have web sites as of the date of writing. Dietitians have also organized informal groupings to communicate via the Internet. *Dietetics Online* is the web site of "a worldwide networking organization of nutrition and dietetic professionals". First developed to provide information on and archives from a long-standing dietitians' bulletin board on the proprietary America Online service, it has grown to include recipes, marketplace information on dietetic-related products, job information, and a reference point for state dietetic associations and several dietitian special interest groups. There are numerous sites belonging to individual dietitians, which generally feature some lay nutrition information along with advertisements for the owner's dietetic services. *Cyberdiet* is a particularly good example of a web site originally set up by an individual dietitian, which has now become part of a larger Internet health network. *Ask the Dietitian* is another of the best dietitian-authored web sites, with a particularly good Q&A section for the lay visitor.

Clinical Nutrition

There are many Internet resources of relevance to clinical nutritionists, but they are scattered amongst many web sites, including university medical school curricularn and course notes, medical CME sites, and organizational home pages. Once again, a good portal site can help to locate what the user is seeking. *The American Society for Parenteral and Enteral Nutrition* web site is one of the better examples, providing news, clinical updates, and lay information on nutrition support. The Arbor Nutrition Guide has a series of articles tailored for family physicians.

Nutrition elements of specific diseases are often best covered by medical organizations concerned with those diseases. For example, useful nutrition resources are available at the

American Diabetes Association, American Heart Association, National Heart, Lung, and Blood Institute, International Association of Physicians in AIDS Care, and *Canadian Pediatric Society.* The *American Association of Clinical Endocrinologists* web site has quite a bit of nutrition material related to osteoporosis and growth disorders, whereas the *Heinz Institute of Nutritional Sciences (HINS)* web site has health professional material on pediatrics.

Food and Food Science

The interactive nature of the Internet is particularly suited to accessing food tables and other forms of nutrition software. The *USDA Nutrient Database* food composition data can be accessed through a simple search engine, in which users can specify food and food quantity in several ways. *Cyberdiet* offers food label food composition data; it is particularly suited to lay users. This site will also calculate a person's recommended dietary intake based on age, gender, anthropometric measures, and activity level. More specialized food composition sites include: *Drive Thru Diet* from Wake Forest University, which provides nutritional information on foods from the seven largest fast-food chains in the United States, and *You Are What You Eat,* an imaginative site originally designed by nutrition students offering nutrient and food label information, food counter and planner, and the ability to report intake data in relation to an individualized nutrient profile. Other types of interactive nutrition calculators are available on-line, particularly for calculating body mass index, energy requirements, compliance with dietary pyramid recommendations (66), and intake of various specific nutrients such as calcium. The food industry is well covered by web sites: generic, industry-wide, and those dealing with individual companies. The range and number of such sites is, however, larger than can be accommodated in this chapter; this is even more the case in relation to sites devoted to food and cooking. Once again, reference to a good nutrition portal site is recommended.

WHAT DOES IT MEAN?

What does all this mean to the nutrition professional and what of the future? The Internet has already boosted the accessibility of nutrition information and fostered communication between nutrition professionals to a degree that few would have anticipated even four or five years ago. The nutritionist who wants to keep up with the latest developments in their field has unprecedented opportunity to do so. Knowledge has also been democratized—our patients and clients are increasingly likely to come to us with information they looked up on the Internet.

Essentially, anyone with an inquiring mind and the determination to do so can now access the same information base as any health professional. To a large extent this is a good thing, but it is also important to remember that there is no quality control on the Internet. We will have to help our patients to learn to distinguish reliable science based information from the unreliable non-science-based. Quite how the Internet will develop over the next five years is hard to tell. It is likely that there will be a major growth in the degree of multimedia richness

and interactivity in nutrition sites—both are relatively primitive in most current nutrition sites (compared to what is commonly seen in web sites dedicated to youth culture and media, for example). We are likely to see an increasing number of interactive web-based nutrition courses available, both reputable and less reputable.

Major web sites devoted to the ever-popular topic of weight loss are already in development, and we will soon feel their influence. Indeed, the Internet will ensure that the spread of the "latest diet"— whether related to weight loss or some other fad—will be faster than ever before. Those who give professional nutrition advice will therefore need to find ways of keeping up with these fads so as to wisely advise their patients and clients. Whatever the direction will be, one thing is certain. With billions of dollars of investment funds pouring into it, and a potential market of hundreds of millions of people, this medium, which has developed so fast in such a short time, has only just begun to unfold its wings.

CHAPTER 3 The Water-Soluble Vitamins

Included in the group of water-soluble vitamins are the well-established members of the B complex. Many writers include choline, inositol, and p-aminobenzoic acid with the B vitamins. Ascorbic acid (vitamin C) and compounds with related activity, such as citrin ("vitamin P"), are also water-soluble vitamins.

The Vitamin B Complex

It is general practice to include in the so-called B complex any water-soluble vitamin found in yeast, liver, or other good source of the well-established members of this group, such as thiamine, riboflavin, and niacin. Though such a practice has justification, it has made the B complex truly complex. The terminology is at present somewhat improved, though the literature of a few years ago and more is often baffling to the uninitiated. As an example, some 15 different names have been applied to the vitamin now known as thiamine. This factor is also called vitamin B_1 by many, although the former term is preferable since it has been adopted by the Council on Pharmacy and Chemistry of the American.

Medical Association, the American Association of Biological Chemists, and the American Institute of Nutrition. Riboflavin, the presently accepted name for another member of this group, was previously referred to as vitamin B_2, vitamin G, and the P-P factor (pellagra preventive). A third factor, niacin (accepted name) was only a few years ago almost universally referred to as nicotinic acid. The various scientific groups, including those just referred to, have done a great deal to unify the terminology of the vitamins. After these organizations officially adopt a name, other synonyms slowly disappear from the literature, and a certain degree of clarification results.

The present trend is to adopt a name that implies in some degree at least the chemical nature of the compound in question. For example, the name pteroylglutamic acid has replaced the name folic acid (another member of the B complex). As a result of the official adoption of vitamin names and discriminating editorial service on the part of the proper authorities, the situation regarding' vitamin terminology is much improved, and there is a continued gain in unification.

Thiamine

A thiamine deficiency in man leads to the condition known as beriberi; in animals the syndrome is referred to as polyneuritis. Reference has been made to the studies by Takaki, who demonstrated that through the proper dietary reforms the incidence of beriberi in the Japanese Navy could be practically eliminated. Also, the important discovery of Eijkman that chickens develop symptoms resembling beriberi when fed a diet of polished rice and that the symptoms disappear on adding rice polishings to the diet has been discussed in the introductory remarks of the preceding section.

The work of Funk was also mentioned. He attempted to isolate the active principle of rice polishings responsible for preventing or curing the symptoms in fowl which develop when these animals are fed diets of polished rice. It was this work that led him to propose the term "vitamine" (now vitamin) for the basic material he isolated that was effective prophylactically or curatively in very small amounts. At about this time'progress in the field was markedly aided by the studies of McCollum and co-workers. In 1915 McCollum and Davis reported that growth of young rats on purified diets was poor when dextrin constituted the carbohydrate of the ration, but if lactose was used, growth was fairly good. They also showed that heating the lactose destroyed the growth-promoting principle and that the active substance was water and alcohol soluble.

The term "water-soluble B" was employed by McCollum to differentiate the factor from his earlier designated growth factor, "fat-soluble A." Yeast was found to be an excellent source of the water-soluble B, and in 1920 Emmett and Luros reported that autoclaved yeast no longer contained the antiberiberi principle of Funk but did contain a substance that promoted growth of rats on certain synthetic diets. This was the first clear-cut demonstration of the dual nature of water-soluble B in yeast. We now know that these workers were dealing with thiamine (water-soluble B_1 or antiberiberi principle of Funk), which was destroyed in the yeast by autoclaving, and riboflavin, which promoted rat growth with the rations employed and is not destroyed in yeast by the heat treatment used. In 1926 the antiberiberi vitamin (thiamine) was isolated from rice polishings by Jansen and Donath. The synthesis of thiamine was accomplished in 1936 by Williams and co-workers.

Chemistry of Thiamine

Soon after the purification and crystallization of thiamine it was evident that a pyrimidine nucleus was a part of the molecule and that there also was present a substituted thiazole ring. In the synthesis of Williams and co-workers, the substituted pyrimidine (2 methyl-5-bromomethyl-6-aminopy-rimidine hydrobromide) was reacted with 4 methyl-5-β-hydroxyethylthiazolc to yield the bromidehydrobromide of the vitamin. The naturally occurring thiamine is a chloride-hydrochloride. After Williams converted the bromide derivative to the chloride-hydrochloride, it was identical with thiamine isolated from natural sources.

The work required for the synthesis of this molecule is, of course, not depicted by the reaction. Years of investigation by many scientists all over the world culminated in success. One gram of thiamine can be dissolved in 1 ml of water; this gives an acid solution due to the hydrbchloride on the amino group. It is soluble to about 1 per cent in alcohol but rather insoluble in other common organic solvents. Thiamine exhibits, absorption bands in the ultraviolet spectrum. At pH 7 (aqueous solution) the maxima are at 235 and 267 mμ. At lower pH values the absorption spectrum changes. This influence of hydrogen ion on absorption is characteristic of many basic nitrogen compounds. The molecule reacts with platinum chloride to form microcrystalline rosettes. It also forms insoluble compounds with phosphotungstic and tannic acids. Thiamine picrate was one of the early derivatives to be crystallized.

Sulfates, nitrates, etc., have also been prepared. Many of these derivatives have vitamin activity. The melting point is between 248 and 250°C. This is for the hemihydrate, which crystallizes as white monoclinic needles. The odor of thiamine is highly characteristic and very like that of yeast. The characteristic odor of yeast is due in large part to the contained thiamine. Thiamine is destroyed at elevated temperatures unless the pH is low. In alkaline solution complete destruction results from boiling for short periods. At a pH of 3.5 boiling results in little destruction.

In yeast, autoclaving at 120°C for short periods does' not destroy the vitamin, but after two or three hours nearly complete destruction is accomplished. Thiamine is readily oxidized; under controlled conditions thiochrome is formed, and this is the basis for a quantitative determination of the vitamin. On reduction, hydrogen is added to the thiamine molecule, and vitamin activity disappears. Schultz prepared a series of 39 compounds chemically related to thiamine; 16 of these showed some vitamin activity. The structure of thiamine as the free base is shown in the following. The naturally occurring molecule and the synthetic vitamin contain a hydrochloride on the amino group and a chloride ion neutralizing the positive charge on the nitrogen atom of the thiazole ring.

Coenzyme Activity of Thiamine

An important derivative of thiamine is the pyrophosphate. This molecule is known as cocarboxylase and is the coenzyme or prosthetic group of the enzyme decarboxylase, which is involved in the decarboxylation of α-keto acids in the body. The structure of thiamine pyrophosphate is shown herewith. Thiaminokinase has been prepared from rat liver and from yeast. This enzyme in the presence of ATP and Mg^{++} synthesizes cocarboxylase from thiamine by transfer of pyrophosphate from ATP to thiamine.

$$\underset{\text{thiamine}}{R{-}CH_2{-}CH_2OH} + ATP \xrightarrow[\text{enz}]{Mg^{++}} \underset{\text{thiamine pyrophosphate (cocarboxylase)}}{R{-}CH_2{-}CH_2O{-}\overset{O}{\overset{\|}{\underset{OH}{\underset{|}{P}}}}{-}O{-}\overset{O}{\overset{\|}{\underset{OH}{\underset{|}{P}}}}{-}OH} + AMP$$

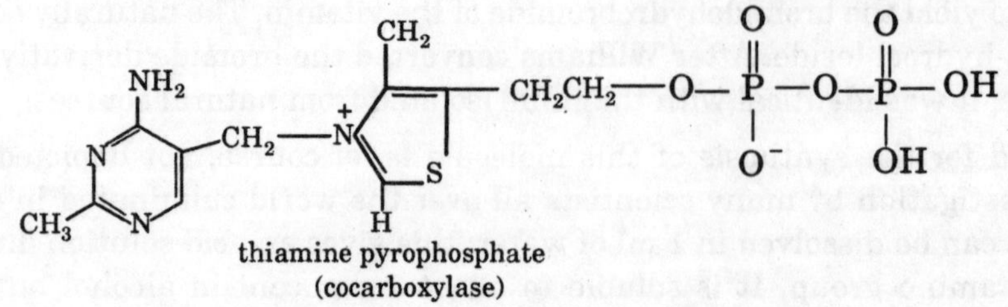

thiamine pyrophosphate
(cocarboxylase)

Barron and associates demonstrated that the utilization of α-ketoglutarate is increased by tissues of thiamine-deficient rats after the addition to the tissue of phosphorylated thiamine. Green and others showed that various animal tissues including rabbit muscle and pig heart contain a thiamine-containing enzyme capable of catalyzing the decarboxylation not only of pyruvate but also of α-ketoglutarate and α-ketobutyrate. The nonoxidative decarboxylation of pyruvic acid (plants, microorganisms) by the thiamine pyrophosphate (TPP) enzyme from yeast produces acetaldehyde. Such a simple decarboxylation of pyruvic acid forming acetaldehyde does not occur in mammalian tissues. Lipoic acid (6, 8-dithiooctanoic acid, or 6, 8-thioctic acid) functions as a co-enzyme in α-keto acid decarboxylation.

Lipoic acid, also called pyruvate oxidation factor, protogen, and acetate replacement factor, is a growth substance for certain protozoa; hence the name protogen. It also can replace acetate for growth in *L. casei,* and so the name acetate replacement factor. The factor is required for normal oxidation of pyruvate by *S. faecalis.* The term lipoic acid was used because the substance is fat-soluble, and the term thioctic acid has been applied because it is an 8-carbon acid containing sulfur groups.

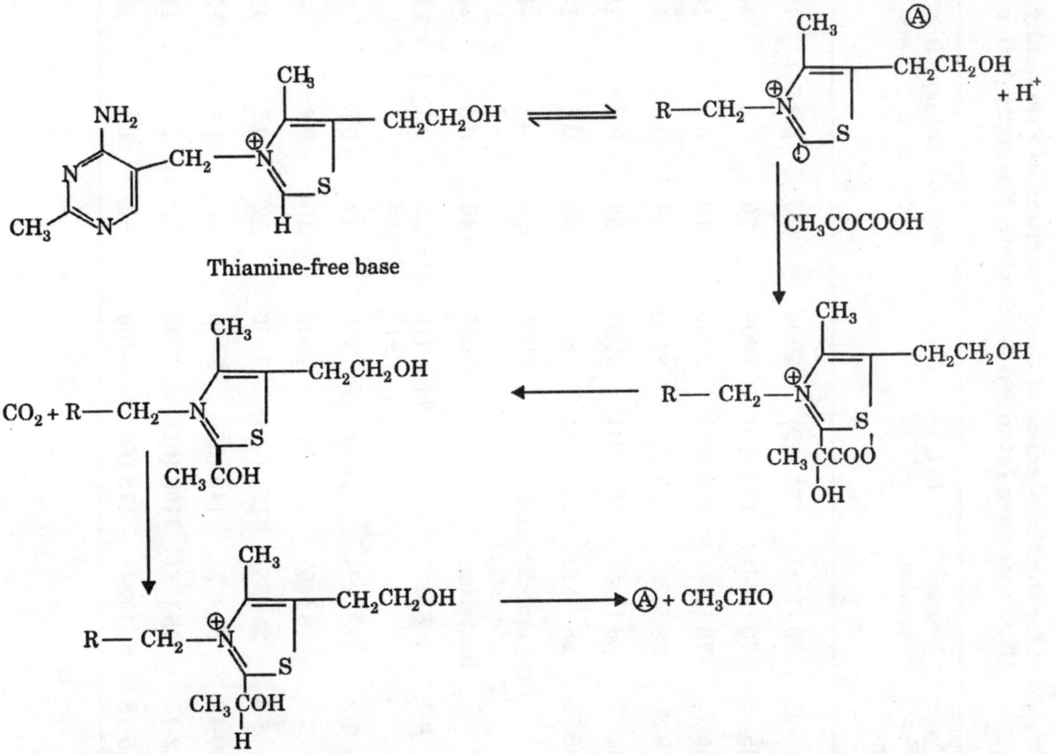

Thiamine-free base

Fig. 3.1. Proposed mechanism for nonoxidative decarboxylation of pyruvic acid.

In mammalian metabolism the decarboxylation of pyruvic acid is an oxidative process and involves the removal of CO_2 with the formation of the transient intermediate, α-hydroxyethyl thiamine pyrophosphate. This was referred to as "active acetaldehyde," and Carlson and Brown

Table. 3.1. Food and Nutrition Board, National Academy of Sciences—National Research Council Recommended Daily Dietary Allowances

Designed for the Maintenace of Good Nutrition of Practically all Healthy Persons in the USA (Allowances are Intended for Persons Normally Active in a Temperate Climate)

	Age + Years From to	*Weight kg (lbs)*	*Height cm (in)*	*Calories*	*Protein g*	*Calcium g*	*Iron mg*	*Vitamin A Value IU*	*Thia-mine mg*	*Ribo-flavin mg*	*Niacin Equlv.‡ mg*	*Aco-rbic Acid mg*	*Vita-min D IU*
Men	18-35	70 (154)	175 (69)	2900	70	0.8	10	5000[||]	1.2	1.7	19	70	
	35-55	70 (154)	175 (69)	2600	70	0.8	10	5000	1.0	1.6	17	70	
	55-75	70 (154)	175 (69)	2200	70	0.8	10	5000	0.9	1.3	15	70	
Women	18-35	58 (128)	163 (64)	2100	58	0.8	15	5000	0.8	1.3	14	70	
	35-55	58 (128)	163 (64)	1900	58	0.8	15	5000	0.8	1.2	13	70	
	55-75	58 (128)	163 (64)	1600	58	0.8	10	5000	0.8	1.2	13	70	
		Pregnant (2nd and 3rd trimester)		+ 200	+0.5	+5	+5	+1000	+0.2	+0.3	+3	+30	400
		Lactating		+1000	+40	+0.5	+5	+3000	+0.4	+0.6	+7	+30	400
Infants§	0-1	8 (18)		kg × 115 ±15	kg × 2.5 ±0.5	0.7	kg × 1.0	1500	0.4	0.6	6	30	400
Children	1-3	13 (29)	87(34)	1300	32	0.8	8	2000	0.5	0.8	9	40	400
	3-6	18 (40)	107 (42)	1600	40	0.8	10	2500	0.6	1.0	11	50	400
	6-9	24 (53)	124 (49)	2100	52	0.8	12	3500	0.8	1.3	14	60	400
Boys	9-12	33 (72)	140 (55)	2400	60	1.1	15	4500	1.0	1.4	16	70	400
	12-15	45 (98)	156 (61)	3000	75	1.4	15	5000	1.2	1.8	20	80	400
	15-18	61 (134)	172 (68)	3400	85	1.4	15	5000	1.4	2.0	22	80	400

	Age + Years From to	*Weight kg (lbs)*	*Height cm (in)*	*Calories*	*Protein g*	*Calcium g*	*Iron mg*	*Vitamin A Value IU*	*Thia-mine mg*	*Ribo-flavin mg*	*Niacin Equlv.‡ mg*	*Aco-rbic Acid mg*	*Vita-min D IU*
Girls	9-12	33 (72)	140 (55)	2200	55	1.1	15	4500	0.9	1.3	15	80	400
	12-15	47 (103)	158 (62)	2500	62	1.3	15	5000	1.0	1.5	17	80	400
	15-18	53 (117)	163 (64)	2300	58	1.3	15	5000	0.9	1.3	15	70	400

About the Recommended Daily Allowance:

In 1940 the Food and Nutrition Board of the National Research Council undertook the development of a dietary standard for the United States. The nutrient allowances were formulated, as for as possible and practical, to be "suitable for the maintenance of good nutrition in essentially the total population rather than those allowance which would fulfill minimal needs for the average persons."

The board has relied on published data and the compined judgment of a large group of nutrition authorities.

Implicit in the chosen term "recommended allowances" is the feeling of the board that the set standards are neither final nor minial nor optimal requirements. The values given in the table represent "levels of nutrient intake oppearing desirable for use in planning diets and food supplies." In general they are higher than the average requirements but lower than the amounts needed in deficiency states.

It is pointed out by the board that if these allowances are used in dietary evaluations, it should be appreciated that most persons whose consumption equals or exceeds the standards are presumably adequately nourished but not all who fail to reach these goals are malnourished. Involved in such a situation are the individual variations in needs among people, differences in degree of intestinal synthesis and intestinal obsorption, variations in metabolk patterns or degree of vitalization of certain nutrients, and other factors.

With the exception of vitamin D, it is safe to assume that a good mixed diet of common foods will supply the recommended allowances, and this would hold with many combination of foods.

* The allowance levels are intended to cover individual variations among persons as they live the United States under usual envirommental stresses. The recommended allowances con be attained with a variety of common foods, providing other nutrients for which human requirements have been less well defined. See text for more detailed discussion of allowances and of nutrients not tabulated.

† Entries on lines for age range 18-35 years represent the 25-years represent the 25-year age. All other entries represent allowances for the specified age periods, *i.e.,* line for children 1-3 is for age 2 years (24 months), 3-6 is for age 41/2 years (54 months), etc.

‡ Niacin equivalents include dietary sources of the preformed vitamn and the precursor, tryptophan, 60 mg typtophan represents 1 mg niacin.

§ The calorie and protein allowances per kg for infants are considered to decrease progressively from birth. Allowances for calcium, thiamine, riboflavin, and niocin increase proportionately with colories to the maximum valves shown.

‖ 1000 IU from preformed vitamin A and 4000 IU beta-carotene.

showed that the reaction product between thiamine pyrophosphate and either pyruvate or acetaldehyde with a wheat gorm carboxylase is α-hydroxyethyl thiamine pyrophosphate. This compound was isolated from microorganisms, and the material was capable of reacting enzymatically.

"Active acetaldehyde" appears to be α-hydroxyethyl thiamine pyrophosphate. The active acetaldehyde is then transferred to the oxidized form of lipoic acid forming acetyl lipoic acid (now reduced lipoic). As the acetyl is transferred to coenzyme A, the reduced lipoic acid is oxidized to the original form, and acetyl CoA may enter one of its various metabolic reactions. Acetaldehyde in the free form is not a product of pyruvic acid decarboxylation in mammalian tissues. The oxidative decarboxylation of α-ketobutyric acid follows a similar scheme. The succinyl semialdehyde formed exists transiently on the coenzyme and then is converted into succinyl lipoic acid and then into succinyl CoA. The enzymatic formation of acetoin in both plants and animals involves TPP as coenzyme. Two molecules of acetaldehyde or a molecule of pyruvic acid and a molecule of acetal dehyde form acetoin as follows:

$$CH_3—CHO + OHC—CH_3 \rightarrow CH_3—CHOH—CH_3$$

$$CH_3—CO—COOH + CH_3—CHO \rightarrow CH_3—CHOH—CO—CH_3 + CO_2$$

The significance of these reactions is obscure, since no role of acetoin is known. Transketolase is an enzyme found in plant and animal tissues. TPP is the coenzyme. The reactions involving this enzyme are detailed. An example of a transketolase reaction involves the transfer of a 2-carbon unit from the 2-keto compound xylulose-5-phosphate to the aldehyde ribose-5-phosphate forming sedoheptulose-7-phosphate and glyceraldehyde-3-phosphate. An "active glycoaldehyde" is an intermediate (glycoaldehyde-thiamine pyrophosphate complex) in this reaction, and the situation is analogous to "active acetaldehyde" discussed previously.

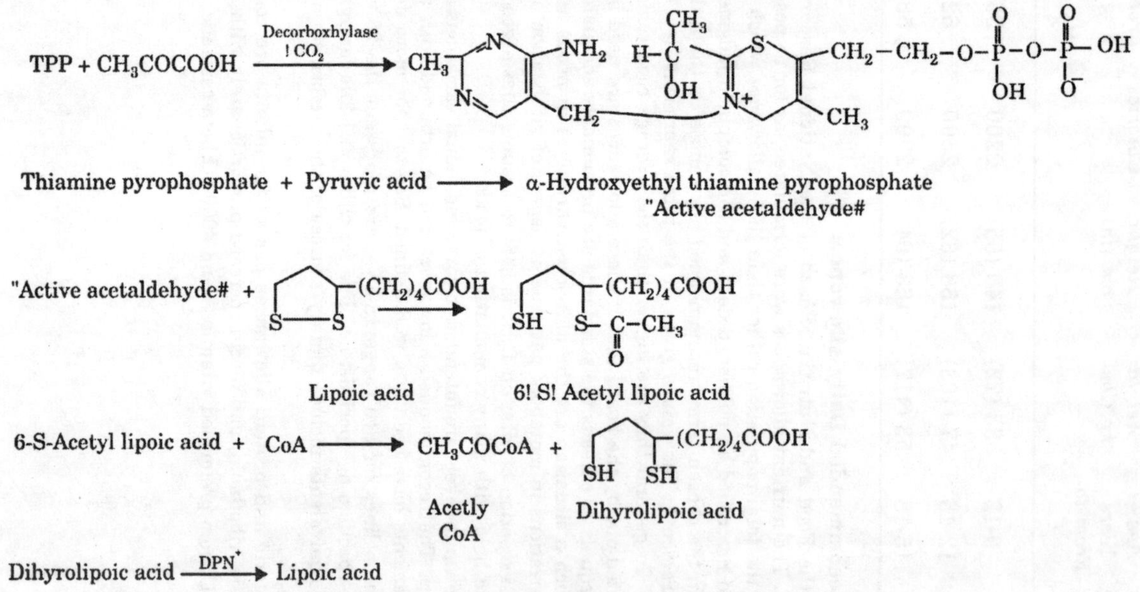

TPP + $CH_3COCOOH$ —Decorboxhylase, ! CO_2→ CH_3 N NH$_2$ H–C(CH_3)(OH) S N+ CH_2 CH_3 –CH_2–CH_2–O–P(=O)(OH)–O–P(=O)(O$^-$)–OH

Thiamine pyrophosphate + Pyruvic acid ⟶ α-Hydroxyethyl thiamine pyrophosphate "Active acetaldehyde#

"Active acetaldehyde# + [S—S ring]–$(CH_2)_4COOH$ ⟶ SH S–C(=O)–CH_3 –$(CH_2)_4COOH$

Lipoic acid 6! S! Acetyl lipoic acid

6-S-Acetyl lipoic acid + CoA ⟶ CH_3COCoA + SH SH –$(CH_2)_4COOH$

Acetly CoA Dihyrolipoic acid

Dihyrolipoic acid —DPN$^+$→ Lipoic acid

Fig. 3.2. Oxidative decarboxylation of pyruvic acid.

TPP is also involved as coenzyme in a phosphoroclastic cleavage of α-keto acids. In such a reaction the acyl of the acid is converted through acyl CoA to acylphosphate, and the carboxyl group may become formate:

$$CH_3COCOO^- + Pi \xrightarrow{-CO_2} CH_3COCOOP + HCOO^-$$

$$\text{Pyruvate + inorganic phosphate} \longrightarrow \text{acetylphosphate + formate}$$

Thiamine Deficiency

In humans the condition known as beriberi results from thiamine deficiency. The condition is rare in the Western world because of food habits and food enrichment. Anorexia (loss of appetite), weight loss, fatigue, and gastrointestinal upsets are noted. Later in the disease cardiac impairment is seen. The administration of the vitamin or foods rich in it is usually followed by dramatic disappearance of symptoms. Most thiamine deficiency states in humans are complicated by deficiencies of other B vitamins. A book by Williams is informative in relation to beriberi. In 1957 Brožek reported the results of a comprehensive study from Keys' laboratory.

Psychological changes and biochemical data are presented in human thiamine deprivation. The study consisted of four experimental periods: (*a*) a month for the standardization and collection of control values; (*b*) partial restriction of the vitamin for 168 days, with daily intakes in three groups of men of 0.61, 1.01, or 1.81 mg, and energy expenditure of about 3300 Cal; (*c*) an acute thiamine deprivation period of 15 to 27 days on "thiamine-free" diet; and (*d*) a thiamine supplementation period, 5 mg daily for 9 to 21 days. Oral administration of thiamine was started for each man when the investigators considered that the deficiency had progressed as far as was safe. At this time nausea and vomiting were present, general weakness was pronounced, anorexia was extreme, and blood pyruvic acid levels were elevated.

Thiamine supplementation restored appetites and resulted in a dramatic change in attitude of the subjects. Performance tests on intelligence were not affected adversely within limits of the experimental conditions. Neurasthenia (nervous exhaustion with many symptoms), emotional deterioration, impaired coordination, and lowering of the pressure-pain threshold all responded dramatically to thiamine administration. During the period of partial restriction, with the other B vitamins supplied in adequate amounts, the intake of 0.2 mg thiamine per 1000 Cal was at the borderline of deficiency. Elevated blood pyruvic acid, especially after glucose ingestion, has been observed in advanced thiamine deficiency by various workers. This undoubtedly results from the diminished supply of thiamine pyrophosphate, required as coenzyme in α-keto acid decarboxylation. Blood pyruvic acid or the pyruvic-lactic ratio, following glucose administration, has been of limited value in studying thiamine status.

Blood thiamine or thiamine pyrophosphate levels may prove to be more informative. On a high-carbohydrate diet supplying only 0.11 to 0.18 mg of thiamine per day, eight growing men developed clinical thiamine-deficiency symptoms in 9 to 27 days. General malaise, headache, nausea, constipation, vomiting and generalized muscle ache were common symptoms. The urinary excretion of thiamine decreased to a point that it was undetectable by the eighteenth day. Transketolase activity of red cells also dropped as deficiency developed and increased during the repletion period.

The excretion of metabolites of the vitamin, the thiazole and pyrimidine moieties, increased above levels found when the vitamin intake was normal and was reduced after adding thiamine to the diet. The authors suggest that there is a body store of thiamine which can be used during insufficient intake. The proceedings of a conference on beriberi held in 1958 contain a great deal of information on the clinical, nutritional, and biochemical problems in this disease. Since the advent of synthetic diets capable of supporting good growth and reproduction in laboratory animals, the development of a thiamine deficiency has been followed closely from both the biochemical and neurological standpoint in a number of species. In pigeons on a severe thiamine deficiency early symptoms are loss of appetite, lassitude, and weight loss.

Paralysis and head retraction (opisthotonus) are followed by 'death in a matter of a few weeks. If the deficiency is less severe, the birds survive for long periods and develop many of the symptoms seen in humans. Rats lose their appetite after a short period on a deficient diet. If the diet contains small amounts of the vitamin, rather typical symptoms may ensue. A spasticity and stiffness of the hind legs is seen, and later the limbs may become paralyzed. It is obvious that a large part of the symptomatology of thiamine deficiency is related to pathological changes in specific parts of the nervous system. It is impractical to discuss here detailed findings of experimental workers in this specialized field.

Rinehart, Greenberg, and Friedman made careful studies of the lesions of the nervous system of rhesus monkeys subjected to one or more periods of thiamine depletion. Outstanding were the lesions in the nuclear structures of the central nervous system. In the animals depleted two or more times the lesions were more severe and extensive. Certain myocardial changes were observed, and according to some investigators, they are similar to those of the human heart in beriberi. North and Sinclair, on the other hand, subjected rats to three to six successive periods of acute thiamine deficiency over an average survival period of 127 days. Other rats were subjected to a combined thiamine and pantothenic acid deficiency. In both studies there was no evidence of degenerative changes in the distal segments of the sciatic and posterior tibial nerves, in the lumbar dorsal-root ganglia, or in the lumbar cord.

Thiamine Antagonists

Pyrithiamine is a synthetic pyridihe analogue of thiamine. In this compound a —C = C— takes the place of the —S— in the thiazole group of thiamine. Woolley and White showed that mice develop a thiamine deficiency when pyrithiamine is administered. The process is reversible when sufficient extra thiamine is included in the diet. The symptoms are similar to those produced by a thiamine deficiency in other species. Pyrithiamine, also called "neopyrithiamine," was shown to inhibit the biosynthesis of cocarboxylase by chicken blood but not to lessen the combination of cocarboxylase with the apoenzyme to produce the decarboxylating enzyme. It was also suggested by Woolley and others that pyrithiamine may affect other thiamine systems than cocarboxylase, since animals dying of typical thiamin-deficiency symptoms as a result of administration of the antagonist showed normal levels of liver cocarboxylase.

The administration of the antagonist oxythiamine (—OH for —NH_2 in thiamine), however, produced low levels of cocarboxylase. Also, De Caro and workers have demonstrated that the normal increases in heat production and in basal metabolic rate (BMR) in the rat following glucose administration are inhibited by pyrithiamine but not by oxythiamine. It seems clear that distinct differences exist in the mechanisms of thiamine inhibition by these two antagonists. The 2-*n*-butyl homologue of thiamine (*n*-butyl group replacing the methyl group on the pyrimidine ring) is a vitamin inhibitor for rats. Emerson and South-wick demonstrated decreased growth rate and certain polyneuritic symptoms in rats fed this compound. The adverse effects were largely obviated by adding thiamine to the diet. These instances are probably examples of the so-called competition reactions. It is thought that the inhibitor molecule may replace the natural vitamin molecule in an enzyme system, or in some other essential metabolic component, and that such a component is not physiologically active.

On the addition of extra, vitamin, the normal system, by mass law action perhaps, may again be formed. An interesting-observation by Gubler and associates regards the action of sorbitol (sugar alcohol of glucose) in rat thiamine deficiency. Rats were made deficient by either thiamine deprivation or by injections of either of the antagonists, oxythiamine or pyrithiamine. Half of the deficient rats were fed a basal diet, and the other half this diet containing. 20 per cent sorbitol.

In all cases where the deficiency caused a decrease in growth or in the degree of oxidative decarboxylation of pyruvic or α-ketoglutaric acid, the feeding of the 20 per cent sorbitol diet reversed the growth effect and also resulted in about normal tissue levels of the keto acids. However, increased blood pyruvate was not lowered by sorbitol feeding. The authors suggest that disturbances in pyruvate metabolism may play only a secondary role in the development of thiamine deficiency. It was generally held some years ago that the metabolism of alcohol required thiamine-containing enzymes. Westerfeld and co-workers were especially active in this field. More recent evidence indicates that alcohol ingestion may actually decrease the thiamine requirement.

In thiamine deficiency pyruvic acid accumulates in the body as a result of impaired carbohydrate metabolism (pyruvic acid decarboxylation). If thiamine were required for alcohol metabolism, one would expect a decreased ratp of removal of this molecule from the body in a deficiency of the vitamin. Westerfeld and co-workers found no decreased rate of alcohol metabolism in pigeons during acute thiamine deficiency. Further work by Westerfeld and Doisy demonstrated that in the pigeon alcohol exerts a sparing action on thiamine. When alcohol was substituted isocalorically for part of the carbohydrate of the diet; the thiamine requirement was lessened. The possibility was pointed out that the metabolism of alcohol and carbohydrate together requires less thiamine than the metabolism of carbohydrate alone. Fat and protein likewise exert a thiamine-sparing action.

The neurological findings in alcoholism, it would appear, cannot be explained on the basis of thiamine deficiency resulting from increased demand, but rather as a deficiency resulting simply in lowered intake on the generally inadequate diet of the alcoholic. The specific role of thiamine in relation to neurological findings in thiamine deficiency in animals or in man cannot be stated. In thiamine-deficient rats the intraperitoneal injection of 1 mg of thiamine is followed by a threefold increase in liver total cholesterol and twentyfold elevation of liver neutral glycerides.

Miller and co-workers also report exceedingly high levels of these components following a second injection of the vitamin. Low levels of liver phospholipids also return to normal with the first injection. The ratio of free to total cholesterol is high during deficiency, indicating faulty esterification or lack of fatty acids. The ratio falls to below normal following thiamine repletion. Oral thiamine produced similar effects, but the parenteral administration caused a much greater effect on the total cholesterol.

Metabolism of Thiamine

McCarthy and others injected thiamine-S into normal rats and recovered about 65 per cent of the injected activity in the urine collected during ten days following administration. Thiamine-deficient rats excreted 52 per cent of the original radioactive sulfur in the time period. In each case most of the radioactive sulfur appeared in the neutral sulfur fraction and a small amount as sulfate.

Thiamine Requirement

One USP unit of thiamine is equal to µg of thiamine hydrochloride. This is identical to the 1U Because of the well-established relationship of thiamine to carbohydrate metabolism, the requirement of this vitamin for man is most accurately predicted on the basis of the caloric intake and primarily the carbohydrate intake. This, however, is not the most practical approach from a nutritional standpoint. Since individual diets vary to such a great extent and because of other individual variations, exact figures for human thiamine requirement are not highly significant.

The general trend in the past few years has been to revise downward the estimated daily thiamine need. Thus the recommended dietary allowance adopted by the Food and Nutrition Board of the National Research Council was revised from 0.6 mg to 0.5 mg per 1000 Cal in 1953 and remained essentially the same in the 1958 revision. A further downward revision to 0.4 mg per 1000 Cal per day was adopted in the 1963 revision. For a 3000 Cal diet the daily thiamine requirement would be approximately 1.2mg. Such allowances are not to be confused with requirements.

The allowances were designed to allow a fair margin above actual requirements. Still lower -intakes are apparently able to cover the needs. In a recent study the thiamine requirement of adults was calculated to be 0.27 to 0.33 mg per 1000 Cal of diet per day. The recommended daily allowance then would be twice this, or 0.54 to 0.66 mg per 1000 Cal per day. The use of massive doses of thiamine for various clinical entities, such as nausea of pregnancy and "nervousness," seems to be successful only in isolated cases.

Distribution of Thiamine

The vitamin is widely distributed in both plant and animal tissues. In plants the seeds generally contain the highest concentration. Dry peas, beas, and soyabeans are excellent sources. The vitamin is concentrated in the outer layers of the grain kernels. Bran and rice polishings are thus rich sources. Whole wheat bread and white bread made from enriched flour contain a good supply of thiamine. Yeast is an outstanding source of thiamine and of certain other members of the B complex.

Many nuts have a high concentration of the vitamin. Peanuts contain a good supply, and pecans and brazil nuts have still more. Milk contains only one tenth the amount of thiamine that is found in peanuts but because of the comparative quantities consumed, milk is by far a more important dietary source. Many cheeses contain thiamine in amounts comparable to that of milk. Avocados, gooseberries, and dried prunes contain the highest thiamine content among the fruits. Oranges have around half the quantity found in figs or prunes. Many animal tissues are excellent sources of the vitamin. Pork chops and fat ham have exceptionally high contents. Many beef cuts are also excellent source.

It is of interest that though live yeast contains a high concentration of thiamine not only is the vitamin unavailable to humans, but the yeast also appears to complete with the body in the gastrointestinal tract for other dietary thiamine. Parsons and co-workers fed live yeast to individuals and noted a marked drop in urinary thiamine excretion to less than the basal level (without yeast). On removal of the yeast from the diet, urinary excretion rose. The thiamine of killed yeast, on the other hand, is almost completely available.

Various workers have demonstrated that the ingestion of viable yeast leads to increased loss of free thiamine in the feces, primarily in the cells of yeast and other organisms, and lessened urinary excretion, which is a good indication of decreased absorption from the intestine.

Table. 3.2. Approximate Thiamine and Riboflavin Content of Some Common Foods

Food	*Thiamine (μg) per 100 g*	*Riboflavin (μg) per 100 g*
Milk, cow	40	150
Cheese, American	40	500
Orange juice	60	55
Peanuts	400	400
Ham	1,200	300
Veal chop	290	220
Beefsteak	200	300
Beef liver	250	2,000
Eggs	170	300
Cornmeal, yellow	400	200
Soyabean flour	500	400
Rolled oats	800	150
Whole wheat flour	500	100
White bread	80	100
White bread, enriched	400	300
Wheat germ	3,000	600
White rice	50	20
Brown rice	300	50
Dry peas	800	300
Potato	150	40
Green beans	90	120
Dry beans	500	300
Beets	40	70
Celery	30	30
Yeast, fresh	3,000	1,500
Yeast, dried brewers	10,000	3,000

Some bacteria synthesize thiamine. There has been considerable controversy over the question of availability to the body of the vitamin from this source. Alexander and Landwehr presented evidence which indicates that neither free nor bound (diphosphate) thiamine of the feces is of nutritional value to the individual. They found that most of the vitamin exists in the bacterial cells and that the greatest part was present as bound thiamine, which is not absorbed until it is dephosphorylated. They point out that probably no enzyme capable of bringing about this hydrolysis exists in the large intestine. They administered both forms of thiamine by retention enema to an individual and found that the urinary thiamine excretion was not increased and that all the thiamine and cocarboxylase administered were recovered in the next 24-hour stool. By the use of C^{14}-labeled thiamine it was shown that rats did not absorb thiamine from sources other than that supplied in the diet. This would indicate that under the conditions employed in these experiments intestinal microorganisms did not add to the total vitamin supply available to the host. Mickelsen has reviewed the subject of intestinal synthesis of vitamins.

Determination of Thiamine

Biological methods for thiamine determination have largely been replaced by physicochemical techniques. Of the former, the rat curative procedure has been widely used. The test material is administered to acutely polyneuritic rats. The duration of cure of rats given various levels of the unknown is compared to the duration after giving known amounts of thiamine.

From such data the approximate vitamin content can be calculated. Yeast fermentation methods and microbiological methods have been developed and widely used. The chemical procedures are increasing in popularity because of ease and rapidity compared to animal assays.

The most widely employed method involves the isolation of thiamine from the test material, the oxidation of the vitamin to thiochrome, and the comparison of the fluorescence of this compound and the fluorescence produced by oxidizing a known amount of pure thiamine. Alkaline ferricyanide is used to oxidize thiamine. The blue fluorescing thiochrome can be separated from many interfering substances by isobutyl alcohol extraction. Special equipment is required to make a quantitative comparison of the fluorescence in the sample and in a standard solution. Fluorescence attachments are available for various types of spectrophotometers:

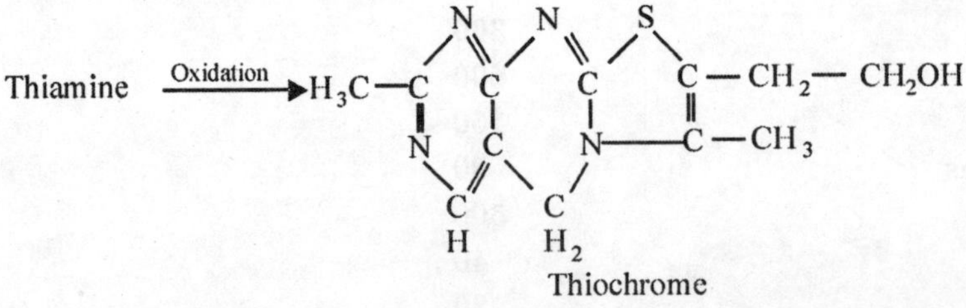

Thiamine $\xrightarrow{\text{Oxidation}}$ Thiochrome

Developments in colorimetric, polarographic, and chromatographic techniques for thiamine, as well as separation procedures for mono-, di-, and tri-phosphoric esters are summarized by Horwitt. Mickelsen and Yamamoto have reviewed analytical methods including animal, physicochemical, enzymatic, and microbiological techniques developed by many investigators over the years.

Riboflavin

The suggestion that water-soluble B contained more than one active principle was given impetus by the work of Emmett and McKim. These workers indicated the presence of two water-soluble vitamins in rice polishings, one that cured rat polyneuritis and another that produced weight gain in rats on specific rations. We now know that the second factor contained riboflavin, and at that time this principle was called vitamin B_2 or vitamin G. Various investigators had studied naturally occurring pigments in foods. Bleyer and Kallmann isolated a crude yellow pigment from whey, and Blooher presented evidence pointing to the probable identity of vitamin G (B_2) and the water-soluble yellow, green fluorescent pigment of whey.

In 1932 Warburg and Christian isolated a yellow respiratory ferment (Warburg's yellow enzyme) from bottom yeast. This was later separated into a protein fraction and a small pigment molecule (riboflavin), neither of which alone possessed enzyme activity. Pure riboflavin was isolated from milk and other foods in 1933 by Kuhn and co-workers. In the early phases of this work 100 mg of flavin were obtained from a quantity of egg white corresponding to 33,000 eggs. This represents a yield of about 7 per cent based on the content of this pigment now known to be present. Lactoflavin (milk), hepatoflavin (liver), ovoflavin (eggs), and verdoflavin (grass) proved to be chemically identical with riboflavin, and it was recognized that riboflavin is actually what had been called vitamin B_2. The synthesis of riboflavin was accomplished by two groups of German workers in 1935. In nature the vitamin is synthesized by all green plants and by most microorganisms, although some bacteria are without this capacity. A review summarizes developments on the biosynthesis and degradation of riboflavin.

Chemistry of Riboflavin

The flavins are widely distributed in nature. Many derivatives have been studied. After the elucidation of the structure of lumiflavin (photolysis product of riboflavin), the relation of this compound to the known alloxazines was established. Shortly after this it was predicted that riboflavin differed from lumiflavin in that a D-ribityl group was present in riboflavin, replacing the methyl group at position 9 in lumiflavin. Total synthesis confirmed this prediction.

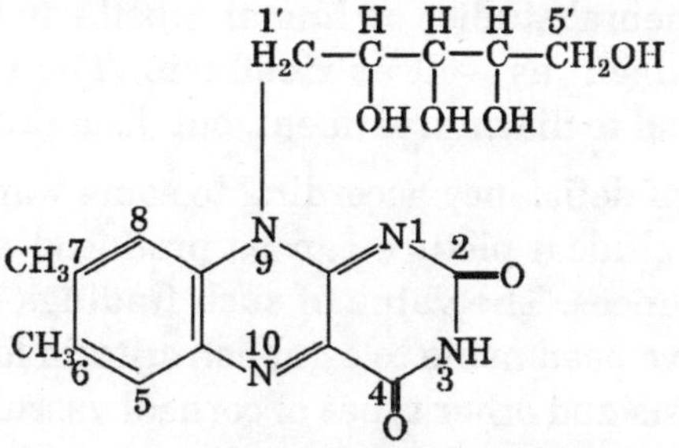

riboflavin
6, 7-dimethyl-9-(D-1'ribityl) isoalloxazine

Riboflovin phosphate has ortho phosphate ester on primary alcohol group

The D in the chemical name of riboflavin indicates that the ribityl group is related to the D series of sugars. The one prime (1′) indicates attachment of the ribityl group at the first carbon atom, and the 9 that this attachment is to position 9 of the isoalloxazine ring system. A number of syntheses are available for the dimethyl compound.

The vitamin is not soluble in ordinary fat solvents but is soluble to a limited degree in water. More soluble derivatives, such as the acetate and the phosphate, have been used medicinally. Riboflavin is reversibly reduced by thiosulfate, H_2S, and H_2 to form the colorless leucoriboflavin. Strong oxidizing agents decompose it, but bromine water and H_2O_2 have only a slight effect. The molecule is in general acid-stable and alkali-labile. Either visible or ultraviolet light irreversibly decompose riboflavin.

The molecule exhibits absorption maxima at wavelengths of 220, 267, 336, and 446 mμ., and a water solution is yellow and has an intense green fluorescence. The orange-yellow needles of the pure compound melt at 282°C. A number of analogues and other derivatives have vitamin activity. The 6-ethyl, 7-methyl compound and arabinoflavin (L-arabityl in place of D-ribityl) are active. The biosynthetic pathways for riboflavin, purines and pteridines are in some respects quite analogous.

Glycine carbon atoms are incorporated into all three classes of compounds by plants and microorganisms. A preformed purine can be converted into riboflavin, although it is not certain whether purines are normally direct precursors. A scheme for part of the biosynthesis of riboflavin has been presented in a review of this subject by Brown and Reynolds.

Riboflavin deficiency

The early studies of human vitamin deficiencies were complicated by the presence in the subjects of multiple deficiencies. Only recently have the symptoms of a frank ariboflavinosis in man been noted, and it must be said that considerable controversy exists regarding the relation of certain of the symptoms to riboflavin deficiency. In general, however, the deficiency is associated in man with abnormal ectodermal tissue maintenance. The usual signs of this are (*a*) inflammation of the tongue (glossitis)—the tongue assumes a magenta color, and there is flattening of the papillae; (*b*) cheilosis, or fissuring at the corners of the mouth and of the lips; and (*c*) a seborrheic dermatitis, often manifested by a sharklike appearance of the skin in the nasolabial folds and elsewhere.

One of the early experimental studies on human riboflavin deficiency was reported in 1938. The diet was deficient in niacin as well as riboflavin. The cheilosis and seborrheic lesions that developed did not respond to niacin treatment, but disappeared following riboflavin therapy.

Another manifestation of deficiency according to some workers is corneal vascularization. It is known that a similar clinical picture can be produced by a variety of deficiencies and even by certain toxic substances. The value of such findings in diagnosing ariboflavinosis is indeed limited. Attempts have been made to establish criteria for differentiating between ocular disturbances of ariboflavinosis and other types of corneal vascularization, and between glossitis of riboflavin deficiency and that found in other diseases. The situation is still complicated, however, and many reports indicate that observers have difficulty .in attaching a specific symptom to a specific avitaminosis.

In many instances one of these symptoms in humans has responded to some other member of the B complex but not to riboflavin. For instance, Cayer and others made a rather intensive biochemical study of a number of patients with glossitis and cheilosis. Excretion and load tests of various B complex vitamins were studied in these patients and it normal individuals.

Interesting among their findings are (*a*) similar riboflavin levels in some patients with and some without cheilosis, indicating the possibility that riboflavin deficiency may not have been responsible for the eheilosis; (*b*) frequent failure in response to riboflavin and (*c*) frequent response in patients with cheilosis to niacin (nicotinic acid, another member of the B complex). Some patients responded only after the administration of yeast or liver extract, sources of many B complex vitamins. In man on controlled low intake of riboflavin over extended periods, Horwitt and co-workers observed erythema of the tongue, angular stomatitis, and scrotal dermatitis as prevalent symptoms. Bessey and co-workers studied the riboflavin content (and the coenzymes containing the vitamin: see further) in blood of men on restricted intakes and of men on liberal intakes of the vitamin.

The content in red blood cells was found to be a reasonably sensitive and practical index for the evaluation of nutritional status with respect to riboflavin. A critical review of pathological changes in riboflavin deficiency can be found in the book by Follis. An informative history of pellagra covering the recognition of the nutritional aspect and the conquest of the disease was published by Sydenstricker. With animals long-continued and well-controlled deficiency experiments are practical.

Rats exhibit decreased growth rate; eye changes, including a keratitis rather typical cataracts; dermatitis; possibly a type of anemia; some nerve degeneration; and poor reproduction. Congenital malformation and a cessation of the estrous cycle have been reported. Definite symptoms of ariboflavinosis in calves have been reported. Wertman and various co-workers have studied a number of vitamins in relation to deficiency and factors in resistance to infection in rats. Paper No. IV of the series concerns a riboflavin deficiency.

Coenzyme Activity

The enzymes containing riboflavin are called flavoproteins. Two coenzymes are known—riboflavin phosphate, or flavin mononucleotide (FMN), and flavin adenine dinuclcotide (FAD). The first flavoprotein isolated was Warburg's yellow enzyme; it is composed of FMN and apoenzyme (specific protein). The enzyme can be separated into protein and FMN by dialyzing against dilute HC1. No enzyme activity resides in the individual constituents, but upon recombining them activity reappears. Highly purified yellow enzyme was found to contain two FMN molecules per molecular weight of 104,000, which amounts to 0.877 per cent of the nuclcotide in the enzyme.

Flavin adenine dinucleotide has been isolated from yeast and animal tissues and various plants. Both FAD and FMN have been synthesized. FAD and FMN combine with different apocnzymes to form a large number of oxidation-reduction enzymes. FAD, for example, is found in xanthine oxidase, D-amino acid oxidase aldehyde oxidase, and fumaric dehydrogenase. FMN is associated with Warburg's yellow enzyme, cytochrome reductase, L-amino acid oxidase, and others.

These enzymes operate in electron transport systems. The structure of the two coenzymes are shown herewith.

riboflavin phosphate,
flavin monophosphate, FMN

flavin adenine dinucleotide, FAD

The mononucleotide (riboflavin-5′-phosphate) is synthesized from the vitamin and ATP under the influence of a flavokinase enzyme:

$$\text{riboflavin} + \text{ATP} \xrightarrow{Mg^{++}} \text{FMN} + \text{ADP}$$

ATP reacts with the mononucleotide to form FAD. The enzyme mediating this reaction has been called flavin nucleotide pyrophosphorylase or FAD synthetase:

$$\text{FMN} \xrightarrow{Mg^{++}} \text{FAD} + \text{PP}$$

It was shown by Tong and others that FMN has a stimulating action in the conversion of iodide into iodoprotein by mitochondrial-microsomal preparations of sheep thyroid glands. The evidence indicated that the increased conversion is enzymatic and is inhibited by thiouracil and some other goitrogenic molecules.

With the accumulation of knowledge regarding riboflavin, one might expect that a deficiency of some specific enzyme system in the body might be correlated with a specific deficiency symptom. No such correlation is known at present although specific enzyme deficiencies are said to be demonstrable in ammals on low riboflavin intakes.

In moderate riboflavin deficiency rat tissues showed no decrease in DPNH dehydrogenase activity (flavoenzymc) according to Burch and co-workers. In severe deficiency the activity dropped off somewhat. In the ever of the deficient rats xanthine oxidase activity paralleled the decrease in FAD while the decrease in D-amino acid oxidase activity followed the decrease m FMN.

Riboflavin Antagonists

Galactoflavin is a potent riboflavin antagonist. Emerson and co-workers reported marked growth inhibition in rats on diets, containing this compound and reversal of the effect upon addition of sufficient riboflavin to the food. The diethyl derivative acts as a riboflavin antagonist in rats but can replace the vitamin for growth of *L. casei*. Egami and others showed that flavin monosulfate inhibits D-amino acidI oxidase and acts as a competitive inhibitor of the vitamin in growth of *L. casei.* Fall and Petering synthesized a number of analogues of riboflavin in which the ribityl group is replaced by other groups. The formylmethyl analogue was found to be a reversible riboflavin antagonist in the rat. The hydroxycthyl analogue proved to be a potent competitive antagonist of the vitamin in rats and with *L. casei*. The compound also exhibits antifungal and antibacterial activity.

Distribution of Ribofiavin

Riboflavin is one of the most widely distributed B vitamins, but there are very few common sources that contain high concentrations All cells of both plants and animals presumably contain the vitamin. Plant seeds synthesize the vitamin during germination. A few bacteria and the yeasts are rich sources. Muscle tissue is low in riboflavin, but certain internal organs, such as the liver and kidney, may contain appreciable amounts. Table 3.2 Lists the approximate riboflavin content of some common foods. Note the high concentration in live, wheat germ, and yeast, and the very low content in celery, polished rice, and potatoes. It should be kept in mind that even though potatoes are low in the vitamin, the continued intake at one or even two meals a day by, the majority of many races makes the potato an important contributor to the total dietary intake It was pointed out previously that the thiamine of live yeast is not available to the body. A similar situation exists with riboflavin.

Parsons and co-workers showed that the riboflavin in live bakers' yeast is only slightly available to humans but that the vitamin is completely available from samples of the same yeast after has been specially dried (dead cells). Other workers have reported similar findings. The methods used involve dearmnining the urinary and fecal riboflavin excretions after the oral administration of a known amount of pure riboflavin and comparing these with excretions after known amounts of the vitamin in yeast before and after special drying and heating treatments. Other foods do not yeld their total riboflavin content to the body. For instance, it was reported that in normal women the availability of riboflavin in ice cream was 90 per cent, whereas in green peas and almonds only 41 and 39 per cent, respectively, were available. The vitamin exists within many plant and animal cells, primarily in combination as one of the two coenzyme forms. The retina and urine and milk contain principally the free vitamin. The nucieotides may in part be combined with specific proteins to form enzymes. In serum of 13 normal adults the sum of the free riboflavin and riboflavin phosphate was found to average 0.8 µg per cent, and the riboflavin dinucleotide, 2.4 µg per cent. The white cells plus platelets contained 252 fig per cent of total riboflavin (both derivatives plus the free vitamin) and the red cells only 22. In a later'study Bessey and others found somewhat similar values in plasma and cells, although the levels were lower in individuals on restricted vitamin intake.

Determination of Riboflavin

Rats and chicks have served as test animals in the biological methods. In the rat growth method, young animals are fed a special riboflavtin-deficient diet for two or three weeks, after which time growth ceases. Then groups of the depleted rats arc given supplements of known amounts of the vitamin, and other groups are supplemented with graded quantities of the test material. Comparisons of growth response over four weeks or more allow the calculation of approximate vitamin content of the substance under examination. Such methods have largely been replaced by shorter and less expensive procedures. The original microbiological method of Strung and Snell using *L. casei* has been modified many times and is the basis of the official USP method. This organism requires riboflavin for growth and lactic acid production.

The culture medium can be prepared so that riboflavin is the limiting factor. With such a culture medium the organism responds to the addition of graded doses of the vitamin by producing corresponding increments of acid. In the test the acid production in response to supplements of known amounts of pure vitamin is compared to the acid production produced by adding graded amounts of extracts of tissues or other substance under examination. From such data an estimate of the riboflavin content of the unknown can be made. Marked variations in technique are required for the preparation of different types of materials for assay. This is especially true in the fluorometric methods, since substances other than riboflavin present in plant and animal tissues are known to fluoresce under the conditions of the test.

The intensity of fluorescence of a solution of riboflavin (under specific conditions) is proportional to the concentration of the vitamin. Light of wavelengths in the region 400 to 500 mm induce the fluorescence, and the intensity of this fluorescence can be measured by one of the various fluorophotometers on the market. In practice riboflavin is determined in an extract of the test substance in terms of the intensity of fluorescence before and after destruction of the vitamin. This is compared with the fluorescence of a known sample of riboflavin or with a solution of a substance such as fluorescein that has been standardized against the vitamin. Yagi described separation procedures for the various forms of riboflavin employing paper chromatography and paper electrophoresis, and a simplified technique for their determination. Murphy and others described procedures for liberating riboflavin from natural sources. An enzymatic method for FAD was developed by DeLuca and co-workers.

Riboflavin Requirement

There is no USP unit or IU for riboflavin. It is becoming common, and logically so, to refer to quantities of all vitamins of known structure on a weight basis. As is the case with most, of the vitamins, the human requirement of riboflavin is not known with any certainty. There are too many complicating factors, and the experimental clinical approach is impractical. Various estimates have been made for average requirements of a large group of people, but these figures cannot be translated to represent the required intake of any one in-dividual. Horwitt and co-workers set the need of an adult on 2200 Cal at between 1.1 and 1.6 mg per day. A riboflavin-depleted individual was able to assimilate at least 6 mg of the vitamin at one time.

The Food and Nutrition Board, National Research Council, 1963 revision, gives as the recommended allowance from 1.0 to 2.0 mg of riboflavin daily depending upon body weight. During pregnancy and lactation, the values given are higher. Bro-Rasmussen has published a comprehensive treatment of available work on riboflavin requirements of man and animals. In animals more exact data concerning requirements are known. The practical value in agriculture of such information is at once obvious. The establishment of required intakes for optimum growth rate, etc., in various species has been of great value in animal industry.

There are reviews on various aspects of ribofiavin.

Niacin—Nicotinic Acid, Niacinamide

The terms niacin and niacinamide have replaced the older terms nicotinic acid and nicotinic acid amide (nicotinamide). The latter names were found to be undesirable because of confusion and the unwarranted belief by many that they were physiologically related to nicotine. Niacin deficiency in humans leads to serious consequences. It is perhaps the most deleterious of the vitamin deficiencies in North America. Although such a deficiency is involved in pellagra (from the Italian *pelle agra,* rough skin), the disease is perhaps never, except experimentally, uncomplicated by other dietary deficiencies. Early in this century the incidence of pellagra, especially in many of our southern states, grew at an alarming rate.

As late as 1941 there were nearly 2000 reported deaths in the United States from this disease. How many more unrecognized pellagra deaths occurred can scarcely be estimated. Funk had isolate nicotinic acid from rice polishings as early as 1914, but he did not realize that it was a vitamin. He was studying polyneuritis primarily at this time, and nicotinic acid did not cure this disease, although he noticed a beneficial response when it was given with antineuritic substances. The earliest demonstration that pellagra results from nutritional deficiency was given by Goldberger and others in 1915. It is difficult to overemphasize the significance of this report.

It was the forerunner of the clinical experimental approach to the study of pellagra control. These workers studied the effect of a marked change in diet at two orphanages, one in Mississippi and one in Georgia, on the incidence of pellagra among the inmates. The change of diet to one with marked increase in "animal and leguminous protein foods" almost completely eliminated a reoccurrence of pellagra in the high percentage of individuals who had demonstrated the typical syndrome the previous year. Canine blacklongue, a deficiency syndrome in dogs described by Chittendon and Under Hill in 1917, has much in common with human pellagra.

It is readily produced by maintaining young dogs on the Goldberger diet. It was observed shortly after this that many of the foods and preparations effective in curing blacktongue were likewise curative for pellagra and that, whatever the active substances might be, they always occurred together in natural foods. In 1935 Warburg and Christian showed that nicotinic acid amide (niacinamide) is an essential constituent of a coenzyme concerned in hydrogen transport (oxidation-reduction system). Now two enzyme prosthetic groups containing niacinamide, coenzyme I and coenzyme II, are known. Niacinamide was isolated from a liver concentrate by Elvehjem and co-workers and they demonstrated that the or the free acid were effective cures for blacktongue in dogs and also that dogs could be maintained in a normal condition on the basal blacktongue-producing by the addition to the diet of small amounts of either compound. Very soon after publication of the initial paper on this subject by Elvehjem an co-workers the first report concerning the use of niacin in pellagrins appeared.

The success of the treatment heralded a new era in the clinical approach to treatment of pellagra. Nicotinic acid had been known to chemists since 1867, and many laboratories had supplies of the product during the many years that thousands and thousands of individuals were suffering from a lack of it. It is an ironical situation. By no logical reasoning, however, one can cast aspersions for the finding not having been made earlier.

Chemistry of Niacin

Niacin or nicotinic acid is pyridine β-carboxylic acid. Niacinamide or nicotinic acid amide is the acid amide. The structures of these and some related compounds referred to in the following pages are given herewith:

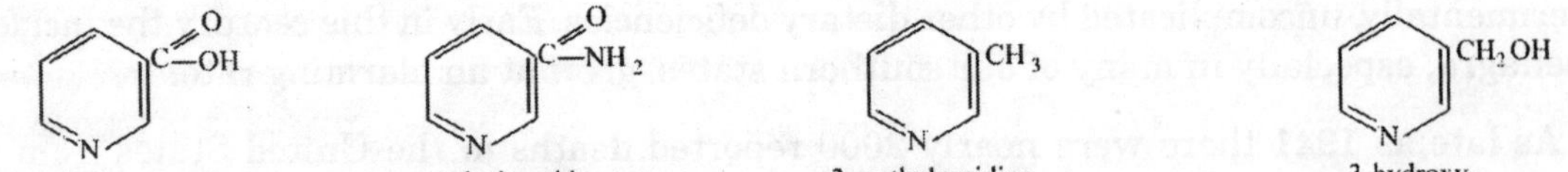

niacin (nicotinic acid) — niacinamide (nicotinic acid amide) — 3-methylpyridine (α-Picoline) — 3-hydroxy-methylpyridine

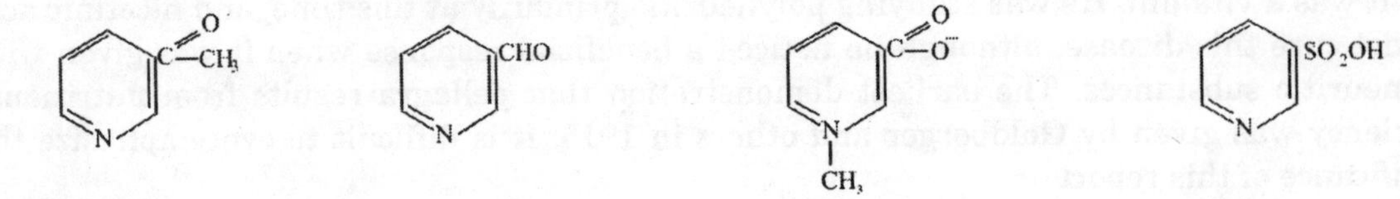

3-acetylpyridine — pyridyl-3-aldehyde — Trignelline (N-methylnicotinic acid betaine) — pyridine-3-sulfonic acid

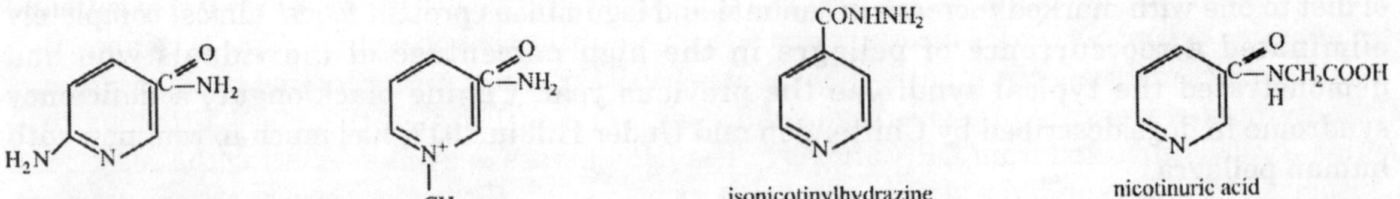

6-aminoicotinomide — N-methyl-nicotinamide — isonicotinylhydrazine (Isoniazid) — nicotinuric acid (nicotinylglycine)

Niacin and structurally related compounds

Niacin is readily prepared by oxidation of nicotine with strong oxidizing agents, such as permanganate or fuming nitric acid. Many other compounds, such as 3-ethylpyridine, can be oxidized to nicotinic acid. A synthesis from pyridine involves sulfonation of pyridine, distillation of the sodium salt of the 3-pyridine sulfonic acid with KCN to give the nitrile, and saponification of nicotinonitrile to yield nicotinic acid.

Bromination of pyridine previous to sulfonation markedly increases the yield. The amide is prepared by forcing NH_3 through a solution of the acid at elevated temperature. Niacin crystallizes as white needles. It is soluble in water to the extent of only about 1 per cent, but quite soluble in alkali (salt formation), alcohol, and glycerin. It forms such salts as the hydrochloride with the basic nitrogen, as well as metallic salts with the carboxyl group. The pure compound melts at about 236°C. In the human body niacin undergoes various changes.

Some of the vitamin is excreted unchanged in the urine, and a substantial portion (depending somewhat on intake) is methylated to N^1-melhylnicotinamide. Part of the latter is oxidized to the corresponding 6-pyridone (N^1-methyl-6-pyridone-3-carboxamide). Dogs excrete mainly N^1-methyl nicotinic acid and the betaine. Using C^{14} carboxy-labeled nicotinamide, Chang and Johnson identified the following metabolites in chickens: β-nicotinylglucuronic acid, nicotinuric acid (nicotinylglycine), both the 2- and the 5-nicotinyl ornithine, 2, 5-dinicotinylornithine, and nicotinic acid and the amide. Some decarboxylation of niacin occurs in animals also.

Coenzyme Activity

Nicotinamide is the form in which the vitamin is found in its physiologically active combinations. Two well-defined coenzymes contain nicotinamide. These act with a large group of hydrogen-transport enzymes and arc discussed specifically. The first is diphosphopyridine nucleotide (DPN), coenzyme I (Col), or NAD for nicotinamide adenine dinucleotide. It contains nicotinamide—ribose—phosphate—phosphate—ribose—adenine. Preiss and Handler consider the following reactions involved in the biosynthesis of DPN by enzymes in yeast, erythrocytes, or liver.

1. Nicotinic acid + 5-phosphoribosylpyrophosphate ⟶ nicotinic acid nucleotide + PP
2. Nicotinic acid nucleotide + ATP ⇌ desamido—DPN + PP
3. Desamido—DPN + glutamine + ATP + H_2O ⟶ DPN + glutamatc + AMP + PP

It should be noticed that amide formation (—NH_2 from glutamine) occurs following the completion of the dinucleotide structure. No enzyme capable of direct conversion of nicotinic acid to the amide has been found.

The other coenzyme containing niacin is triphosphopyridjne nucleotide (TPN), also called coenzyme II (Coll) or NADP for nicotinamide adenine dinucleotide phosphate. This is composed of nicotinamide—ribose—phosphate—phosphate—ribose-2′-phosphate adenine: It differs from DPN only in the presence of a third phosphate group on carbon 2 of the ribose attached to adenine. It is synthesized from DPN and ATP. The DPN kinase enzyme mediating the reaction was isolated from pigeon liver by Wang and others:

$$DPN + ATP \xrightarrow{Mg^{++}} TPN + ADP$$

The primary action of the two coenzymes is to remove hydrogen from substrates as part of dehydrogenase enzymes and transfer hydrogen and/or electrons to the next coenzyme in the chain or to another substrate which then becomes reduced. These enzymes are thus alternately oxidized and reduced. DPNH represents the reduced form. Many metabolic processes utilize one or other of these coenzymes. Discussion of the mechanism by which they operate can be found. The structure of DPN and TPN are shown herewith:

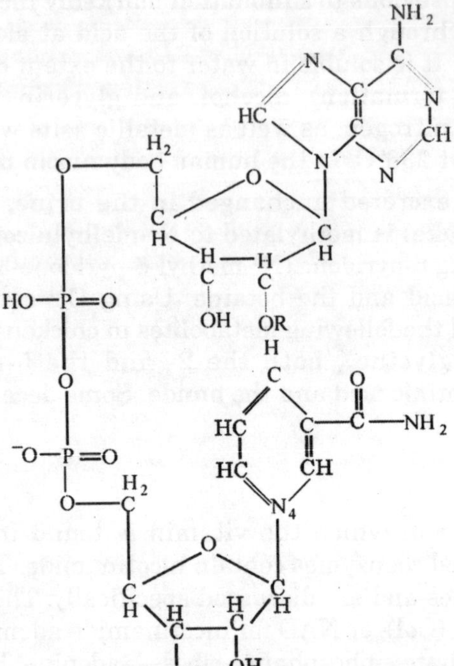

diphosphopyridine nucleotide, coenzyme I in which $R = H$ (NAD)

triphosphopyridine nucleotide, coenzyme II in which $R = PO(OH)_2$ (NADP)

Niacin Deficiency

In human pellagra the deficiency is generally not one of niacin alone. The response of pellagrins to niacin treatment is a partial one. The cheilosis, if present, may not be altered, and neurological symptoms may remain. The term pellagra can scarcely be thought of as indicating a specific set of symptoms. In other words, the symptoms in one individual case may be very different from those of another, and the response to various members of the B complex may show little resemblance in the two cases.

Variations in the pellagrous syndrome seen in different parts of the world are well known. Other B vitamins and general inanition are certainly involved. The symptom's generally associated with the condition are (*a*) dermatitis, accompanied by pigmentation (exaggerated by sunlight), and erythema of the tongue, which often becomes smooth and atrophic; (*b*) nervous lesions, especially in late stages, which involve varying regions of the cerebrum and spinal cord fibers (myelin degeneration); (*c*) gastrointestinal lesions characterized by cyst formation in the colon and atrophy in the later stages; and (*d*) mottled liver with fatty degeneration. The nervous symptoms especially are thought to arise from a deficiency other than that of niacin. They are similar, except in distribution, to the findings in thiamine deficiency. Studies of various tongue lesions in man have led to the conclusion that other members of the B complex are often deficient and that niacin frequently has little or no beneficial effect.

Goldsmith and others, however, produced clinical pellegra symptoms in humans and reported that all the lesions responded to nicotinamide or to tryptophan. Horwitt and others studied a group of men on low niacin and tryptophan intake over long periods of time. Other men received the same diet with supplements of the vitamin and tryptophan.

Physical changes other than those attributed to the accompanying riboflavin deficiency were not found in either group. No cutaneous lesions such as found in pellegra were noted. It was found from the studies on urinary excretion of niacin metabolites that the amount of N^2-methylnicotinamide in the urine correlated best with the niacin and tryptophan content of the diet. Pellegra is especially frequent among peoples eating corn as a large part of the diet. In some degree this may be explained by the low tryptophan content of corn, although there is more to the story than is known at present. This problem is discussed in some detail by Goldsmith. The flushing syndrome is well known in human beings taking nicotinic acid either by mouth or parenterally.

The amide does not produce these symptoms. It is the latter compound that is found in natural sources of the vitamin. Some individuals flush with only a few milligrams of niacin, while others are able to take 100 mg without untoward effects. Many people flush on the intravenous administration of 5 or 10 mg of niacin. In animals experimental niacin deficiency is readily produced. Blacktongue in the dog has been studied intensively, and this syndrome has a close resemblance to that seen in humans, especially with regard to the oral cavity. Special dietary precautions are required in rats, for example, to show disorders which respond to niacin administration. Diets high in corn or otherwise low in tryptophan predispose to this condition. Corn is low in tryptophan, and it has been suggested that this amino acid is required in the gastrointestinal tract for bacterial synthesis of niacin—an important source of the vitamin in this species.

The addition of tryptophan-containing protein, or the amino acid, to the diet of rats cures or prevents the symptoms. It is now established that in some species tryptophan can substitute in the diet for niacin. The conversion of the amino acid to the vitamin has been studied in great detail. A number of compounds arising in the body from tryptophan metabolism have been isolated. Several of these have niacin activity in rats and support growth in *neurospora*. The following scheme has been proposed by Hayaishi and others for the conversion of tryptophan into niacin in mammalian liver.

Tryptophan → kynurenine → 3-hydroxykynurenine → 3-hydroxyanthranilic acid →

2-acroleyl-3-aminofumaric acid → quinolinic acid → nicotinic acid (niacin)

Actually the quinolinic acid is converted to a nucleotide structure and then loses CO_2 to become niacin ribonucleotide, which is phosphorylated by ATP yielding deamido DNP. This reacts with glutamine forming DPN and thus the series of events constitutes the synthesis of the coenzyme from the amino acid.

The last steps are identical with the conversion of niacin into the coenzyme. The structures of these compounds can be found under tryptophan metabolism. In the rat the site of this transformation was thought to be the gastrointestinal tract. However, it is obvious now that other tissues participate in catalyzing this reaction. Henderson and Hanks removed the gastrointestinal tracts of rats from the duodenum to the anus and then gave them sterile mixtures of amino acids, carbohydrate, salts, etc., by subcutaneous injection.

Some of the animals (and unoperated controls) were given tryptophan in the food mixture. Analyses demonstrated the presence of increased amounts of N^1-methylnicotinamide. "free," and total niacin in the urine of both control and enterectomized rats given the amino acid. The N-methyl derivative is known to be formed from niacin in the body. Handley and Bond offered a clear-cut demonstration of this by feeding rats niacin containing C'^3 in the carboxyl group. Results of examination of the isolated and purified N-methylnicotinamide from the urine indicated that over 95 per cent of this compound had arisen from the administered niacin. These results establish the activity of body tissues other than the gastrointestinal tract in the conversion of this amino acid to the vitamin. Serum cholesterol is lowered in humans after the ingestion of large doses of nicotinic acid; the amide is without effect. In young adults the basal metabolism is increased also.

The mechanism by which blood cholesterol is lowered has not been resolved. In rats the blood levels of cholesterol are also decreased by massive intakes (up to 4 per cent of the diet). This effect was attributed to the great demand for methyl groups used in nicotinic acid detoxication, with a resulting decreased cholesterol synthesis.

Niacin Analogues

Ethyl nicotinate, nicotinic acid N-methylamide, β-picoline β-methyl pyridine) and 3-hydroxymethylpyridine have vitamin activity. The evidence indicates that these compounds are active on the basis of their conversion into nicotinic acid. Pyridyl-3-aldehyde added to the diet was effective in niacin-deficient ducks, but pyridyl-3-carbinol and β-picoline were inactive when fed with the diet. They were active given per, and all three were active upon injection.

Niacin Antagonists

A number of compounds structurally related to niacin act as antagonists in animals and/or in microorganisms. Pyridine-3-sulfonic acid, and 3-acetylpyridine are examples. It is probable that such antagonists are used in the synthesis of DPN or TPN analogues and that these combinations are not physiologically active. Thus the organism is deprived of its normal complement of these coenzymes. In most cases additional niacin overcomes the antagonism. The compound 6-aminonicotinamide produces niacin deficiency symptoms in rats at a level of 15 to 30 mg per kg of diet. The symptoms are prevented by the further addition of 150 to 300 mg of niacinamide per kg of ration. The 6-amino analogue is quite active against some tumors in rats, and this tumor inhibition is reversed by niacin-amide.

Occurrence of Niacin

In natural sources the vitamin occurs as free nicotinic acid, bound nicotinic acid, or nicotinamide in coenzyme form. Ghosh and others determined the content of these forms in a variety of natural food products. About 40 per cent of the nicotinic acid of oil seeds and most of it in cereals is present in bound form. Yeast, crustacea, fish, animal tissues, and milk contain principally free and amide forms of nicotinic acid. The vitamin is rather widely distributed in both plant and animal tissues. Certain animal organs, including liver, kidney, and heart, as well as lean meat and some fish flesh are outstanding sources. Yeasts, peanuts, wheat germ, and dried legumes are in the same category. Fresh legumes and a few green vegetables are good sources, whereas milk, eggs, and most fruits are generally classed as poor sources.

The niacin content of 160 Indian foodstuffs, many of which are common constituents of the American diet, has been reported. According to the data of this report, dried yeast may contain from 20 to 62, rice polishings around 28, wheat germ 7, sheep liver 15, and pork muscle 2.8, all in milligrams per cent. Milk and eggs contain only 0.1 mg per cent. Elvehjem and co-workers also gave an extensive list of foods and their niacin content. Niacin, like other B vitamins, resides in wheat primarily in the bran and middling fraction. These workers list the niacin-content of patent flour 0.80, fancy first clear flour 1.68, second clear flour 5.55, clean shorts 5.44, and wheat feed and screenings 19.2 mg per cent.

Niacin Requirement

It should be noted that in the table of recommended dietary allowances, the niacin allowance is stated in milligram equivalents. It has been established that in man about 60 mg of food tryptophan are equivalent to 1 mg niacin (due to conversion), so that the required niacin intake varies with the tryptophan content of the diet. A diet providing 60 g of mixed protein contains around 600 mg of tryptophan or 10 niacin equivalents from the amino acid.

On the basis of other studies and allowing for a liberal safety factor, the recommended daily allowance (1963 revision) has been set at 19 mg niacin equivalents for a 70 kg adult and up to 22 for boys 15 to 18 years old (68 kg). In a controlled study in four women the mean urinary excretions of N^1-methylnicotinamide and its 6-pyrone were 5.8 and 7.3 mg per day, respectively. With a diet providing 8.7 mg nicotinamide and 770 mg of tryptophan from 60 g of protein daily, these excretions indicate the adequacy of the 21 niacin equivalents ingested. More recent studies upheld the original value of about 60 mg tryptophan equivalent to 1 mg niacin in humans.

Determination of Niacin and Related Compounds

Dogs and chickens have been used for the biological assay of the vitamin. In these procedures total vitamin activity is determined. Microbiological methods are far more practical from the standpoint of time and expense. The present USP procedure employs *Lactobacillus arabinosus* as the test organism. The culture medium contains all essentials for growth except niacinactive compounds.

By graded addition of test material, the amounts added can be estimated by determining acid production and comparing this to acid produced upon addition of niacin. A microbiological method useful in the differential determination of nicotinic acid, bound nicotinic acid, and nicotinamide in natural materials was reported by Ghosh and others. Chemical methods are principally refinements of the original cyanogen bromide procedure adapted to niacin. After extraction of niacin-active substances, hydrolysis converts to niacin, which is reacted with CNBr and a reducing agent to yield a color.

The color density is compared to that produced from a known amount of niacin. Methods for a number of niacin metabolites can be found in the paper by Chang and co-workers. A sensitive spectrophotometric method for DPN and TPN in the oxidized and reduced forms was described by Clock and McLean.

Pantothenic Acid

In 1933 Williams and co-workers demonstrated the widespread distribution of a substance that acted as a growth factor for a yeast and other microorganisms. At that time these workers applied the name pantothenic acid (Greek, from everywhere) to the active principle. Later Williams group obtained 3 g of a crude pantothenic acid concentrate (40 per cent pure) from 250 kg of liver. Further purification yielded an amorphous product with over 11,000 times the activity (yeast growth) of a standard rice bran extract.

The presence of the vitamin was detectable in concentration of 5 parts in 10 billion parts of culture medium. Other investigators were studying concentrates of plant and animal tissues which showed various physiological activity. After some years the so-called chick antidermatitis factor from liver, also called the filtrate factor, was shown to be identical with Williams pantothenic acid by Woolley, Waisman, and Eivehem and also by Jukes.

The former investigators showed that the antidermatitis factor, like pantothenic acid, is alkali and heat-labile and is a hydroxy acid derivative of β-aianme. They were able to split β-alanine from the active compound and reactivate by coupling the remaining part of the molecule with synthetic β-alanine. In 1940 pantothenic acid was crystallized in pure form, its structure determined, and its synthesis accomplished by investigators of the Merck Laboratories.

Chemistry of Pantothenic Acid

Chemically, pantothenic acid is α, γ-dihydroxy-β, β-dimethylbutyryl- β′-alanide. The molecule is perhaps more easily comprehended if it is thought of as a dihydroxydimethylbutyric acid in peptide bond formation with β-alanine. The β′ indicates that the amino group is on the β carbon atom rather than on the ∝ carbon, as found in ordinary alanine. Many syntheses have been developed for this molecule.

The direct condensation of β-alanine with the lactone of the substituted butyric acid gives excellent yields and the desired product is obtained directly. Kagan and co-workers reported the synthesis of calcium pantothenate in high yield and in a high state of purity:

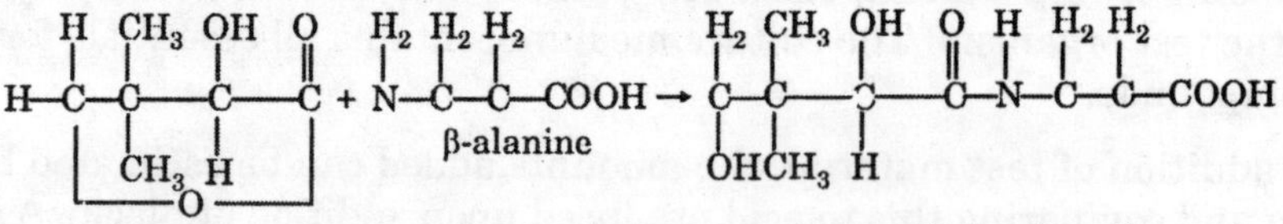

```
   H  CH3 OH  O    H2 H2 H2              H2 CH3  OH  O  H  H2 H2
   |   |   |  ||   |  |  |               |   |    |  ||  |  |  |
H--C---C---C--C + N--C--C--COOH  →      C---C----C--C--N--C--C--COOH
   |   |   |                             |   |    |
   |  CH3  H       β-alanine             OH CH3   H
   |_______O_____|
      lactone                                 pantothenic acid
```

$(CH_2OHC(CH_3)_2CHOHCONHCH_2CH_2CONHCH_2CH_2S)_2$

pantothine

$CH_3CHCHC(CH_3)_2CHOHCONHCH_2CH_2COOH$

ω-methylpantothenic acid, N-(α, gdihydroxy-b, β-dimethylvaleryl)-b-alanine

$CH_2CHC(CH_3)_2CHOHCONHCH_2CH_2S)_2$

bis (β-pantoylaminoethyl) disulfide

Pantothenic acid and related structures

The free acid is a viscous yellow oil, soluble in water and ethyl acetate, but insoluble in $CHC1_3$. It is acid, alkali, and heat-labile. It is readily available as either the sodium or calcium salt. The latter salt is crystalline material fairly soluble in water (7 g per 100ml) and insoluble in alcohol. It is quite stable, although autoclaving destroys the activity. $[\alpha]_D^{26}$ and for the free acid is +37.5° and for the calcium salt, +24.3°.

Coenzyme Activity

Pantothenic acid is a part of coenzyme A of Lippmann, characterized in 1951 by Lynen and co-workers. The CoA molecule is a nucleotide of the following composition:

adenine—phosphorylated ribose(C-3)— phosphate—phosphate—pantothenic acid—β-mercaptoethylamine.

The sulfhydryl group (—SH) is readily acylated to form a great variety of acyl derivatives for the purpose of transferring these groups in synthesizing reaction: *i.e.,* acetyl CoA + choline ⟶ acetylcholine. Also acyl groups are accepted by CoA in metabolism of substrate molecules: *i.e.,* pyruvate ⟶ ⟶ acetyl CoA + CO_2. CoA enters into metabolic pathways involving most classes of compounds, and discussion and reactions involving it can be found especially in the chapters on metabolism.

The biosynthesis of CoA is reviewed by Brown and Reynolds and by Cheldelin and Baich. Brown presented the scheme shown herewith. All of the enzymes required can be found in microorganism and in rat liver and kidney.

pantothenic acid + ATP ⟶ 4′-phosphopantothenic acid

4′-phosphopantothenic acid + cysteine ⟶ 4′-phosphopantothenylcysteine

4′-phosphopantothenylcysteine $\xrightarrow{-CO_2}$ 4′-phosphopantotheine

4′-phosphopantotheme + ATP ⟶ dephospho coenzyme A + PPi

dephospho coenzyme A + ATP ⟶ coenzyme A + ADP

A partial alternate route involves the reaction of 4′-phosphopantothenic acid with β-mercaptoethylamine to form 4′-phosphopantotheine. The CoA structure is shown herewith.

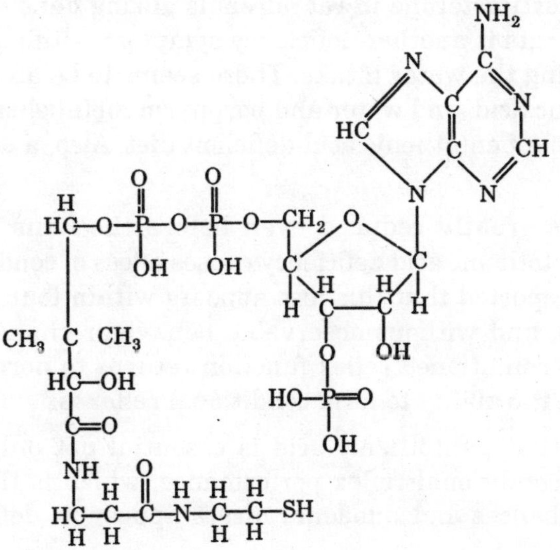

```
                                              NH2
                                              |
                                          N   C
                                            /   \\
                                          C       N
                                     HC   ||      |
                                          C       CH
                                            \\   /
           O       O                      N       N
    H      ||      ||                     |
    HC—O—P—O—P—O—CH2   O
    |      |       |       \\   /  \\
    |      OH      OH       C       C
    |                       |H     H|
    |                     H C —— C H
 CH3 C CH3                  |      |
    |                       O      OH
    HC—OH                   |
    |                   HO—P=O
    C=O                     |
    |                       OH
    NH     O
    |  H   ||  H  H  H
    HC—C—C—N—C—C—SH
    H  H       H  H
```

The dephospho CoA lacks a phosphate group on carbon 3 of the ribose residue. Pantotheine is the part of the coenzyme A molecule composed of pantothenic acid in combination with β-mercaptoethylamine. This structure is also called *Lactobacillus bulgaricus* factor (LBF) because of its relation to growth of this organism. Pantothine is the disulfide compound and bears the same relation to pantotheine that cystine does to cysteine.

The structure and synthesis were reported in 1950. LBF is a growth factor for various microorganisms and can replace pantothenic acid in the diet of various animals. It is widely distributed in nature and constitutes a second bound form of the vitamin.

Pantothenic Acid Deficiency

All species of animals investigated require pantothenic acid. The fox, mouse, monkey, rat, pig, dog, chick, and other species have been studied in this respect. Microorganisms also need the vitamin, although some are able to synthesize it and thus do not need an external supply. In the rat certain fairly well-defined symptoms of deficiency are observed. Retarded growth in young animals is seen.

Reproduction is seriously affected. It was demonstrated that institution of a pantothenic acid deficiency as late as the day of mating resulted in failure of implantation, resorption, or defective litters. Achromotrichia, or graying of hair, in black rats (and foxes) is related to but not due solely to a pantothenic acid deficiency. Other deficiencies, perhaps including inositol, are involved. The condition has not been produced in rats after adrenalectomy. Adrenal cortical necrosis results from continued deprivation of the vitamin. Rats suffering such a pathological condition exhibit serious abnormalities in salt and water metabolism. The cholesterol content of the adrenals is markedly lowered, indicating decreased ability on the part of the rat to synthesize this compound under the stress of pantothenic acid deficiency.

Cholesterol is the likely precursor of adrenal cortical hormones, and coenzyme A is known to participate in the utilization of acetate in cholesterol formation. Adrenal cortical secretion is decreased in pantothenic acid deficiency, and Langwell and co-workers demonstrated a reduced production of corticosterone in rat adrenals during deficiency. The porphyrin-caked or "bloody" whiskers in rats is another deficiency symptom, although a similar condition can be produced by restricting the water intake. There seems to be an important interrelation of the adrenals, pantothenic acid, and water and porphyrin metabolism. In chickens growth and appetite are retarded on a pantothenic acid-deficient diet. Also, a dermatitis and a fatty liver condition develop.

Egg hatchability is greatly reduced even before the hens demonstrate any visible symptoms. In dogs a pantothenic acid deficiency causes a loss of conditional reflex performance. Gantt and co-workers reported that this loss appears within four to ten days prior to other neurological symptoms, and without observable behaviour changes. Upon addition of the vitamin to the diet, the conditioned reflex function returns to normal. In rats a pantothenic acid deficiency lessened the ability to form conditional reflexes.

The authors state that pantothenic acid is essential not only for physical health but apparently for normal conditional reflex performance, which is the basis of mental health. Gastrointestinal disturbances and duodenal ulcers appear in deficient rats, and in severe deficiency there is degeneration of testicular interstitial tissue. A reduction of 30 to 40 per cent in coenzyme A content was found in tissues of pantothenic acid-deficient rats.

The utilization of pyruvate by liver tissue was also decreased. In deficient ducks the injection of pantothenate markedly increased the coenzyme. A content of liver and the ability of liver slices to utilize pyruvate. Deficient rats also show a decreased ability to acetylate administered β-amino-benzoic acid. However, it was reported that in such animals the ability to acetylate a test dose of an aromatic amine was restored in as short a time as three hours after injection of pantothenate. Pantothenyl alcohol (alcohol analogue) is growth promoting and prevents achromotrichia in rats. In man it is just as efficient as the vitamin in increasing pantothenic acid excretion in urine after a test dose. These facts indicate that the alcohol is converted to the acid (vitamin) in the body. The alcohol, however, does not support growth of *Lactobacillus arabinosus,* and so it seems that this organism cannot carry out the conversion.

Deficient rats respond to pantotheine, pantothine, pantothenylcysteine, and pantothenylcystine. Animal deficiency symptoms have been reviewed by Wagner and Folkers. Pantothenic acid deficiency in man is known only in the cases in which it has been produced experimentally. This undoubtedly results from the wide distribution of the vitamin in, plant and animal foods, and from the fact that some microorganisms, such as *E. coli,* synthesize quantities of the vitamin and excrete the excess above their needs. In man this organism is normally found in the gastro-intestinal tract. This is an important contributing factor in the absence of human pantothenic acid deficiency. The extravagant claims for pantothenic acid as an anti—gray hair factor for humans have been adequately disproved in carefully controlled experiments. Hodges and others restricted the pantothenic acid intake of two young men. Two others consumed the same deficient diet plus 1000 mg daily of the antagonist, ω-methylpantothenic acid. A third group of two men received the diet plus 20 mg of pantothenic acid daily. After a few weeks the antagonist group developed serious personality changes, including irritability and restlessness.

Alternate periods of somnolence and insomnia and excessive fatigue after mild exercise were experienced. A little later the men in the deficient group noted similar complaints, and soon the symptoms were indistinguishable in the two groups. They developed a. staggering gait and gastrointestinal symptoms became common. Biochemical changes reflected by laboratory tests were not prominent. The loss of eosinopenic response to adrenocorticotropic hormone (ACTH) was the most consistent. The men in the control group showed none of the foregoing symptoms. Massive doses of pantothenic acid to the four men who were deficient corrected the faulty eosinopenic response and alleviated most of the clinical symptoms.

Pantothenic acid Antagonists

Pantoyltaurine has been studied extensively as an antimetabolite of pantothenic acid. Inclusion of this compound in the diet of animals leads to deficiency symptoms. Twenty-five, derivatives of this compound were prepared and studied for inhibitory activity. Another antagonist that has been studied in various species is ω-methylpantothenic acid. This compound interferes with the formation of CoA and produces typical deficiency symptoms in animals.

This compound was used to produce human pantothenic acid deficiency discussed previously. Antagonists of pantothenic acid probably all act by inhibiting normal functions of CoA. It is reasonable to suppose that of the many functions of this coenzyme, some would be affected to a greater extent than others, due to different degrees of coenzyme-apoenzyme and enzyme-substrate affinity, etc.

Dietrich and Shapiro showed that ω-methylpantathine markedly inhibited sulfanilamide acetylation at concentrations which were ineffective against citrate formation in pigeon liver homogenates. On the other hand, bis (β-pantoylaminoethyl) disulfide was found to inhibit citrate formation at concentrations which did not affect sulfanilamide acetylation. The action of 6-mercaptopurine in blocking nuclear mitosis in animals can be reversed by administration of pantothenic acid or coenzyme A.

Occurrence of Pantothenic Acid

The distribution of this vitamin is so widespread in nature that little comment is required. It is important, however, to understand that in large degree pantothenic acid is found in natural products in bound forms. Yeast, liver, and eggs are outstanding sources. Some meats, skim milk, and sweet potatoes contain moderate amounts of the vitamin, while most other vegetables and fruits have considerably less. Zook and associates reported the pantothenic acid content of a large number of foods using newer methods which release bound forms of the vitamin.

Requirements of Pantothenic Acid

Chicks require a dietary source of pantothenic acid, although laying hens get along and continue to lay eggs on diets containing very little of the vitamin. The eggs produced under these conditions are poor in hatchability and in fact reach a level of zero in this respect. A careful study of the pantothenic acid requirement for laying hens indicates that about 650 μg per 100 g of ration is sufficient for the production of eggs of high hatchabiltiy. It is interesting that for considerable periods of observation only about 150 μg per 100 g of ration is sufficient for maintenance of the adult chicken.

The requirements for various other species are more or less established. The human requirement is largely unknown. Estimates of 10 to 15 mg per day seem to be met easily even with diets classed below optimum. The estimates lack significance because of the many unknown factors, such as the degree of release of the bound vitamin in the intestinal tract and the amount of intestinal bacterial synthesis of the vitamin as well as the availability of this source. Goldsmith and others reported that high-cost 3000 Cal diets contained 5 to 11 mg of free and 13 to 19 mg of total pantothenic acid, while low-cost 2500 Cal diets contained 4 to 12mg free and 9 to 20 mg total vitamin. Normal adults excrete 1 to 7 mg of pantothenic acid in the urine daily.

Assay of Pantothenic Acid

The growth rate of chicks has been used to determine the pantothenic acid content of various products. It is a time-consuming procedure, and although still in use, it has largely been replaced by microbiological methods, especially since the development of more nearly optimal methods for releasing the bound vitamin from its firm combination (such as coenzyme A) in plant and animal materials. Certain organisms show a growth response (vitamin effect) to the β-alanine part of the molecule and others to the pantoic acid methyl. Such organisms synthesize pantothenic acid if supplied with a part of the molecule. An improved technique using *L. casei* was described by Clarke.

A method for estimating CoA in tissues, involving enzymatic liberation of pantotheine and microbiological assay of this substance and free pantothenic acid, was reported by Wolff and others. Brown has devised methods for the estimation of pantothenic acid, phosphopantotheric acid, pantotheine phospho enzyme and coenzyme A in a single sample. The major pantothenic acid-containing found in animal tissues and in microorganisms was CoA, although phosphopamezeme to from 10 to 25 per cent that of CoA in some tissues and was actually higher kidney.

Pyridoxine, Pyridoxal, Pyridoxamine, Vitamin B_6

The complex nature of the B vitamins was further unraveled through the discovery of pyridoxine. Other names applied to the principle rat acrodynia factor, rat antidermatitis factor, included and vitamin H. The was used some in the European literature. Gyorgy showed that an eluate of adsorbate of yeast extract was able to cure a type of dermatitis in rats that developed on synthetic rations containing B_1 (thiamine) and B_2 (lactoflavin).

This condition differ from that found in the absence of the pellagra preventive factor (then thoozyl to be B_2) and was characterized by denuding around the paws, nose, and mouth. Thus at least two antidermatitis factors for the rat became known and the dual nature of called B_2 preparations was established. Various groups of workers isolated the compound from natural sources in 1938. Keresztesy and Stevens, for instance, isolated the pure substance from rice polishings and studied certain properties of the material. Its structure was elucidated by Stiller and others and by Harris and co-worker in this country.

Details of one method of synthesis were described by Harris and Folkers in 1939. The synthesis was also accomplished at about the same time by a group of German workers headed by Kuhn. The name pyridoxine was given to the compound at this time. There are three compounds, interconvertible in the animal body, which possess vitamin B_6 active. In general, pyridoxine is a general term indicating vitamin B_6 active material. Specifically this is not correct, since pyridoxine is a definite compound, as are pyridoxal and pyridoxamine, and all three have B_6 activity. The phosphorylated compounds likewise have vitamin activity in animal.

Chemistry of Pyridoxine

Pyridoxine is a basic substance. The colorless crystals melt at 160°C. The compound is soluble in alcohol and water, but only slightly soluble in ether or chloroform. The vitamin is generally used as the hydrochloride salt which melts at 206° to 208°C with some decomposition. This salt is highly soluble in water (pH about 3), less soluble in alcohol, and insoluble in ether. The vitamin shows marked absorpiion in the ultraviolet range. Such studies were of value in determining the structure of the molecule. The total synthesis is complicated and involves

The total synthesis is complicated and involves many steps. Syntheses from somewhat related and more readily available derivatives are numerous.

The structures of pyridoxine and some related compounds are shown herewith :

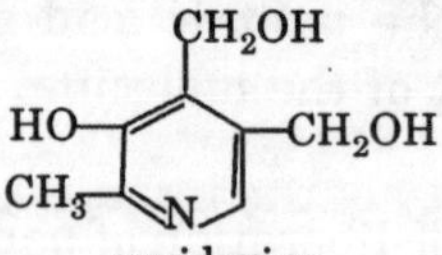

pyridoxine 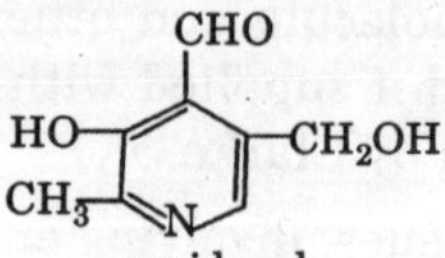 pyridoxal 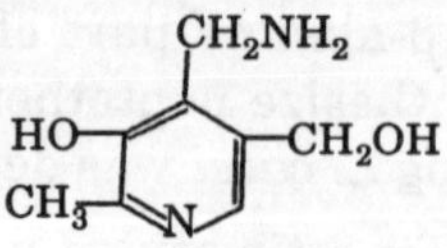 pyridoxamine

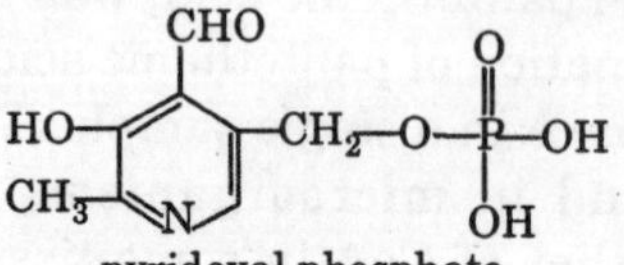

pyridoxal phosphate 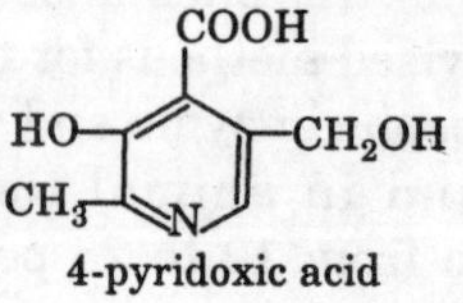 4-pyridoxic acid

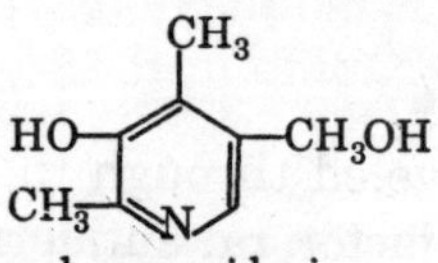

desoxpyridoxine 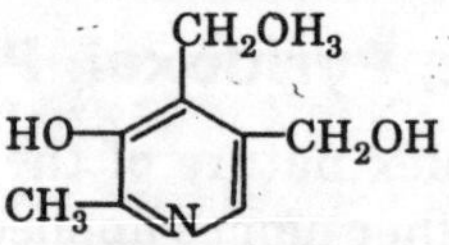 methoxypyridoxine

Pyridoxine and related compounds

Pyridoxine Deficiency

The early description of a pyridoxine deficiency in rats was one of multiple deficiency; other unknown factors were absent from the synthetic diets. At the present there is recognized in rats a specific syndrome representing pyridoxine want. This includes acrodynia, or a typical dermatitis which is generally symmetrical and affects the paws and various parts of the head. Seborrheic lesions are frequent. Edema of the connective tissue layer of the skin is thought to be characteristic. Loss of muscle tonus was observed after long-continued deprivation in rats, and convulsive seizures have been observed. Growth is subnormal on pyridoxine-deficient diets.

All animal species studied show deficiency symptoms, but these vary considerably from one type of animal to another. Mice, for instance, develop fatty livers on a B_6 deficiency, while in pigs a microcytic, hypochromic anemia is a characteristic part of the syndrome.

Dogs develop hypochromic anemia, high serum iron levels, and atherosclerosis. Monkeys develop dermatitis, lymphocytopenia, and atherosclerosis. Many species, including man, excrete extra xanthurenic acid in the urine during a deficiency state following a test dose of tryptophan.

Human B_6 deficiency may be more prevalent than was generally recognized. Hunt reviewed a number of cases of apparent B_6 deficiency. In 1951 an out-break of deficiency in infants was observed in various parts of the country. The cardinal symptom in these infants was convulsions, although only a small per cent of the infants developed the deficiency to such an extent. It was soon determined that the deficiency resulted from the use of a commercial liquid, infant-feeding mixture, which during processing had had the B_6 content reduced to very low levels. A. change in diet or administration of pyridoxine promptly relieved the symptoms.

In normal tryptophan metabolism very small amounts of xanthurenic acid air formed and excreted in the urine. In B_6 deficiency larger amounts are excreted. The determination of urinary xanthurenic acid after administering tryptophan to a human or animal (load test) gives an indication of the nutritional status with respect to vitamin B_6. It has been found by Bessey and co-workers and others that some infants require more pyridoxine than others to prevent xanthurenic acid excretion after a tryptophan load. In Hunt's review it is pointed out that some infants given quantities of B_6 sufficient for the majority of the infantile population are likely to suffer convulsions.

A number of experimental human deficiencies have been reported. Cheslock and McCully maintained eight college students on a diet low in B_6 for 52 days. The blood level of B_6 dropped to zero within four weeks. Lymphocyte counts decreased in five of the subjects, but no dermal symptoms were evident. Xanthurenic acid excretion increased markedly after a tryptophan load.

Vilter and others induced human deficiency with the antagonist deoxypyridoxine. Lymphopenia (reduced,white cell count) was the most common finding in the blood studies. Clinical symptoms included seborrheic dermatitis, usually about the eyes, in the eyebrows, and at the angles of the mouth.

Hodges and co-workers also produced a deficiency state in humans and found a slight impairment in antibody formation. Large amounts of xanthurenic acid were excreted after tryptophan ingestion, but the ability to convert the amino acid into N-methylnicotinamide was not altered. Will and associates showed low transaminase blood levels in such patients. Supplements of other B vitamins had no effect on skin, mucous membranes, or peripheral nerve lesions, whereas all lesions responded promptly to 5 mg of pyridoxine, pyridoxal, or pyridoxamine daily.

A case of hypochromic anemia that did not respond to iron therapy was successfully treated with 20 mg per day of pyridoxine given orally.

Coenzyme Activity

The active form of pyridoxine is the coenzyme pyridoxal phosphate. The phosphates of pyridoxine and of pyridoxamine are also found naturally.

The synthesis of the coenzyme from pyridoxal and ATP was demonstrated by Hurwitz:

$$\text{pyridoxal} + \text{ATPpyridoxal} \rightarrow \text{phosphate} + \text{ADP}$$

Human brain is a very rich source of pyridoxal kinase, the phosphorylating enzyme. The enzyme mediated phosphorylation of all three forms of the vitamin and was activated strongly by Zn^{++}.

The ever increasing number of molecular interconversions mediated by pyridoxal phosphate-dependent enzymes attests to the significant metabolic role of this vitamin.

The coenzyme is required in the decarboxylation of amino acids. In this function it is referred to as codecar boxylase to distinguish it from cocarboxylase, which is thiamine pyrophosphate, involved in α-keto acid decarboxylations. Some of the amino acid decarboxylases have been studied in detail. Histidine decarboxylase, for example, produces histamine and CO_2 from histidinc. The general reaction is

$$RCHNH_2COOH \rightarrow RCH_2NH_2 + CO_2$$

A number of amino acid derivatives not found in proteins are also decarboxylated by enzymes requiring pyridoxal phosphate. An example of this is the demonstration by Weissbach and co-workers of the functional role of the coenzyme in the decarboxylation of 5-hydroxytryptophan with the production of 5-hydroxytryptamine (serotonin) plus CO_2. Another step in tryptophan metabolism involves the pyridoxal phosphate-dependent enzyme kinureninase, which catalyzes the conversion of kynurenine to anthranilic acid. The significance of these reactions is discussed under tryptophan metabolism in Chapter 25 on protein metabolism.

Cysteinesulfinic acid decarboxylase is another, example of a decarboxylase using a nonprotein amino acid as substrate. In general the amino acid decarboxylases are highly specific and can be used in the determination of various amino acids by following CO_2 production in purified systems. The wide scope of amino acid transamination reactions was established by Cammarata and Cohen, who studied 22 transaminase reactions in animal tissues and suggested that each involved a distinct enzyme with pyridoxal phosphate as, coenzyme. Typical of these enzymes is glutamic-oxalacetic transaminase (glutamicaspartic transaminase) which mediates the reaction

$$\text{glutamate} + \text{oxalacetate} \rightleftharpoons \alpha\text{-ketoglufarate} + \text{aspartate}$$

Diamine oxidase brings about oxidative deamination of certain diamines, such as cadaverine. Pig kidney histaminase and diamine oxidase appear to be identical, with pyridoxal phosphate as coenzyme in both cases. Cysteine desulfhydrase is another pyridoxal phosphate enzyme. It converts cysteine to pyruvate and $H_2S + NH_3$.

A number of dehydrases found in animal tissues require pyridoxal phosphate and remove water from hydroxy amino acids. Examples are serine dehydrase (yields pyruvate and NH_3 + H_2O) and threonine dehydrase (yields α-ketobutyric acid and NH_3 + H_2O). The coenzyme is required, among other factors, in the synthesis of δ-amino levulinic acid, a requisite intermediate in porphyrin synthesis. An interesting role of the coenzyme is found in bacterial enzymes responsible for racemization of certain amino acids. For instance, *S. faecalis* requires D-alanine for cell wall construction.

The organism can convert L-alanine to the D form, providing pyridoxal phosphate is supplied, since this is the coenzyme of alanine racemase. The presence of 4 mols of pyridoxal phosphate per mol of phosphorylase-a and 2 mols per mol of phosphorylase-b was reported by Cori and Illingsworth . The significance of the presence of the coenzyme in this system is not clear. Human muscle phosphorylase-b contains 2 mols of pyridoxal phosphate and phosphorylase-a 4 mols on the basis of a molecular weight of 250,000 for the b and 500,000 for the a. From the foregoing it is clear that the pyridoxine-active compounds are associated with most of the nonoxidative metabolic changes of amino acids, and with some pathways not involving amino acids.

Further details and proposed mechanisms of action of pyridoxal phosphate enzymes can be found in. Pyridoxine and pyridoxamine can be converted in the body to pyridoxal. An enzyme system in rat liver catalyzes the oxidation of pyridoxamine or its phosphate to pyridoxal or the phosphate. An example of a biochemical deficiency related to a functional role of a vitamin or coenzyme is the decreased production of taurocholic acid in B_6 deficient rats, Doisy and co-workers showed that dietary correction resulted in increased production of this bile salt. Pyridoxal phosphate is required in the enzyme system catalyzing the decarbcxylation of cysteinesulfonic acid which is a source of taurine. The latter compound reacts with cholic acid to form taurocholic acid. A decreased activity of specific transaminase enzymes in a B_6 deficiency would be expected. This has been demonstrated in various species. Glutamicaspartic transaminase was decreased in deficient rats. In a deficiency state the decreased transaminase activity of tissues shunts greater amounts of amino acids into other metabolic pathways, such as deamination, accounting for the increased urea production under these conditions.

Vitamin B_6 Antagonists

Desoxypyridoxine (2, 4-dimethyl-3-hydroxy-5-hydroxymethylpyridine) has a potent B_6 antagonistic action. Its use in animal diets is of value in accentuating B_6 deficiency symptoms and decreasing the time required to develop them. The compound is a strong tyrosine decarboxylase inhibitor, for example, but only in the phosphorylated form. Methoxypyridoxine is another antagonist that has been studied in detail. Toxopyrimidine (2-methyl-4-amino-5-hydroxymethyl pyrimidine) administration lowers the B_6 levels of rat tissues and produces liver damage.

Glutamic acid decarboxylase is inhibited by toxopyrimidine, and the product of the reaction of this enzyme, γ-aminobutyric acid, is lowered in brain tissue of rats just prior to convulsions. Injection of β-hydroxy-γ-aminobutyric acid, but not of γ-aminobutyric acid, suppressed the convulsions resulting from toxopyrimidine. The compound isonicotinic acid hrydrazide (INH) in combination with other drugs has been used successfully in the treatment of tuberculosis. Daily doses of 300 to 900 mg caused increased excretion of kynurenine and xanthurenic aeid in tryptophan load tests, according to Price and co-workers. The structure of this drug is given with those of B_6 and related compounds. It is chemically similar to pyridoxine and was thought to opeiate as α-vitamin antagonist.

It is known now that it reacts with pyridoxal forming a hydrazone (aldehyde hydrazone) combining the aldehyde form of the vitamin, RCHO, and the free amino group of the hydrazine, forming RCH=N—NHR', the hydrazone, which is excreted. Heilbron considers that such a reaction cannot be the only mechanism involved in this B_6 enzyme, inhibition.

Fate of Vitamin B_6

In man the B_6 active compounds are largely converted to pyridoxic acid and excreted as such in the urine. In women 75 to 85 per cent of an average daily intake of 0.86 mg (range 0.52 to 1.2 mg) of B_8 was excreted as pyridoxic acid.

Pyridoxine Requirements

All animal species so far studied require B_6. The dietary requirement of mice is increased by feeding rations high in protein. After depletion mice respond better to pyridoxine than to the aldehyde or the amine. This is puzzling, since only the latter two have been found in the coenzymes. Many microorganisms required a source of B_6, although in some instances the presence of a specific amino acid in the culture medium may abolish this requirement. The human requirement is not firmly established. Infants apparently require in the neighbourhood of 0.3 mg per day. In adults 0.5 mg daily was found to be insufficient.

A figure of around 1.5 to 2.0 mg per day is a reasonable estimate for adults and readily obtainable from most diets, according to the Food and Nutrition Board, National Research Council, 1963 revision. It is still questionable whether man makes use of B_6 compounds synthesized by intestinal microorganisms. Reports of excretion of B_6 and metabolites in greater quantity than the intake are numerous.

Occurrence of Pyridoxine

The best natural sources of pyridoxine are those foods containing other members of the B complex. Yeast, rice polishings, the germ of various grains and seeds and egg yolk are outstanding sources. The vitamin is widely distributed in other foods of both plant and animal nature. In many of its sources pyridoxine is chemically bound to protein and is present in liver, for example, mainly as pyridoxal and pyridoxamine protein complexes.

Determination of Pyridoxine

Both rats and chicks have been used in the biological methods. The rat growth method is used more than other biological methods. It consists of a depletion period on a specific ration and then a comparison of weight gain of groups fed this ration, supplemented with various levels of pure pyridoxine, and other groups fed the basal ration, plus at least two levels of material under test. Microbiological methods are rapid and accurate. Cheslock reported 100 per cent recovery of pyridoxine added to blood nitrates in her modification of the method of Parrish and co-workers.

Some microorganisms do not respond equally to the different forms of B_6, and in some cases they show no response to one or two of the forms; thus methods for the different molecules are available. An enzymatic procedure for pyridoxal phosphate capable of estimating as little as 0.03 μg was developed by Wada and others. Pyridoxal, pyridoxamine, and pyridoxine are not active in the test. A sensitive fluorometric method has been used for pyridoxic acid, the main excretory product of pyridoxine.

Biotin

Biotin, one of the members of the group of B vitamins, has been known by a variety of names including bios, vitamin H, and coenzyme R. As early as 1916 the toxicity of diets high in egg white was observed. Some years later Boas described egg white injury in rats fed diets containing raw egg white as the source of protein. She described muscle incoordination, dermatitis, loss of hair, and nervous manifestations as symptoms of this syndrome. She observed that cooked egg white was not toxic and that liver, yeast, and certain other foods apparently contained a substance that protected rats against the toxicity of the raw egg protein.

The protective substance was called vitamin H by Gyorgy. Lease and Parsons showed that chicks were subject to egg white injury. Williams and co-workers some six years later demonstrated that egg white injury in rats and in chicks was actually due to an antivitamin in egg white. This substance, known as avidin, is a basic protein, and its ability to inactivate biotin was confirmed in 1941. Previous to this time a potent growth stimulant for yeast had bern isolated from dried egg yolk by Kögl and Tonnis and named biotin. Coenzyme R was the name given another growth factor isolated in 1933. It was in 1940 that György. du Vigneaud, and their co-workers announced the identity of vitamin H (anti-egg white injury factor) coenzyme R, and biotin. Du Vigneaud and co-workers characterized biotin and published its structure in 1942. Harris and others announced the synthesis of *d-biotin* in 1943. Improvements in the synthesis have been numerous, and at present biotin is readily available to the research worker. The elucidation of the nature of biotin is unique among vitamin studies.

Real progress developed after the demonstration of the toxic effect of egg white, and then it was shown that foods contained a protective substance, vitamin-like in nature. Later it was demonstrated that avidin of egg white combines with and inactivates biotin of foods and perhaps that produced by bacteria in the intestine leading to a dietary deficiency.

Chemistry of Biotin

Biotin *(d* or natural isomer) is a monocarboxylic acid, only slightly soluble in water (0.03 to 0.04 g per 100 ml at 25°C and 1 g per 100 ml at 100°C) and alcohol (0.06 g per 100 ml at 25°C). Salts of the acid are quite soluble; the sodium salt can be prepared in 20 per cent aqueous solution. The free acid is practically insoluble in acetone and ether. The colorless crystalline needles melt at 231° to 232°C. Water solutions (pH 4-9) are stable at 100°C, and the dry material is also thermostable and photostable.

The vitamin is destroyed by acids and alkalies only on rigorous treatment and by oxidizing agents such as peroxide and permanganate. The specific rotation, $[\alpha]_D^{26}$, is 91.0° in 0.1 *N* NaOH, and it shows maximum absorption in the ultraviolet at a wavelength of 234 mμ.

The structures of biotin and some related compounds are shown herewith:

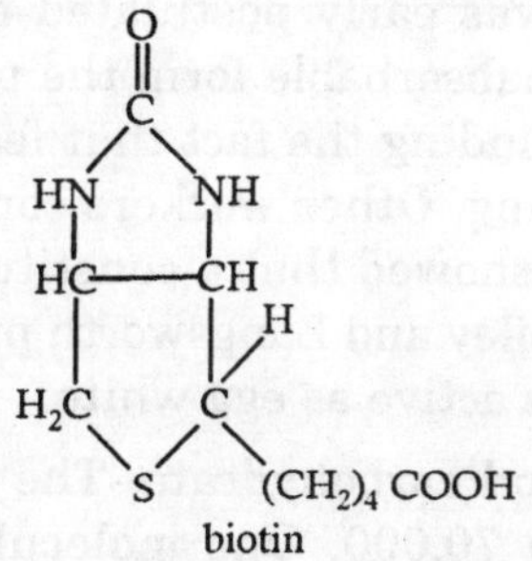

biotin

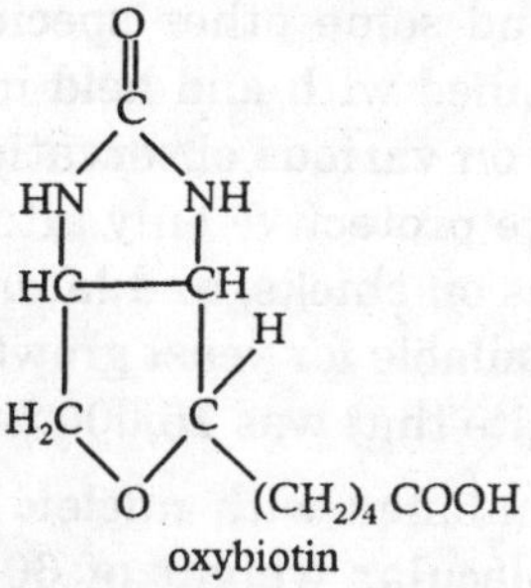

oxybiotin

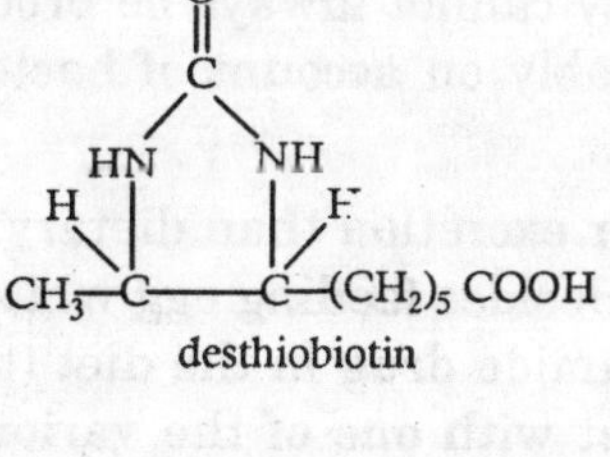

desthiobiotin

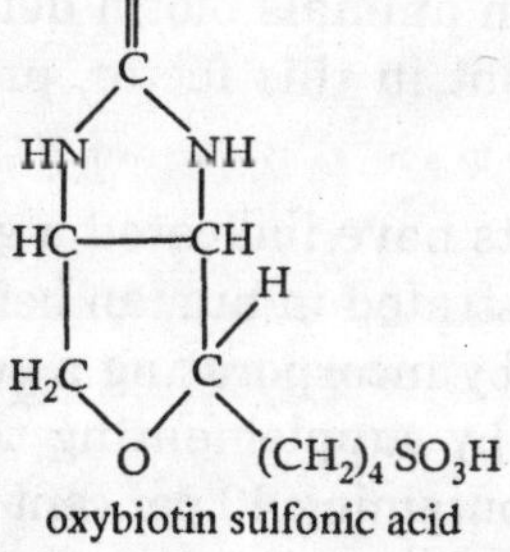

oxybiotin sulfonic acid

O=C(O⁻)–N / NH, HC–CH, H₂C, C, S, $(CH_2)_4COOH$

biotin-CO_2 combination

biotin coenzyme [12]

Biotin and related compounds

Biocytin is a term designating a bound form of biotin first isolated from yeast by Wright and co-workers. It was identified as ∈-N-biotinyl-Iysine and later synthesized. It occurs in plant and animal tissues also. The synthetic material reacts in the same way as naturally occurring biocytin with respect to microbiological activity, combination with avidin, hydrolysis rates, infrared absorption, and other criteria.

In biocytin the ∈-amino group of lysine and the carboxyl of biotin are combined in a manner similar to that in a peptide bond. Another bound form of the vitamin—soluble bound biotin—contained in peptic digests of hog liver and other tissues was found to be convertible to free biotin by an enzyme-from liver named biotinidase. Biotin has been synthesized by various procedures. There are several isomers of biotin, but only the one designated (+)-biotin is biologically active.

Biotin Deficiency

Various microorganisms, including bacteria, yeasts, and molds, require biotin; it is a vitamin for these organisms. It is also a required dietary vitamin for man, monkey, dog, rat, rabbit, turkeys, chickens, and some other species. It was early postulated that egg white in the intestinal tract combined with and held in an unabsorbable form the protective factor of the diet. This was based on various observations including the fact that feces from rats showing egg white injury were protective only after heating. Other workers concluded somewhat the same thing in studies on chicks, and later they showed that a constituent of rave-egg white rendered biotin unavailable for yeast growth. Wooiley and Langsworth prepared an amorphous fraction from egg white that was 15,000 times ,as active as egg white.

Avidin occurs combined with nucleic acid and carbohydrate. The free protein is highly basic and has a molecular weight of 60,000 to 70,000. One molecule combines with two molecules of biotin. In animals biotin deficiency cannot always be produced by maintaining them on diets deficient in this factor, presumably on account of bacterial synthesis in the intestinal tract.

Many experiments have indicated a greater excretion than dietary intake of biotin. This has also been demonstrated in human beings. Besides feeding egg white diets, the symptoms have been produced by incorporating a sulfonamide drug in the diet (to minimize intestinal bacterial growth) or by supplementing the diet with one of the various biotin antagonists. Hamsters fed a diet containing 40 per cent egg white supplemented with sulfaguanidine showed extreme deficiency in six weeks. Weight gain was far below normal, and the animals were irritable and nervous. They showed jerky movements and dragged their hind legs. The eyes were sealed shut with incrustations, and the mouth and nose were swollen. They became nearly hairless.

Symptoms disappeared rapidly following daily injections of 4 μg of biotin. Boas fed rats on diets high in raw egg white and observed a dermatitis, loss of hair, and loss of muscular control. Specific studies on nervous tissue and muscle of deficient animals were reported by Sullivan and others. Experimental studies on human biotin deficiency have been reported. Sydenstricker and associates fed egg white-containing diets to adult volunteers. After three to four weeks a transient dermatitis was seen, and shortly thereafter lassitude, anorexia, muscle pains, and hyperesthesia were observed.

The administration of 150 to 300 μg daily of a biotin concentrate relieved these symptoms in a few days. During the course of the experiment the urinary excretion of biotin decreased and rose abruptly upon therapy with the vitamin. Other investigators have not confirmed the production of human biotin deficiency symptoms by feeding egg white or avidin concentrates. Follis has reviewed the symptomatology of deficiency in animals and in man.

Coenzyme Activity

The action of biotin as a required component in some specific carboxylations and decarboxylations is well known. Oxalacetate stimulates the growth of *L. arabinosus* in a medium deficient in both aspartic acid and biotin, This probably results from the failure of the deficient organism to condense pyruvic acid with CO_2 to form oxaiacetate, which could then be transaminated, forming aspartic acid. It has been demonstrated that $C^{14}O_2$ from $NaHC^{14}O_3$ was incorporated into aspartic acid to a much smaller extent by livers from biotin-deficient chicks than by livers from normal chicks. Biotin is known to constitute the coenzyme of a variety of carboxylating (CO_2 fixation) enzymes.

Propionyl CoA carboxylase reacts with CO_2 and ATP; in this form the enzyme incorporates CO_2 into propionic acid (as propionyl CoA) to form methyl malonyl CoA. The CO_2 reacts with biotin to form a biotin-CO_2 combination (an active CO_2 molecule). The point of attachment of the CO_2 is somewhat unsettled, but according to Lynen and co-workers the CO_2 combines with the nitrogen atom as indicated in the structure shown. Another carboxylation reaction results in malonic acid from acetic. In these carboxylation reactions the process is visualized as a two-step mechanism :

$$CO_2 + \text{biotin-enzyme} + ATP \rightarrow CO_2\text{-biotin-enzyme} + ADP + Pi$$

$$CO_2\text{-biotin-enzyme} + \text{acetyl CoA} \rightarrow \text{malonyl CoA} + \text{biotin-enzyme}$$

The point of attachment of CO_2 in oxalacetic transcarboxylase is also on the nitrogen of biotin indicated, according to Wood and others. In the production of urea in the body, one step involves the reaction of carbamyl phosphate and ornithine to yield citrulline. MacCleod and others showed that livers from biotin-deficient rats were capable of carrying out this conversion at a rate of only 50 per cent that of normal livers. In microorganisms the role of biotin in purine synthesis is well established. One step in the synthesis involves CO_2 fixation (5-aminoimidazole ribotide (AI)- + -$CO_2 \rightarrow$ 5-amino-4-imidazolecarboxylic acid ribotide (AICA)).

It is at this point that biotin is involved, since AI accumulates in *Saccharomyces cerevisiae* under conditions of biotin deficiency and is utilized in the formation of AICA upon supplying the vitamin.

Interestingly enough, biotin also appears to be involved in a subsequent step, the formation of the carboxamide derivative. The details of these and other reactions concerned with purine synthesis can be found. Some interesting relations of biotin to enzyme and other protein synthesis in animals has been reviewed. Specific roles of biotin enzymes in lipid synthesis in animals are indicated in a review by Vagelos.

Fatty acid synthesis is decreased in animals during biotin deficiency. As might be expected, CO_2 transfer is the significant reaction here involving biotin enzymes. Oxybiotin can be utilized by various microorganisms, and it has been demonstrated that higher animals (chicken) can also utilize this molecule without conversion into biotin. Biocytin and desthiooiotin are all used in place of biotin by various microorganisms. Pimelic acid ($HOOC(CH_2)_5COOH$) stimulates biotin synthesis in some microorganisms and is growth-promoting, in others.

Biotin Antimetabolites

Many compounds more or less chemically related to biotin are known to inhibit the growth-promoting effect of biotin on various microorganisms. Each of a series of biotin homologues (various lengths of the side chain) was found to be capable of inhibiting the growth of *Lactobacillus casei and Saccharomyces cerevisiae* in the presence of biotin. It was pointed out that in this series of compounds the antibiotin activity was similar to the avidin-combining power. The mode of action of antimetabolites in general is not clear. However, important observations in this field have been reported.

It is well established that the fermentation rate of biotin-deficient yeast is markedly increased upon the addition of biotin to the medium. These authors showed that if certain biotin analogues (oxybiotinsulfonic acid and others) were added prior to the addition of biotin, the stimulatory effect of biotin was lost. However, if biotin had established the increased rate of fermentation, then the addition of the antimetabolites did not directly inhibit fermentation. From this it appears that the inhibitors act by preventing biotin from incorporating into a required coenzyme and not by inhibiting the action of the coenzyme in fermentation processes.

Fate of Biotin

The fate of biotin in the body has been studied by Mistry and co-workers. Rats and chicks were sacrificed one, two, or three hours following an injection of carboxyl-C^{14}-biotin. A maximum of 14 per cent of the dose was found in the liver. After fractionation of liver homogenates, 39 to 53 per cent of the activity-there was found in the supernatant, 18 to 29 per cent in the mitochondria, 15 to 23 per cent in nuclear material, and only 2 per cent in the microsomes. It was established that about 90 per cent of the biotin in the cellular fractions of normal rat liver exists in bound forms, except in microsomes, where 80 per cent is, free or as biocytin.

Distribution of Biotin

This vitamin enjoys an ubiquitous distribution in plant and animal tissue. Foods rich in biotin include liver, kidney, molasses, yeast, milk, and egg yolk. The high content of biotin in egg yolk and of avidin in egg white is an interesting point for speculation. Vegetables, nuts, and grains also contain biotin in both the free and the combined form.

For man and for some other animals the biotin produced by the bacteria of the large intestine may be a more important source (quantitatively) than that of the diet. A study of the distribution of a near physiological dose of carboxyl labeled (C^{14}) biotin in deficient chicks and rats showed that about 16 per cent of the radio-activity was found in the liver, and around half of this was in the pH 5-2 fraction of the liver cell supernatant. Microsomes contained only a small amount, and very small amounts were oxidized to CO_2. Most of the tissue biotin was found in bound form, except in microsomes. Only about one tenth of the normal quantity of the vitamin was found in the liver of the depleted animals.

Requirement of Biotin

The biotin requirement of microorganisms is well established. In animals there is no practical method now of establishing the quantitative need for this vitamin, since large amounts of it are supplied through intestinal bacterial synthesis. In some of the earlier balance studies in man it was found that the urinary excretion of biotin often exceeded the dietary intake, and the fecal excretion was greater than the intake in every case. At any rate, daily needs appear to be met by diets containing 150-300 μg of the vitamin.

Determination of Biotin

The assay of biotin is complicated by the presence in natural materials of biotin analogues, some of which may have varying degrees of biotin activity or even antibiotin activity for some of the test organisms used. Animal assays have largely been replaced by microbiological procedures. Some materials, such as urine and milk, may be assayed without previous hydrolysis, but in many substances the combined biotin must be released by acid hydrolysis. Glick and others presented an extremely sensitive method for biotin using *L. arabinosus*.

Folic Acid (Pteroylglutamic Acid) and Related Substances

The development of our present knowledge of folic acid, called pteroylglutamic acid (PGA), resulted from many studies of the nutritional needs of animals, on the one hand, and of bacterial requirements, on the other. It is no wonder that this compound (and closely related molecules) has been assigned a wide variety of designations, since so many test animals and different bacteria have been employed under diverse experimental conditions in the elucidation of its nature. Also, a variety of symptoms were used as deficiency criteria in animals. Further complications surely resulted from the fact that the vitamin occurs in several chemical forms.

A few of the names previously applied to this vitamin include vitamin M (a hematopoietic factor for monkeys), vitamin B_c (chick growth factor), factor R (bacterial growth), norite eluate factor *(Lactobacillus casei), L. casei* factor, vitamin B_{10}, and vitamin B_{11}. In 1941 the name folic acid (Latin *folium,* leaf) was assigned to a principle required by *Streptococcus lactis* R by Mitchell, Snell, and Williams. The term folacin is commonly used now as a synonym for folic acid. In no other instance has the correlation of the work of the bacteriologist and the nutritionist been so important and productive in the success of elucidating our knowledge of a vitamin. An important link between the two fields of investigation was formed in 1941, when investigators at the University of Wisconsin pointed out certain similarities between a bacterial growth factor and a substance required for chick growth.

Activity for *L. casei* and for the chick and loss of activity on storage for both were shown to be similar in a preparation from liver. Mitchell, Snell, and Williams developed a method for concentrating the vitamin from spinach to 137,000 times the activity of their standard - liver preparation. A crystalline product was obtained by several groups of investigators in 1944 from liver, yeast and other natural sources. About this time investigators recognized a difference between fermentation *L. casei* factor and liver *L. casei* factor, although the products of chemical degradation were the same.

It developed that the fermentation factor contains three molecules of glutamic acid, one molecule of *p*-aminobenzoic acid, and a substituted pteridine, while the liver factor is chemically similar save for the fact that only one molecule of glutamic acid is present per molecule of vitamin. In 1946 the isolation, proof of structure, and synthesis of folic acid were described by a group of workers at the Lederle Laboratories.

Chemistry of Folic Acid

Folic acid is a yellow crystalline material soluble in water to an extent of only about 0.1 per cent. It is soluble in dilute alcohol and can be precipitated from solution as the barium or lead salts or with basic precipitants, such as phosphotungstic acid. It shows characteristic absorption bands in the ultraviolet and in the infrared portions of the spectrum. In one of two methods of synthesis used by Angier and others equimolecular amounts of 2,4,5-triamino-6-bydroxypyrimidine; *p*-aminobenzoylglutamic acid and 2, 3-dibromopropi-onaldehyde were allowed to react together in the presence of an acetate buffer.

The reaction mixture contained about 15 per cent of active material (microbiological assay). Purification procedures and final recrystallization from hot water yielded pteroylglutamic acid. At present the identity of vitamin M, factor U, factors R and S, vitamin B_c, norite eluate factor, and liver *Lactobacillus casei* factor with pteroylglutamic acid (synthetic folic acid) is assured.

The fermentation *L, casei* factor differs in that three glutamic acid residues are present per molecule, and vitamin B_c conjugate contains seven glutamic acid residues. Other related molecules known to be active for one or more species of microorganisms include xanthopterin, thymine, and pyracin. A naturally occurring enzyme, vitamin B_c conjugase, hydroiyzes folic acid-like compounds with several glutamic acid residues to pteroylglutamic acid and glutamic acid. This enzyme is widely distributed in animal tissues and may be of importance in converting pteroylglutamates to PGA. although pteroylglutamic acid, pteroyltriglutamic acid, and pteroylheptagiutamic acid are active as hematopoietic agents for man. Two diglutamic acid derivatives have been synthesized.

Pteroyl-α-glutamylglutamic acid (Diopterin) is inactive for *L. casei* and *Streptococcus faecalis* R but active in the types of human blood discrasias in which it has been studied. Pteroyl-γ-glutamylglutamic acid is active in certain types of human anemias. There are five triglutamic acid derivatives. The γ, γ-derivative is identical with naturally occurring fermentation *L. casei* factor. The α in these designations indicates that the carboxyl of glutamic

acid adjacent to the carbon holding the amino group is involved in the peptide bond, while the γ indicates that the other carboxyl (γ from the carbon holding the amino group) is thus involved. A two-stage synthesis of PGA was reported by Sletzinger and others.

pteroylglutomic acid (PGA)
folic acid

petroic acid
Folic acid and some related structures

rhizopterin (N^{10}-formylpteroic acid)

folinic acid (citrovorum factor), N^5-formly FH_4,
N^5-formly-5, 6, 7, 8-tetrahydrocteroylgutamic acid

xanthopterin

aminopterin (4-amino PGA)

N^{10}-formyltetrahydro PGA (partial structure)

N^{5}-formyltetrahydro PGA (partial structure)

$N^{5,10}$-methenyltetrahydro PGA (partial structure)

Folic acid and some related structures (continued)

N^{5}-methyltetrahydro PG (partial structure)

folic acid and some related structures (continued)

Coenzyme Activity

Folinic acid, citrovorum factor (CF), and leucovorin are names for a naturally occurring derivative of folic acid. Unlike folic acid, this molecule supports the rapid growth of *Leuconostoc citrovorum.* It has been synthesized and is one of the active coenzyme forms of the vitamin involved in one-carbon (C_1) transfer mechanisms mentioned later. It is N^5-formyl-5, 6, 7, 8-tetrahydropteroylglutamic acid (reduced and formylated PGA). The natural material is about twice as active as the synthetic substance, since the latter is a D, L mixture. The L isomer only is utilized by the organisms studied and appears to have the same activity as natural CF. Active enzyme systems have been prepared from liver and kidney which are capable of converting PGA into CF. It is well established that ascorbic acid is involved in this conversion. The administration of ascorbic acid to rats, or to humans given PGA markedly increases the urinary excretion of CF. In monkeys citrovorum factor content of liver was found to be higher with additional ascorbic acid intake but no change in the amount of folic acid ingested. Thus,

Misra and co-workers found that in monkeys on a daily intake of 200 μg of folic acid the CF content of liver increased from around 568 to 2375 mμg per gram of tissue as the ascorbic acid intake was increased from 0 to 50 mg daily. An important, function of ascorbic acid, then, is to maintain folic acid in the reduced form required for coenzyme formation.

It has been proposed that ascorbic acid protects the folic acid reductase and that folic acid antagonists, such as aminopterin and amethopterin, inhibit this protection and inactivate the reductase. The different coenzyme forms of folic acid participate in one-carbon metabolism—a number of specific metabolic reactions in which one-carbon units are transferred for the synthesis of a variety of molecules. Other coenzyme forms of the vitamin active in Q metabolism are N^{10}-formyl FH_4, N^{5-10}-methylene FH_4, N^5- or N^{10}-formimino FH_4, N^5-methyl FH_4, and N^5-hydroxymethyl FH_4.

Structures of these molecules and a discussion of one-carbon metabolism are given in Chapter 25 (one-carbon metabolism), and these will not be repeated here. Among the metabolic reactions discussed under this heading are (1) serineglycine interconversion, (2) purine and pyrimidine synthesis, (3) methionine-homocysteine relations (methyl synthesis), (4.) histidine synthesis, (5) formiminoglutamate formation, and other one-carbon reactions. Methylation actions in general are discussed under a separate heading.

Folic acid Deficiency

A deficiency of folic acid is difficult to produce in most animals unless intestinal bacterial growth is inhibited (by feeding a sulfonamide drug or antibiotic, for example). The use of folic acid antagonists has also been widely used to produce deficiency symptoms. In monkeys a folic acid deficiency leads to a characteristic type of anemia. Rats develop anemia and leukopenia which disappear following the administration of a folic acid active substance.

Formic acid and formiminoglutamic acid excretion are increased in a deficiency state in this species. In man there results a macrocytic anemia which resembles pernicious anemia except that the nervous involvement of the latter condition is absent. Vilter and co-workers pointed out that in a deficiency of folic acid coenzymes the ability of the erythrocyte precursor cells to produce DNA is impaired. Thus the cell nucleus remains young, and division is delayed while the cell grows large.

By such abnormality, there results the megaloblast, a cell with a limited supply of DNA precursors due to a deficiency of folic acid, vitamin B_{12} or ascorbic acid. Vitamin B_{12} and ascorbic acid are both required for the formation of the folic acid coenzymes. A patient with megaloblastic anemia associated with ascorbic acid deficiency was maintained on a vitamin-free diet for two weeks during which time the anemia progressed in severity. Ascorbic acid failed to prevent the progression of the condition. However, 50 μg daily of folic acid intramuscularly initiated hematological improvement. This experiment shows that ascorbic acid has no effect on the course of the anemia unless folic acid is present. Diarrhea, gastrointestinal lesions, and other symptoms are present. Folic acid is highly effective in the treatment.

The vitamin is also effective in treating the anemia of pernicious anemia, although it is without effect on the nervous symptoms, which respond to vitamin B_{12}. The sprue syndrome and some other types of anemia in humans respond to folic acid therapy. What part of the symptomatology of deficiency may be related to abnormal interconversion of glycine and serine, or to the impaired handling of formiminoglutamic acid (from histidine catabolism), or to the decreased methylation of homocysteine to yield methionine cannot be defined at this time. Folic acid labeled with tritium is given by mouth and radioactivity determined in the serum three hours later as a measure of folic acid absorption.

Folic acid Antagonists

Much has been learned regarding pathways and coenzyme activity through the use of antagonists in animals and in microorganisms. One of the most potent inhibitors is aminopterin (4-amino PGA). The marked potency of the inhibitor aminopterin is attested to by the fact that when the compound was fed at a level of only 1 mg per kilogram of diet, mice died in a few days. It produces anemia and leukopenia in guinea pigs and in rats. This antagonist has been used successfully to bring about remissions (not a cure) of acute leukemia in children.

Amethopterin (4-amino-N^{10}-methyl PGA) has also been used to produce experimental deficiency and in the treatment of leukemia. Goldin and co-authors have reviewed the clinical use of such antagonists. Amethopterin markedly inhibited nucleic acid synthesis in rats following partial bepatectomy to induce mitosis. This is probably explained on the basis that the 4-amino antagonist suppresses formylcoenzyme synthesis, resulting in inadequate purines and/or pyrimidine production, with the consequent inability of the organism to fabricate nucleic acids. Timmis has reviewed many antifolic acid compounds with real or potential value in cancer chemotherapy.

Occurrence of Folic acid Compounds

Green leafy vegetables and yeast are good sources. Many microorganisms contain appreciable quantities and may supply man and other animals, through intestinal synthesis, with significant amounts. Liver contains more folic acid activity than other materials studied (around 300 μg per 100 g). Grains and green leafy vegetables vary from 20 or 30 up to 100 μg per 100 g. Some of the activity may be free in natural products but a larger portion is in a bound form.

Assay of Folic Acid Activity

The microbiological methods of assay are preferred, since they are more practical than rat or chick assay procedures from the standpoint of time and cost. Even at this time the assay methods are not adequate, especially for differentiating the various active molecules. *Lactobacillus casei* (acid production) and *Streptococcus faecalis* (growth) have been employed for total activity toward these organisms. The Association of Official Agricultural Chemists adopted revised procedures in 1958. One of the problems involves hydrolysis of the bound forms of the vitamin in foods. Newer enzymatic procedures for hydrolysis offer promise.

Blood serum folic acid may be determined by the method of Herbert, and the results of this determination gives an indication of the state of folic acid nutrition of an individual. Chemical methods are developing around the fluorescence character of folic acid compounds. Duggan and co-workers surveyed the potential applicability of such methods for the determination of many biological compounds and list the wavelength of maximum activation and of maximum fluorescence for a variety of molecules including folic acid and folinic acid (CF).

Requirements of Folic Acid

A study of food content showed, in adequate high-cost and low-cost diets, 0.193 and 0.157 mg total folic acid, respectively, of which 0.065 mg was free. In deficient diets 0.047 total and 0.015 mg free folic acid were found. These dietary levels should meet the requirement for the vitamin, with some increase during pregnancy through increased amounts of food rich in foiic acid. The actual amounts required are not known. As little as 0.025 mg of crystalline folic acid daily was found to initiate a hematological response in sprue. A daily intake of 0.05 mg prevents a decrease in serum folic acid, but 0.025 mg daily does not. Reviews on folic acid are listed.

Vitamin B_{12}, Cobalamins

After many years of intensive work the antipernicious anemia factor of liver extract (used in clinical treatment) was isolated in 1948 by Rickes and co-workers and by Smith. Various workers had been studying the same or a similar substance found in other material. The "animal protein factor," which promotes the growth of animals on diets containing vegetable proteins and is found in such materials as fish solubles and cow manure, is similar to the factor isolated from liver extract.

The name B_{12} was given to the vitamin. Rapid progress in isolation of B_{12} followed the finding that the active principle could be separated chromatographically. Activity during purification was followed by microbiological assay. A pink color seemed to be associated with activity. Red crystals of the pure material were finally isolated. The crystals decompose without melting at 300°C.

Chemistry of Vitamin B_{12}

Cyanocobalamin crystals are tasteless and odorless. One gram dissolves in about 80 ml of water at room temperature, forming a neutral solution. The pure material is quite soluble in alcohol and insoluble in ether and acetone. In aqueous solution crystalline cyanocobalamin has three absorption maxima at 278, 361, and 550 mμ, with extinction coefficients ($E^{1\%}_{1cm}$) of 115, 107, and 64, respectively. It is remarkable that cobalamin contains about 4.35 per cent cobalt. The molecular weight is 1355 figured on the ensuing structure. Cyanocobalamin has a net charge of one. The cobalt has a coordination number of six.

It has one coordinate linked cyanide group, one coordinate pyrrole nitrogen, and a coordinate link to a nitrogen of the 5, 6-dimethylbenzimidazole moiety. Other B_{12} active compounds are known in which the cyanide radical is replaced by various groups forming other cobalamins, such as hydroxycobalamin, chlorocobalamin, nitrocobalamin, and thiocyanatocobalamin. Treatment with cyanide converts these molecules into cyanocobalamin. The similarity of the cyanocobalamin molecule and the porphyrins is of in-terest. In B_{12} two pyrrole rings are joined directly rather than through the methene (—CH=) bridge, as in other porphyrins.

Coenzyme Activity, Physiological Action of Vitamin B_{12}

Many microorganisms require B_{12} for growth. Lactic acid bacteria are highly sensitive to deficiency. The B_{12} requirement of some bacteria is met by methionine, or one of the pseudocobalamins. Other microorganisms synthesize quantities of B_{12} above their requirements. Animal protein factor (APF) was known to be an essential nutrient for growth in experimental animals, and after the discovery of B_{12}, it was established that the vitamin possessed high APF activity in the chick.

The vitamin is known to be essential for the growth of other laboratory animals and nonruminant farm animals. It is likewise essential in human nutrition. More specifically, B_{12} has been implicated in several metabolic reactions. The evidence is not clear-cut in all cases, and considerable controversy exists with respect to certain of the proposed functions. One coenzyme form of B_{12}, 5,6-dimethylbenzimidazole cobamide coenzyme, is identical with the vitamin except that in place of the—CN radical there is attached to the cobalt an adenine nucleoside in which the sugar is 5′-deoxyribose and the attachment is cobalt to the 5′-carbon atom.

The structure of this molecule is given with the discussion of coenzymes. The coenzyme acts in the enzyme system which catalyzes the reversible isomerization of methylmalonyl CoA to succinyl CoA in mammalian and mtcrobial metabolism and in the isomerization of methylaspartate to glutamate in microorganisms. Another coenzyme of B_{12} has a methyl group attached directly to cobalt in place of the deoxyadenosine nucleoside. This coenzyme participates in the formation of methionine from homocysteine in the presence of N^5-methyl FH_4 and other factors.

One idea is that methyl Bi_2 coenzyme is an intermediate in the transfer of the methyl group. This is discussed under one-carbon metabolism. This coenzyme also acts in the methylation of the purine ring in thymine synthesis. B_{12} has been implicated in cholesterol metabolism, porphyrin biosynthesis, and other phases of metabolism. Another coenzyme function of cobamide coenzymes involves the metabolism of 1,2-glycols. Abeles and others showed that cell-free extracts of *Aerobacter aerogenese* treated with charcoal lose their ability to mediate conversion of ethylene-glycol to acetaldehyde and propanediol to propionaldehyde. Full activity was re-stored to the treated extract after addition of cobamide coenzyme.

Further studies by these investigators suggested that the enzymatic reaction involves a transfer of hydrogen from C-1 to C-2 of the diol and a loss of the elements of water:

$$\overset{(2)}{CH_2OH}—\overset{(1)}{CH_2OH} \rightarrow CH_3CHO + H_2O$$

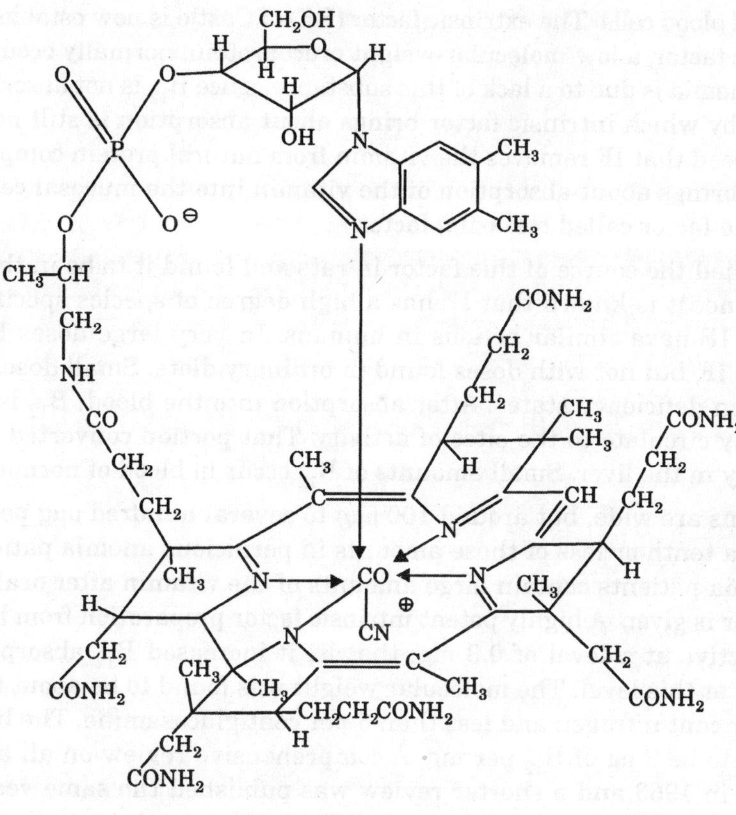

vitamin B_{12}
(cobalamin)

Further purification studies of the enzyme called dioldehydrase showed that hydroxocobalamin and cyanocobalamin are inhibitors of the reaction. Certain interactions of enzyme and coenzyme were postulated and thought to be related to catalytic activity.

Deficiency of B_{12}

Johnson has reviewed the methods used to produce deficiency of B_{12} in pigs, chicks, and rats. On a "soy-protein synthetic milk" ration pigs develop B_{12} deficiency. Growth is poor and death ensues unless B_{12} is administered. Development of formed elements of blood is abnormal. The ease with which uncomplicated B_{12} deficiency can be produced in the young pig has made this animal a valuable tool in studying metabolic pathways involving the vitamin. Rats and chicks are susceptible to lack of dietary B_{12}.

Growth is poor, and rats also develop porphyrin whiskers and scaly feet. Vitamin B_{12} relations to cobalt nutrition in ruminants are discussed with cobalt. In humans a dietary B_{12} deficiency is rare. In a review B_{12} deficiency symptoms of humans subsisting on diets devoid of animal products are discussed. Of more interest is the relation of the vitamin to human pernicious anemia and other diseases. Castle and others some years ago suggested that in pernicious anemia there is a deficiency of an intrinsic factor (stomach factor) and an extrinsic factor (food factor). The two factors were thought to react to form something required for the maturation of red blood cells. The extrinsic factor (EF) of Castle is now established to be vitamin B_{12}. The intrinsic factor, a low-molecular-weight mucoprotein, normally occurs in gastric juice; and pernicious anemia is due to a lack of this substance, since B_{12} is not absorbed in its absence. The mechanism by which intrinsic factor brings about absorption is still not clear. However, it has been proposed that IF removes the vitamin from natural protein complexes with animal proteins. It also brings about absorption of the vitamin into the mucosal cells with the aid of an intestinal juice factor called releasing factor.

Herbert studied the source of this factor in rats and found it to be in the proximal end of the small intestine. It is known that IF has a high degree of species specificity. Fortunately hog and human IF have similar actions in humans. In very large doses B_{12} is absorbed in humans without IF, but not with doses found in ordinary diets. Small doses parenterally are highly effective in deficiency states. After absorption into the blood, B_{12} is bound to plasma proteins and may circulate to the sites of activity. That portion converted into coenzymes is stored principally in the liver. Small amounts of B_{12} occur in blood of normal individuals.

The variations are wide, but around 100 μμ*g* to several hundred μμg per ml of blood have been found and a tenth or less of these amounts in pernicious anemia patients. The stools of pernicious anemia patients contain large amounts of the vitamin after oral administration if no intrinsic factor is given. A highly potent intrinsic factor preparation from hog pyloric mucosa was clinically active at a level of 0.3 mg; that is, it increased B_{12} absorption in pernicious anemia patients at this level. The molecular weight was found to be about 5000; the material contained 10 per cent nitrogen and less than 3 per cent glucosamine. The binding capacity *in vitro* was found to be 3 μg of B_{12} per mg. A comprehensive review on all aspects of intrinsic factor appeared in 1963 and a shorter review was published the same year. Another role of intrinsic factor appears to involve retention of B_{12} by tissues, Active preparations increased B_{12} uptake of liver slices.

In liver perfusion experiments B_{12}-Co^{60} was taken up to a slight extent, and the uptake was tremendously enhanced in the presence of intrinsic factor. Sorbitol (sugar alcohol analogue of glucose) enhances B_{12} absorption in man. In rats this compound, manuitol, sorbose, and xylose also increased absorption. In normal persons, studies with Co^{60}-vitamin B_{12} show that over 70 per cent of a 0.5 μg dose is absorbed and that increasing the dose results in lower percentage absorption. With a 5.0 μg dose only 30 per cent was absorbed. Vitamin B_{12} is used clinically in diseases other than pernicious anemia. It is effective in various megaloblastic anemias and in neurological disturbances accompanying various other conditions. Much of the information in this connection has been brought together and summarized.

Assay of B_{12}

The most sensitive methods depend upon growth stimulation in microorganisms. Special procedures are required to liberate the vitamin from bound forms as it is extracted from crude samples. *Lactobacilus leichmannii* has been used for cobalamins with either growth or acid production as end points. *E. coli* and other organisms have been employed in special instances. The organism *Ochromonas malhamensis* is said to be almost specific for cyanocobalamin. Spectrophotomctric methods are sensitive and applicable to pure substances.

The USP method involves the determination of absorbance of a solution of the vitamin at 361 mμ. Details of a radioisotope assay technique for cobalamins and a study of the adaptability of the method to crude samples was reported by Bruening and others. The assay of the cobamide coenzyme was reviewed by Heinrich and others.

Sources of B_{12}

Plants do not contain vitamin B_{12}. Microorganisms synthesize cobalamins, especially those normally present in the rumen of herbivorous animals. Some bacterial synthesis may occur in humans, but not in sufficient quantity to meet the needs, since man is dependent upon dietary sources. Liver and kidney are excellent sources; animal and fish muscle contain moderate quantities; while vegetables and grains contain little or none of the vitamin. Goldsmith and co-workers found 31, 16, and 2.7 μg of B_{12} in high-cost, low-cost, and poor daily diets, respectively.

Requirement of B_1

The human requirement of B_2 is not established but probably is not over 1.5 μg per day and very likely is less than this. In the absence of intrinsic factor (following gastrectomy) patients may show no signs of pernicious anemia for several years. This is in agreement with the observations of Schloesser and others, who estimated the half-life of cobalamin to be over a year in humans. Thus the rate of destruction and/or excretion of the vitamin is exceedingly slow. Individuals with megaloblastic anemia from B_{12} deficiency or those with pernicious anemia respond to as little as 0.1 μg daily given parenterally.

Inositol

The compound inositol (muscle sugar) was discovered in 1850, and some years later it was shown to be a hexahydroxycyclohexane. It was not until 1928 that Eastcott recognized the nutritional significance of the compound. She isolated a substance from, tea: showed that it was essential for the growth of certain yeasts, and identified it with the long-known "Bios I," a concentrate with growth-promoting properties for various microorganisms. In 1940 Woolley produced a type of alopecia in mice by dietary means and showed that inositol prevented the condition. In the course of these studies some of the animals showed spontaneous cures. It was demonstrated that bacteria from the intestinal tract of such animals were able to synthesize much more inositol than were organisms isolated from the tract of animals that retained the symptoms of deficiency.

Under specified dietary conditions it was shown that inositol was beneficial in the treatment of the so-called spectacled-eye condition, or denuding about the eyes, in rats. From the following formula the possibility of *cis-trans* isomerism in the molecule is evident. Seven optically inactive forms and a pair of active isomers can exist. Only one of the optically inactive forms, myo-inositol or meso-inositol has biological activity:

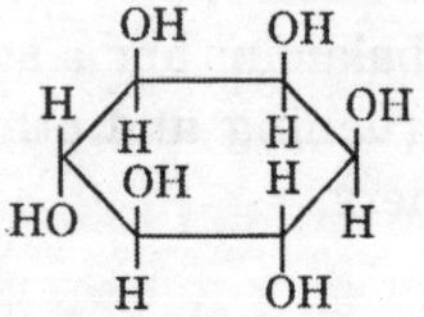

myo-inositol

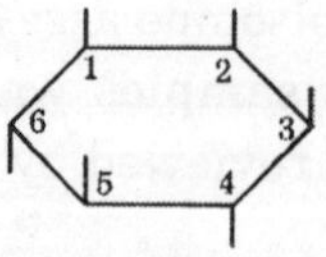

myo-inositol
1235/46

The designation 1235/46 devised by Hornstein indicates that the hydroxyl groups on carbons 1, 2, 3, and 5 are projected in the same direction in space, while those on carbons 4 and 6 are oriented in the opposing plane. Such a system simplifies the nomenclature of the various inositol isomers. *myo*-inositol is a white crystalline material, soluble in water to the extent of about 12.5 per cent at 20°C. It is insoluble in alcohol and ether.

The crystals melt at 225°C. It is an alcohol with the same empirical formula as glucose, $C_6H_{12}O_6$. The term *myo* is used to indicate that the compound occurs in muscle. Synthesis of *myo*-inositol is not practical as a source of the material, primarily because of the relative ease of isolation from natural sources, many of which contain an abundance of it either in the free or in various combined forms. From: the standpoint of proving the structure, the synthesis is highly significant. This was accomplished by various workers. The primary symptoms of deficiency in mice are alopecia and subnormal growth rate. In rats a retarded growth and swelling and loss of hair about the eyes are typical symptoms.

The appearance of the animals led to the designation spectacled-eye condition. Other animals reported to require inositol include the guinea pig, hamster, pig, chicken, and turkey. Under specified dietary conditions rats develop fatty livers which contain large amounts of cholesterol. Choline, a well-established lipotropic agent, does not prevent the disease, since the diet and supplements used contained liberal amounts. The administration of inositol prevented the accumulation of fat and cholesterol in, the livers. This lipotropic action of inositol in rats has been amply confirmed.

Best and co-workers reviewed previous work in this field and presented results of their carefully controlled experiments. They concluded that inositol exerts a limited lipotropic effect in rats on a diet very low in fat but that on diets containing fat this effect is abolished. Abels and co-workers reported that in their cases the livers of patients with gastrointestinal cancer invariably were infiltrated with fat. After the administration of inositol postoperatively, the livers of a number of such patients, with one exception, were found to have normal levels of fat. It is generally agreed that inositol is a lipctropic substance under special conditions and that it may act synergistically with choline or other active molecules giving it more general lipotropic properties. Ordinarily the cholesterol and phospholipid levels of the blood serum of rabbits rise markedly when diets high in cholesterol are fed.

Dotti and others, however, reported that 0.5 g of inositol daily added to the high cholesterol diet eliminated the expected rise in blood cholesterol and phospholipid in these animals. Inositol is in a unique position as far as its status as a vitamin is concerned. Under highly specific dietary conditions a need in several species has been demonstrated.

On the other hand, the synthesis of inositol from glucose in the immature rat and in the chick embryo has been demonstrated by Daughaday and others. This synthesis has been confirmed by Halliday and Anderson who isolated *myo*-inositol-C^{14} from rats that had been injected with glucose-1-C^{14}. Tissue slices from brain, liver, and kidney are able to convert labeled glucose into inositol which was isolated from both water-soluble and lipid soluble-extracts of the tissues, Of interest in this connection are the findings of Eagle and co-workers, who demonstrated that inositol is a required growth factor for each of 18 human cell lines tested in tissue culture. It appears that inositol can be metabolized as carbohydrate by the rat, since it acts antiketogenically. Of a 250-mg dose administered, less than 1 mg was recovered in the urine. Deuterium-labeled inositol fed to a phlorizinized rat led to the excretion of deuterium-labeled glucose in the urine. Following intra-peritoneal injection of C'Mabeled *myo*-inositol in fasting rats, about 25 per cent of the dose was recovered in the respired CO_2 in 12 hours. Tissue phospholipids and liver glyeogen were labeled to a much smaller extent. Diabetics excreted increased amounts of inositol (280 to 850 mg per day in seven patients) is and after insulin treatment the excretion fell to a normal amount of around 30 mg per day. Inisitol occurs in nature in the free form and in many combined forms.

Methyl and also phosphoric acid derivatives have been isolated from a variety of materials. Mono-, di-, and tri-phosphoric acid esters are known to occur naturally. The hexaphosphoric acid ester in the form of mixed calcium and magnesium salts is known as phytin and constitutes one of the abundant sources in nature. Inositol is readily obtained by $Ca(OH)_2$ hydrolysis of this and other phosphate combinations. The high concentration of phytin in many grains, especially in corn, is an important factor in the supply of dietary phosphate from these sources. The occurrence of inositol in phospholipids of animal and plant tissues is discussed. The very high concentration of free inositol in seminal plasma of various species is of interest, although no explanation for its presence there is available at present.

Thus, Hartree found that boar seminal plasma contained 600-700 mg per 100 ml and that in the bull, human, rabbit, ram, and stallion the levels were less than 100 mg. These levels are far higher than are found in other body fluids. The most satisfactory assay method for inositol is one of the microbiological procedures. In these the growth response of *Saccharomyces cerevisiae* or of *Sacch. carlsbergensis* to additions of inositol-containing material to a culture medium devoid of the vitamin is compared to growth with additions of known increments of pure meso-inositol.

LeBaron and others described a micro-spectrophotometric method for the determination of inositol in tissues and lipid hydrolysates. The animal requirements for inositol have not been studied intensively. In rats a 20 mg daily dose was sufficient (probably an excess) to cure inositol deficiency symptoms. It has not been established that humans require a dietary source of inositol.

p-Aminobenzoic Acid

p-Aminobenzoic acid has been known to the chemist for many years. It was not until 1940 that a physiological role was proposed for this compound. Woods reported the interesting observation that *p*-aminobenzoic acid (PAB, PABA, or p-AB) counteracts the bacteriostatic action of sulfanilamide in *vitro.* It was subsequently demonstrated that other bacteria would grow normally in the presence of sulfanilamide (and a variety of other sulfonamide drugs) if sufficient *p*-aminobenzoic acid were present in the culture medium. This antisulfonamide action of PAB was also demonstrated *in vivo.*

Mice infected with various pathogenic organisms were protected against disease by the administration of sulfanilamide, but when PAB was given also, the sulfonamide protection was absent. It was these observations that led to the development of our knowledge in the field of metabolic inhibitors or antagonists—a field which is now a science in itself. It should be pointed out that PAB is used rather routinely in culturing an organism from patients on sulfonamide therapy.

The purpose of culturing an organism in such cases is to identify it, and obviously the bacteriostatic action of the sulfonamide must be counteracted if growth of the organism is to be attained. Some of the rickettsial infections in man were found to respond favorably to large doses of PAB, although presently antibiotics have replaced it in the treatment of the disease. The vitamin status of PAB ($COOH—C_6H_4—NH_2$) is probably restricted to microorganisms where it is utilized in the synthesis of folic acid compounds and possibly in other functions in growth. Further discussion of the relation of sulfonamide inhibition of PGA synthesis in microorganisms can be found. A short review on PAB is available.

Ascorbic Acid (Vitamin C)

Scurvy in man has been known for centuries. The use of fresh vegetables and especially of the juice of lemons and of limes was established as specific cures for the condition centuries ago. The earliest accurate description of the disease and its control was published by Lind in 1757 in his "Treatise on Scurvy." Years later limes became a required article in the diet of the English Navy. The terms limey for English sailors and Limehouse for the wharf area in London stem from this fact. Vitamin C deficiency not only was prevalent on long sea voyages because fresh foods were unavailable, but also was epidemic over parts of the world during times of famine and war.

During the wars of our own times outbreaks of scurvy in parts of the world have not been uncommon. Progress in the development of specific knowledge concerning vitamin *C* was initiated by Hoist and Frolich in 1907. These investigators showed that a scorbutic state could be produced in guinea pigs by restricting the diet of these animals to oats and bran. This development paved the way for intensive research into the properties of a protective substance in certain foods as well as into the biochemical defects associated with the scorbutic state. Zilva and associates provided important information on the concentration of active antiscorbutic material from lemon juice and on certain important chemical properties of the active substance.

They showed, for instance, that following oxidative destruction of the principle, the activity could be restored by reduction and proposed a relation between antiscorbutic activity and reducing power of his preparations. In 1928 Szent-Györgyi isolated a strong reducing substance from adrenal glands of rats and from citrus juices. The name hexuronic acid was given to this material. It was only a matter of a few years before hexuronic acid, and the reducing substance of Zilva were shown to be vitamin *C,* the antiscorbutic vitamin.

Waugh and King reported in 1932 that the vitamin C isolated by them-from lemon juice and the reducing heurbnic acid studied by others were identical because of the similarity in many chemical and physical properties as well as in biological potency in protecting guinea pigs against scurvy. It was well agreed at this time by workers in the field that vitamin C was an acid with the general formula $C_6H_8O_6$. Several proposed structures were offered, but the true configuration of the molecule was established in 1933 in Haworth's laboratory in England. The same year the synthesis of ascorbic acid was achieved by Reichstein and co-workers and also by Haworth's group.

Chemistry of Ascorbic Acid

Pure vitamin C is a white crystalline odorless substance with a sour (acid) taste. It melts at 190°-192°C, and in the crystalline form it is stable for years. It is insoluble in most oragnic solvents, although a 2 per cent solution can be made in alcohol. In water the vitamin is soluble to the extent of 1 gin 3 ml. The strong reducing property of vitamin C depends on the loss of hydrogen atoms from the hydroxyls on the double-bonded (endiol) carbons. At pH 4 and at 35°C the E'_0 +0.166 volt.

A dilute solution of vitamin *C* has a pH of about 3. The acidity is due to ionization of the enol group on carbon atom 3. The first pK_a is 4.17; this indicates that ascorbic acid is considerably more dissociated than acetic, for instance. The second pK_9 is 11.57. Vitamin *C,* or *L*-ascorbic acid, in water has a specific rotation $[\alpha]_D^{20} = +23°$. The ultraviolet absorption maximum of ascorbic acid is at a wavelength of 265 mμ. Ascorbic acid readily forms salts of several metals. It takes up iodine at the double bond and can be reduced here by hydrogenation. Oxidation of ascorbic acid yields dehydroascorbic acid. This is a freely reversible reaction. H_2S, among other things, may be used to reduce the oxidized form in the laboratory.

The dehydro form, except in rather acid solution, undergoes hydrolysis at the lactone ring with the formation of diketogulonic acid. The reverse of this Reaction does not proceed in the body but can be brought about in the laboratory. The greater stability of ascorbic acid in acid solution depends on the decreased tendency toward hydrolysis of the Jactone ring with decreasing pH. In alkaline solution the hydrolysis is fairly rapid, and such solutions lose vitamin activity in a short period of time. The oxidation of ascorbic acid *in vitro* is catalyzed by various substances.

The copper ion is quite active and, of course, the plant ascorbic acid oxidase (a copperprotein enzyme) is highly active. The rate of destructive oxidation is greater with increasing pH. This type of oxidation involves molecular oxygen, and, consequently, in processing vitamin *C*-containing foods, such' as orange juice, the removal of oxygen by nitrogen or CO_2 results in decreased losses of the vitamin during canning or other processing. Low-temperature storage of vegetables before processing, though usually impractical, and a quick preheating (blanching) just previous to canning or freezing also aid in decreasing ascorbic acid destruction. Unfortunately, the vitamin C loss during blanching of many foods may be considerable owing in part to oxidation at the elevated temperature, and perhaps to a greater extent through solubility of the vitamin in the blanching water:

L-ascorbic acid $\underset{+2H}{\overset{-2H}{\rightleftharpoons}}$ L-dehydro ascorbic acid $\xrightarrow{HOH}$ L-diketo-gulonic acid $\rightarrow$ oaxalic acid

Several synthetic compounds have some vitamin-C activity. Thus, 6-desoxy-L-ascorbic acid has about one-third, L-rhamnoascorbic about one-fifth, D-arabo-ascorbic about one-twentieth, and L-glucoascorbic around one-fortieth the activity of L-ascorbic acid. The activity of D-ascorbic acid is nil.

Ascorbic Acid Deficiency

The primates, including man, and the guinea pig are the only animal species that require a dietary source of vitamin C. All other animal species studied need the vitamin for normal metabolic functions, to be sure, but they are able to synthesize all their requirement of this factor. The guinea pig has been the animal of choice in the bulk of the experimental work on the pathological and biochemical defects associated with ascorbic' acid deficiency. This animal is highly susceptible to a lack of vitamin C, and the alterations are known to be similar to those found in human scurvy. Vitamin C deficiency varies in degree from a mild condition scarcely recognizable to a profound state resulting in death.

In human beings the latter condition is not now seen except under extreme stress. Outstanding among the pathological defects found in ascorbic acid deficiency is the failure to deposit intercellular cement substance. The presence of an abnormal collagen leads to a tendency to hemorrhage and to slow wound healing. Roe and others showed that ascorbic acid is essential for the production of connective tissue to insure postoperative healing as well as for maintenance of connective tissue in formed scar tissue. They demonstrated also that the vitamin accumulates in scar tissue quickly after a wound, and remains for a long time.

The connective tissue has the highest concentration of the vitamin. In the teeth there are changes as a result of scurvy in the very young, but they are not prominent features of the disease in adult humans. The loss of teeth and the gum changes are more related to the deficiency effect on soft tissues.

In guinea pigs the poor dentine formation leads to poor tooth development. This was studied and described especially by Wolbach and co-workers. In the bones a deficiency results in the failure of the osteoblasts to form the intercellular substance osteoid. Without osteid, deposition of bone salts is arrested. The resulting scorbutic bone is weak and fractures easily. A detailed description of bone changes in guinea pigs and in man during ascorbic acid deficiency is available.

The advanced stage of scurvy in man is easily recognized. In infants the symptoms include irritability and fretfulness, tenderness and swelling of joints, some degree of apathy and pallor, and a desire to remain quite motionless. In adults the advanced stage is attended by loosening or even loss of teeth, accompanied by sore, spongy gums, internal hemorrhage, subcutaneous hemorrhage upon mild injury, painful joints, dyspnea, edema, and anemia. A loss of weight and a marked pallor are also noted. It is doubtful if all these symptoms are attributable to ascorbic acid deficiency alone; more likely they represent the effects of a multiple deficiency.

In support of this view the important contribution of Crandon and others on experimental human deprivation of vitamin C is cited. Crandon remained on a diet totally deficient in vitamin C but supplemented with the other known vitamins for a period of six months. Some of the pertinent findings are listed herewith. After 41 days the blood plasma ascorbic acid level reached zero.

Since it was many weeks later before the first clinical signs were observed, the authors consider the plasma level of vitamin C as a poor index of the vitamin C status of an individual. They consider the ascorbic acid level in the white-cell-platelet layer of blood a good index, since this remained elevated during most of the experimental period and fell to zero only shortly before the onset of clinical signs. The earliest symptoms noted were hyperkeratotic papules over the buttocks and calves. These began to develop after 132 days on the deficient diet.

According to these workers, this may be the earliest signs of deficiency. Perifollicular hemorrhages appeared after 161 days. Wound healing failed after the subject had remained on the diet for six months, although it was adequate at about the halfway mark in the experiment. At this time the plasma level of vitamin C had been zero for 44 days and the white-cell-platelet level was 4 mg per 100 ml (normal is 25 to 30 mg per 100 ml). No gross changes were seen in the gums or teeth although x-ray pictures showed interruptions in the lamina dura in the early acute phase of scurvy. This was considered likely to be a good diagnostic criteria in early scurvy. Correlation between capillary fragility and vitamin C deficiency was not good, and the authors doubt if the tests are as valuable as they were once thought to be, especially in a subclinical deficiency.

No anemia was found at any time during the experiment, nor was there any evidence of decreased resistance to infection. During the scorbutic state both the glucose and the insulin tolerance tests were found to be normal. Blood lactate disappeared abnormally slowly after exercise. Following the intravenous administration of ascorbic acid all the signs and symptoms of scurvy disappeard rapidly.

The tissue deficiency of this individual is shown by the fact that after the injection of 1 g of the vitamin the plasma level returned to zero after five hours. Even after 3 or 4 g of vitamin C the urinary excretion was subnormal, again demonstrating that the tissues were actively removing the vitamin from the blood stream. Another interesting experiment on experimental human ascorbic acid deprivation employing several volunteers was conducted in England. The findings were somewhat similar to those reported by Crandon and co-workers, although the English workers were attempting to establish requirements as well as to study the pathological and biochemical aspects of severe and moderate deficiency.

Physiological Functions of Ascorbic Acid

Ascorbic acid, is intimately concerned, in an unknown manner, in the normal production of supporting tissues of mesenchymal origin, such as osteoid, denlin, and collagen. The outstanding chemical property of the vitamin is its reversible oxidation-reduction between ascorbic and dehydroascorbic acid. Some of the physiological properties of the vitamin must be related to this redox system although no specific coenzyme function for ascorbic acid has been demonstrated. It may help maintain the oxidation-reduction potential of cells at the proper level.

As a hydrogen acceptor it is thought to act in plants and possibly in animals in a system involving TPN and glutathione in which the metabolite passes H to TPN, and this passes hydrogen to glutathione, which may pass them on to ascorbic acid and finally to oxygen forming water:

$$2H \text{ from } MetH_2 \text{ to } TPN^+ \text{ to } GSSG \text{ to dehydroacorbic to } O_2$$

The significance of such a proposed system in animal tissues is not yet clearly defined.

Guinea pigs lose the ability to maintain normal levels of collagen as they become depleted in ascorbic acid, and the administration of the vitamin promptly corrects this defect. It was felt by Gould some years ago that ascorbic acid promotes collagen formation, in part at least through its action in bringing about hydroxyproline synthesis from proline. Collagen contains large amounts of hydroxyproline. Mitoma and Smith felt that hydroxylation of proline was not affected by ascorbic acid and that collagen synthesis and not the amino acid synthesis is at fault in the deficiency state. Gross has suggested that synthesis and breakdown of collagen may be progressing at somewhat the same rate in scorbutic guinea pigs.

Udenfriend and others more recently showed that scorbutic guinea pigs build collagen-type peptides which incorporate C^{14}-containing proline but are deficient in hydroxyproline. This hydroxyproline-deficient intermediate may require ascorbic acid for the conversion of proline already in peptide formation to hydroxyproline. The collagen intermediate may be short-lived due to its abnormal composition; it is readily attacked by collagenase. This does not establish ascorbic acid as a part of proline hydroxylase, but that may prove to be the case. The fact that the abnormal collagen may not be stable may account for the lack of new collagen formation in the scorbutic state.

Ascorbic acid or a derivative of it has been implicated in certain other hydroxylation reactions such as the hydroxylation reaction in the conversion of tryptophan to serotonin (5-hydroxytryptamine).

The vitamin has been placed in the electron transport chain between DPNH and one of the cytochromes, a reaction that may be coupled with hydroxylation. Transport is thought to proceed from the DPNH to the active ascorbic acid derivative (monbdehydroascorbic acid; see further) which may be a partially oxidized radical form. This reduces the active compound to ascorbic acid, which may be oxidized to the monodehydro form by cytochrome b_5. The transfer may then proceed through the other members of the transport chain to O_2. The reaction product with oxygen may be the active participant in the enzymatically mediated hydroxylation. Baker and others have isolated an ascorbic acid derivative—the monodehydro-ascorbic acid in combination with ascorbic acid. They feel that such a complex may be the active form in hydroxylation reactions.

The foregoing scheme is not accepted by all workers in this field. In the scorbutic state tyrosine metabolism is abnormal. The administration of tyrosine or phenylalanine to man or to guinea pigs in the deficiency state results in the urinary excretion of *p*-hydroxyphenylpyruvic acid, and ascorbic acid medication abolishes the defect.

Normally the enzyme *p*-hydroxyphenylpyruvic acid oxidase mediates the oxidation of *p*-hydro-xyphenylpyruvic acid to homogentisic acid. At one time ascorbic acid was considered to be a specific cofactor for this enzyme, but a number of other compounds—some, entirely unrelated chemically, such as hydrocminone—were found to replace ascorbic acid in tyrosine oxidation *in vitro*. It was later established that ascorbic acid and compounds that can replace the vitamin *in vitro* do so by protecting β-hydroxyphenylpyruvic oxidase from inhibition by its substrate, and Zannoni and LaDu have now demonstrated that ascorbic acid operates *in vivo* through this protection of the liver enzyme in guinea pigs.

It appears that ascorbic acid is required for tyrosine metabolism only when large amounts of the amino acid are ingested, and the vitamin may not be required for tyrosine metabolism under normal conditions. Ascorbic acid enhances iron absorption from the intestine in humans and animals. The mechanism probably involves the reducing property of ascorbic acid, since iron in the ferrous state is prefereally absorbed.

Biosynthesis of Ascorbic Acid

The rat supplies his own ascorbic acid requirement through synthesis, as do most species of animals. Plants also synthesize vitamin C. The rat has been studied extensively to establish the synthetic pathway. Glucose or other hexoses convertible to glucose serve as the starting material. King and associates first established through isotope work that carbon 1 of D-glucose becomes carbon 6 of L-ascorbic acid, while carbon 6 of glucose turns up in the 1 position of ascorbic acid. This work indicated that during the biosynthesis the glucose carbon chain underwent inversion of configuration and that the intact glucose chain was converted into ascorbic acid. Specific enzymes mediating each step in the scheme of synthesis have been demonstrated in rat liver. Much of the work was done with labeled compounds.

A review by Wagner and Folkers gives details of the various studies involved in the deveolpment of the pathway. Preliminary to the inversion of the glucose chain the following reactions apparently occur: glucose ⟶ glucose-6-PO_4 ⟶ glucose-l-PO_4 ⟶ UDP-glucose (uridine diphosphate glucose) ⟶ UDP-glucuronic acid→D-glucuronic acid-l-PO_4 ⟶ D-glucuronic acid. The next step yields L-gulonic acid and then L-gulonolactone, which is converted to L-ascorbic acid through 2-keto L-gulonolactone as an intermediate. The structures involved in the biosynthesis are not presented here, since they are given. Galactose is converted to ascorbic acid by rats through the following sequence of reactions: galactose ⟶ galactose-1-PO_4 ⟶ UDP galactose ⟶ UDP glucose ⟶ as above for glucose. It is further established that in man, monkeys, and guinea pigs there is a deficiency in the enzyme or enzymes responsible for the conversion of gulonolactone to ascorbic acid. Thus, Burns found no detectable conversion of L-gulonolac-tone-1-C^{14} into labeled L-ascorbic acid in homogenates Of human, guinea pig, or monkey liver, while rat liver showed about 8 per cent conversion under similar conditions.

Two other animals in the avian species more recently have been found to be unable to synthesize ascorbic acid and must have a dietary supply of the vitamin. These are the Indian fruit bat and the red vented bulbul.

Metabolism of Ascorbic Acid

In man ascorbic acid is partly excreted unchanged and partly as diketo-L-gulonic acid and oxalic acid. Hellman and Burns using labeled ascorbic acid, found 12 to 24 per cent of the urinary C^{14} activity in L-ascorbic acid, 12 to 18 per cent in diketogulbnic acid, and 24 to 63 per cent as oxalic acid in several experiments with three subjects. No C^{14} was detected in respiratory CO_2. The vitamin exists in the body in an equilibrium between the reduced and the oxidized (dehydro) states, with only a small fraction in the latter form. The oxidized form can be either reduced to ascorbic acid reversibly, or metabolized to diketogulonioacid irreversibly.

Both the oxidized and reduced forms have vitamin activity, while the diketogulonic acid has none. Oxalate and ascorbic acid in urine accounted for about half the total ascorbic acid turnover in an adult man in the experiments of Atkins and others using C_{13}Magged ascorbic acid. Most of the oxalate was shown to arise from ascorbic acid or glycine. Respiratory CO_2 contains carbon atoms from C-l of ascorbic acid. Oxalate arises from carbons 1 and 2 of ascorbic acid. Rats and guinea pigs metabolize the carbon chain of ascorbic acid producing principally CO_2 and oxalic acid, although pathway differences in these species are established. Chan and others found that guinea pig liver preparations formed dehydroascorbic acid which was degraded to-oxalate, CO_2, and L-xylose. No xylose was found from the catabolism of ascorbic acid by rat kidney in studies by Burns and co-workers. The half-life of ascorbic acid in the guinea pig is only a few days, while in man it has been estimated to be about 16 days. This may account for the long period required to produce a deficiency state in man compared to the time involved to produce symptoms in the guinea pig.

Determination of Ascorbic Acid

Two principles have been employed in the important chemical methods for vitamin C determination. In the first the strong and fairly rapid reducing property of the vitamin is determined by titration against a standard oxidizing solution. At present the dye 2, 6-dichlorobenzenoneindophenol (2, 6-dichlorophenolindophenol) is the oxidant of choice. Rather than titrate ascorbic acid, an excess of the dye solution may be added to the ascorbic acid solution, and the loss of color due to reduction of the dye, determined by use of a photoelectric colorimeter.

Many modifications of each of these approaches have been used. The vitamin is generally extracted from animal or vegetable tissues with metaphosphoric acid (HPO_3) or trichloroacetic acid. In the case of vegetable tissue high in ascorbic acid oxidase the enzyme can be destroyed by extracting with hot HPO_3. Dehydroascorbic acid does not reduce the 2,6 dye, and, in plant tissue especially, appreciable amounts of the vitamin may exist in this form. Treatment of the acid extract with H_2S converts the vitamin C to the reduced form. This is an important consideration, since both forms have vitamin activity. The direct titration methods, the photoelectric method and the xylene extraction method each have certain advantages. The latter is especially adapted to turbid or highly colored ascorbic acid-containing solutions. Another chemical principle employed in ascorbic acid determination involves the conversion of the vitamins into a soluble colored complex by treatment with 2, 4-dinitro-phenylhydrazine.

The color intensity can be determined in any of the various instruments for this purpose. A critical review of various methods is available. In the various biological methods the guinea pig is the animal of choice. Practical methods have been developed employing growth, tooth structure, and growth of odontoblasts of incisors as the criteria of the ascorbic acid content of the test material. Details can be found in the sixteenth revision of the *United States Pharmacopoeia*.

Ascorbic Acid Status

Most of the approaches to the study of an individual's nutritional status with respect to ascorbic acid involve an attempt to study the degree of tissue saturation. There is little diffculty in the estimation of blood or urine levels, but such data alone often do not truly reflect the condition in the tissues. It has been pointed out that the level in the white cells may be more significant than that of the plasma for diagnostic purposes. The so-called saturation tests give additional information. The blood, and urine levels may well be affected by the intake during the day previous to the time the samples are tested, but the degree of tissue saturation would not be markedly altered unless the intake were very high. The tissue saturation tests involve a study of urinary ascorbic acid excretion in response to a test dose.

In the individual with low reserves the vitamin is held in the tissues, and the total excretion is below normal. In one of the tests 5 mg of ascorbic acid per pound is given orally, and if 50 per cent is excreted during the ensuing 24 hours, the patient is said to have been saturated. Excretion of less than this amount indicates unsaturation. In a severely deficient patient the excretion may be from almost none up to 20 per cent of the test dose. Human blood plasma ascorbic acid levels of 0.7 to 1.2 mg per 100 ml are considered to be within the normal range. Many investigators now consider that levels of 0.4 to 0.7 mg per cent indicate a mild deficiency, and those below 0.4 mg a severe deficiency. On high dietary intakes the plasma level may rise to 2.0 mg per cent. In controlled studies of men in the Canadian Air Force on known vitamin C intakes good correlation was found between the white cell level and the amount consumed daily. Thus on intakes of 8, 23, and 78 mg ascorbic acid daily the white cell levels were found to be 11.9, 12.9, and 24.2 mg per 100 g white cells.

The plasma level of ascorbic acid except on constant intake probably reflects the recent consumption of the vitamin more than it does the degree of body saturation and thus may tell one less about vitamin C status of an individual. In older people the blood level may decline. Thus, Berlina found the low average level of 0.13 mg per 100ml in 20 healthy aged men and women on intakes of 30 to 35 mg of ascorbic acid per day. Urinary excretion ranged from two thirds to nearly all of the vitamin C intake.

Distribution of Ascorbic Acid

Vitamin C is distributed rather widely in nature. Important dietary sources for man include many vegetables and fruits. Canning, cooking, and other processing result in various degrees of vitamin C loss. Fresh vegetables such as broccoli, kale, parsley, and turnip greens have a high content of the vitamin, but these foods are generally not eaten in the raw state. The citrus fruits are also excellent sources and are consumed largely without processing, although for economic reasons they are not an important vitamin C source for the overall population. Certain vegetables constitute important sources by virtue of the amount eaten rather than due to a high level of the vitamin. In this class are potatoes, beans, and peas. Most animal products contain only small amounts of the vitamin. Because the vitamin of milk is largely destroyed in pasteurization or in the evaporation process, another source of vitamin C is regularly supplied to infants. By definition 1 IU or 1 USP unit is equivalent to 0.05 mg of L-ascorbic acid. This makes 1 mg equal to 20 USP units.

Table 3.4 gives the ascorbic acid content of a number of foods. The values given are, of course, subject to considerable variation due to climatic and soil conditions, etc.

Ascorbic Acid Requirement

The optimum human requirement for vitamin C remains a controversial matter. The recommended intakes set by the Food and Nutrition Board of the National Research Council are not necessarily based on sufficient experimental data, since such data are largely. lacking. This board [1963] recommends daily ascorbic acid intakes of 70 mg per day for adults, 100mg during pregnancy and lactation, and 30 mg for infants increasing to 80 mg in young adults. These intakes are considerably higher than many workers feel are needed. Abt and co-workers

Table 3.4. Ascorbic Acid Content of Some Foods.

Food	*Ascorbic Acid, mg per 100 g*
Strawberries	40-80
Watermelon	5-8
Orange jucie	40-70
Lemon jucie	40-60
Apple	3-10
Pineapple, fresh	20-30
Pineopple, conned	2-10
Potato, white	20-30
Patato, winter stored	5-10
Cabbage, fresh	40-70
Cabbage, cooked	15-20
Sauerkraut	10-20
Lettuce, head	100-150
Beans, green (canned)	2-5
Tomato, fresh	20-30
Tomato, juice	10-20
English walnut's unripe)	500-2000

first reported the exhalation of $C^{14}O_2$ after ingesting labeled ascorbic acid. From studies on excretion patterns of ascorbic acid these investigators set the minimum requirement for man at 1.0 to 3.0 mg daily. This is not necessarily an optimum or recommended intake and is, of course, only a fraction of that recommended by the Nutrition Board. It is certain that a large proportion of our population has an intake considerably lower than the recommended allowances. A study of ascorbic acid intakes by Olympic athletes in London is of interest.

The seven subjects studied in this connection ate duplicate meals, cafeteria style, for a four-day period, and these were assayed for vitamin C as well as for other nutritional constituents. The average intakes were found to be 41, 43, 45, 71, 80, 81, and 98 mg of ascorbic acid per day for these subjects.

"Vitamin P"

After pure ascorbic acid became available, it was noticed that in guinea pigs on scorbutic diets this material, in certain instances, was not as effective in alleviating the tendency to hemorrhage as was one of the natural food sources of ascorbic acid, such as citrus fruits. In 1936 Szent-Gyorgyi and co-workers reported the presence of material in red peppers and in citrus fruits which they claimed was beneficial in the control of hemorrhage in man and in guinea pigs and was chemically different from ascorbic acid.

Since the material was thought to be involved in capillary permeability and was first found in paprika, the name "vitamin P" was given the newly discovered factor. It is difficult to demonstrate a vitamin P deficiency state in animals or in man. It was claimed, however, that guinea pigs on a flavone-free diet supplemented with ascorbic acid developed capillary weakness which responded specifically to vitamin P. Many workers feel that the active substances do not warrant the status of a vitamin, since it is possible to produce symptoms only in the scorbutic guinea pig treated with ascorbic acid. Of the several compounds with vitamin P activity the most effective are rutin from buckwheat and esculin from chestnuts. Rutin is 3, 5, 7, 3', 4'-pentahydroxy-flavone-3-rutinoside. Rutinose is a disaccharide containing glucose and rhamnose. Esculin is 6, 7-dihydroxycoumarin-6-glucoside. The active substances are all of plant origin. In citrus fruits the concentration is higher in the rind than in the juice. It is not established that the "vitamin" is required in man. No figures can be given for a requirement that may not exist.

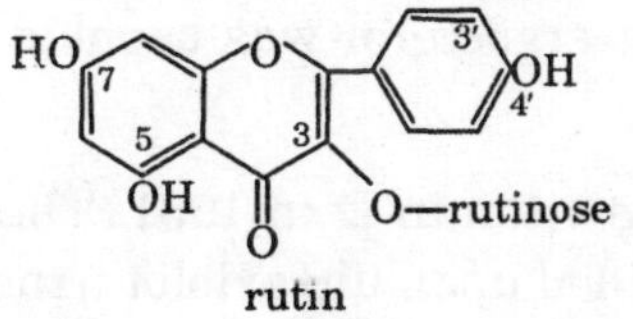

rutin

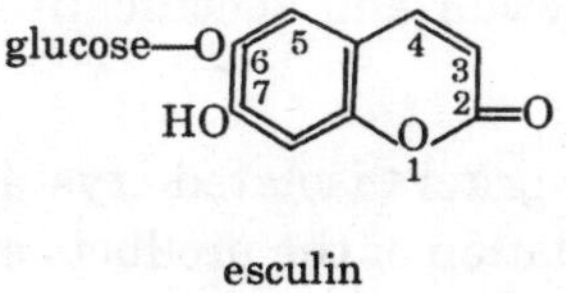

esculin

Vitamin D

The disease rachitis, now commonly called rickets, has apparently plagued mankind since ancient times. In 1650 infant rickets was described by Glisson in England. The disease was rampant especially among the children of the lower classes of people in England (and other sections of the world) for centuries. Infection was early postulated as a cause of the disease. Around London the abundance of fog was held by some to be a contributing factor. It is now known that the latter assumption was correct, although at that time it may have been predicated on erroneous beliefs.

In other words, the lack of sunshine due to the fog was a predisposing factor. The curative value of liver, especially cod liver oil, was known centuries ago. Late in the eighteenth century it was used in parts of England as a therapeutic agent, although its specific value in preventing or curing rickets was apparently unknown until many years later. As late as 1920 Hess and Unger indicated that rickets was a very common disorder of infants living in the temperate zone. After World War I infant rickets in various parts of Europe was widespread. Our present-day knowledge of rickets and the chemical nature of the vitamers D is based primarily on a series of seemingly unrelated findings. Important among these are the following:

1. In 1918 Mellanby produced the first clear-cut experimental rickets in animals. Dogs were fed various diets of milk and porridge or bread and milk. Rickets developed regularly on such diets. He was able to demonstrate the presence of the antirachitic factor in cod liver oil and showed that this substance was far superior to butter fat and peanut, olive, or linseed oils in curing or protecting against rickets in his dogs.

2. Huldschinsky in 1919 demonstrated marked clinical improvement in severely rachitic children by playing ultraviolet light on their bodies. This finding was important not only from the clinical standpoint but also afforded impetus to the theory, doubted by many, that sunlight is beneficial in the treatment of rickets.

3. In 1922 McCollum and co-workers demonstrated that cod liver oil contains a specific substance (vitamin) concerned with calcium deposition in rachitic rats. The presence of "fat-soluble A" in cod liver oil had been known for some time. By running air through the oil at 100°C the destroyed the xeropbthalmia-curing principle, but the oil maintained a high activity in respect to calcium metabolism. Thus the dual vitamin nature of cod liver oil was established.

4. Two groups of investigators reported in 1924 that the irradiation of certain foods with ultraviolet light endowed them with antirachitic (vitamin D) activity. Steenbock and co-workers and Hess reported this remarkable finding at about the same time. Shortly after this it was demonstrated that the sterol fraction of foods actually contained the material that became antirachitic on irradiation and specifically that ergosterol was capable of a high degree of activation.

5. Angus and co-workers isolated crystalline vitamin D in 1931. They accomplished this by high vacuum distillation of the products obtained upon ultraviolet irradiation or ergosterol. This was named calciferol and is referred to as vitamin D_2.

6. Vitamin P_8, activated 7-dehydrocholesterol, was isolated by Wiridaus and co-workers in 1936. They irradiated 7-dehydrocholesterol in benzene and were able to separate crystalline derivatives-of the vitamin. The name vitamin D_3 was proposed for the substance at that time.

It is established now that small amounts of 7-dehydrocholesterol accompany animal cholesterol. The skin contains this precursor, and thus ultraviolet light, from the sun or from artificial sources, is able to activate and yield vitamin D_3 to the body. As a result of the preceding findings and, of course, many other outstanding observations, the value of sunlight and many artificially produced vitamin D products in the treatment of rickets was realized.

In the early 1930's vitamin D milk (ultraviolet irradiated) was widely available in the larger centers of population, and a number of highly concentrated vitamin D preparations were marketed previous to that time. Likewise, the elucidation of the chemistry of the various vitamers D arid the photochemical changes proceeding during the ultraviolet irradiation of the precursors were markedly hastened. It became evident that two forms of vitamin D are of importance in human nutrition: calciferol (vitamin D_2), produced from irradiation of the plant sterol ergosterol, and vitamin D_3, the activation product of the animal sterol 7-dehydrocholesterol. Many other vitamers are known, but they are at present of academic interest only.

Dihydrotachysterol, or AT 10, is a reduction product of tachysterol; the latter is one of the products of irradiation of ergosterol (see further). The parent substances which can be activated are also known as provitamins.

Chemistry of Vitomin D

The compounds with vitamin D activity are closely related to the perhydrophenanthrencyclopentane ring system, and thus to the naturally occurring sterols. Only the two antirachitic substances calciferol or vitamin D_2, and vitamin D_3 will be discussed in any detail.

When ergosterol dissolved in alcohol, benzene, ether, etc., is irradiated with ultraviolet light, a series of photochemical reactions takes place.

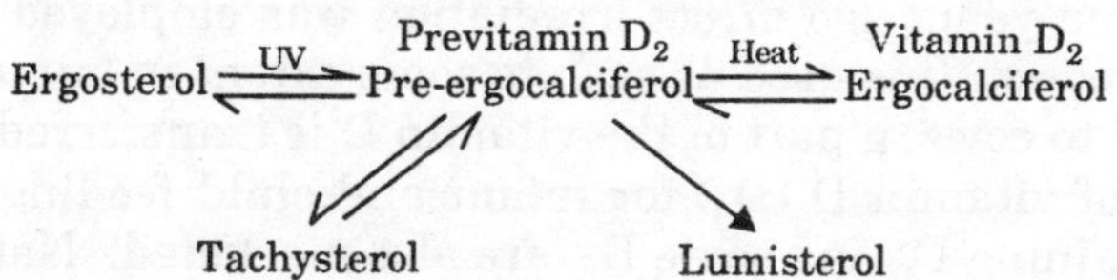

Fig. 3.3. The conversion of ergosterol to vitamin D_2, and side reactions.

Ergosterol does not absorb visible light but does absorb various wavelengths in the ultraviolet spectrum. The absorption maxima are at 260, 270, 282, and 293.5 mμ. The absorption of ultraviolet light or heat supplies energy for the aforementioned transformations. The reactions are photochemical or activated by heat. Nothing is added to or removed from the molecules during these rearrangements; all the compounds are isomeric, and their structures are known. It should not be inferred that all the ergosterol is converted to preergocalciferol (or pre-calciferol) and that this compound is then converted to the next in the series.

Various rearrangements proceed simultaneously so that mixtures of the various products are present at any one time. Ergosterol and lumisterol are not antirachitic, Tachysterol, after reduction to dihydrotachysterol, is used especially in the treatment of some types of human tetany. It was found that following irradiation of ergosterol at 20°C the vitamin D_2 content of the solution increased on heating, without further irradiation. The intermediate compound indicated by these findings was isolated and shown to be an isomer of D_2. This previtamin (preergocalciferol) transforms itself into ergocalciferol, a process that is accelerated by heat. Ring *B* is open in the previtamin, but no double bond is present between carbon atoms 10 and 18. A shift in double bonds is involved in the conversion to calciferol and *cistrans* isomerism (C6) in tachysterol formation. The changes involving the conversion of 7-dehydrocholesterol into vitamin D_3 are known to be similar to those shown for ergosterol, and such a conversion occurs in the body from the ultraviolet of sunshine or other source. There is no vitamin D_1; the original proposal for such a vitamin was later shown to be in error, since the material was found to be a mixture of calciferol and lumisterol. Vitamin D_2 and vitamin D_3 are of importance in nutrition and medicine.

The value of other vitamers D cannot be accurately assessed at present, since small amounts may occur in natural sources and their value, if any, is not established. Vitamin D_4 is activated 22-dehydroergosterol. Vitamin D_5 is activated 7-dehydro-sitosterol. A number of other products also have some degree of antirachitic activity.

Sources of Vitamin D

It is practical to consider the natural and synthetic sources of the antirachitic vitamins at this time, since interesting differences in regard to animal vitamin D (activated 7-dehydrocholesterol) and plant vitamin D (calciferol) content of the different substances arise. At present many commercial vitamin D products are available in concentrated form. Calciferol dissolved in vegetable oil or in propylene glycol is marketed for use primarily in infant and child feeding. High-potency capsules (100,000 units for instance) are also available. These preparations contain vitamin D_2. Vitamin D milk is widely distributed in the United States. Practically all the evaporated milk is fortified with concentrates to increase the vitamin D content.

Either D_2 or D_3 may be used. Fluid milk is now generally fortified with concentrates (D_3 primarily), although some years ago direct irradiation was employed (D_3). Irradiated yeast (D_2) is a high-potency product. It is used as such to some extent in human and more in animal nutrition. When it is fed to cows, a part of the vitamin D is transferred to the milk. Cod liver oil is a popular source of vitamin D (D_3) for infant and child feeding. Many other types of preparations, some containing D_2 and some D_3. are also marketed. Naturally occurring foods have practically no vitamin D activity.

Milk contains insignificant amounts as far as infant and child nutrition is concerned unless vitamin D is added. Grains and vegetables in general have still less. Butter and liver have small quantities. It is interesting that, in contrast to many fish livers, this organ in mammals is not considered a source of the vitamin in nutrition. Reference to the discussion of vitamin A shows the vitamin D content of a number of fish liver oils.

Physiological Action of Vitamin D

Vitamin D is associated with a number of physiological processes in the body. At our present state of knowledge we can say that from a qualitative standpoint the physiological actions of the well-known vitamers are similar, if not the same. Quantitatively, however, marked differences are known. This matter will be discussed later. Unless otherwise specified, the term vitamin D will hereafter refer to either D_2 or D_3 or to other mixtures as they may occur in natural or synthetic preparations.

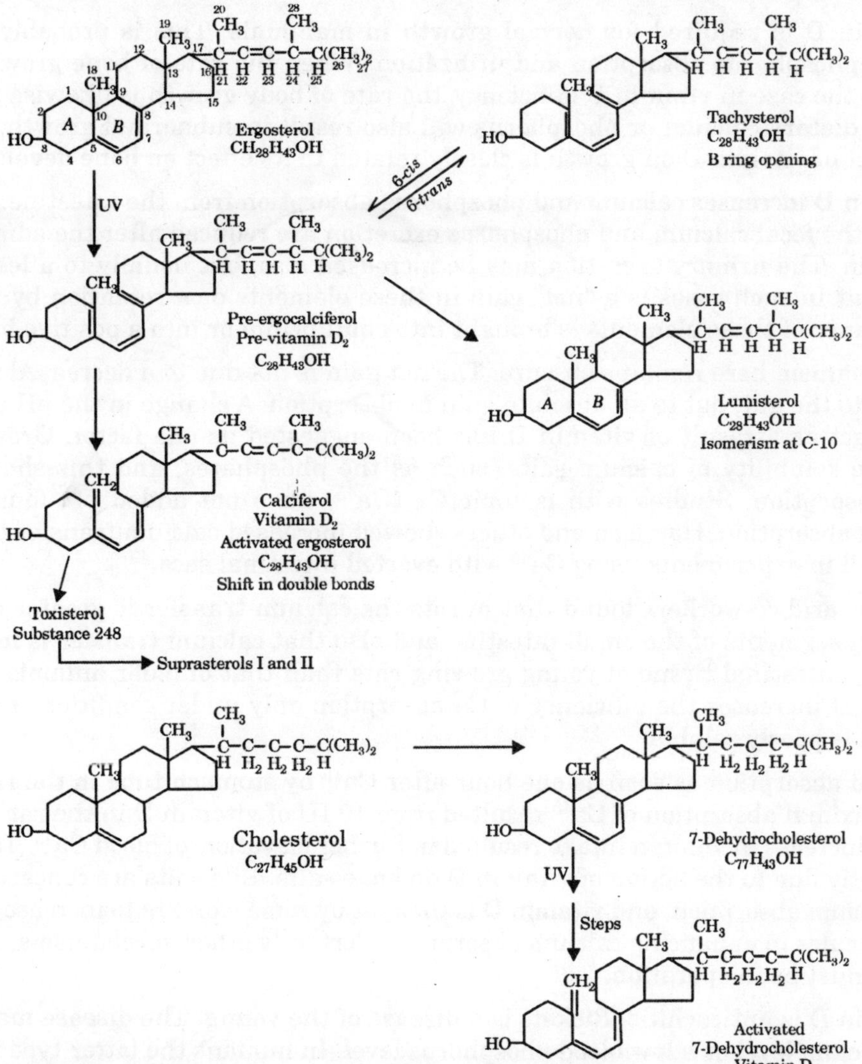

Ergosterol
$CH_{28}H_{43}OH$

UV

Pre-ergocalciferol
Pre-vitamin D_2
$C_{28}H_{43}OH$

6-cis
6-trans

Tachysterol
$C_{28}H_{43}OH$
B ring opening

Lumisterol
$C_{28}H_{43}OH$
Isomerism at C-10

Calciferol
Vitamin D_2
Activated ergosterol
$C_{28}H_{43}OH$
Shift in double bonds

Toxisterol
Substance 248

Suprasterols I and II

Cholesterol
$C_{27}H_{45}OH$

7-Dehydrocholesterol
$C_{77}H_{43}OH$

UV

Steps

Activated
7-Dehydrocholesterol
Vitamin D_3
$C_{77}H_{43}OH$

Fig. 3.4. Scheme of rearrangements in ultraviolet irradiation of ergosterol and of 7-dehydrocholesterol producing vitamins D_2 and D_3.

The complexity of vitamin D physiology can well be appreciated if one considers (*a*) the close interrelations of the actions of the vitamin and of the hormone(s) of the parathyroid gland; (*b*) the several forms of both calcium and phosphorus in foods, blood, tissues, urine, and feces; (*c*) the ill-understood mechanism concerned in bone growth and repair; and (*d*) the apparent variations in calcium and phosphorus metabolism in young and adult life.

It is impossible, of course, to discuss vitamin D without at the same time discussing various aspects of the points *a*, *b*, *c*, and *d* above.

The following are established as specific actions of vitamin D in the body :

1. Vitamin D is required for normal growth in mammals. This is probably related to calcium and phosphorus absorption and utilization. When the rate of bone growth is below normal, as is the case in vitamin D deficiency, the rate of body growth is likewise retarded: A deficiency of dietary calcium or phosphorus will also result in subnormal growth. It appears that the effect of vitamin D on growth is closely related to its effect on bone development.

2. Vitamin D increases calcium and phosphorus absorption from the intestine. In vitamin D deficiency the fecal calcium and phosphorus excretion are reduced after the administration of the vitamin. The urinary excretion may be increased also, but usually to a lesser extent. The end result in such cases is a "net" gain in these elements or a retention by the body. A negative balance of these elements is brought into equilibrium or into a positive balance.

The mechanism here remains obscure. The net gain is not due to a decreased reexcretion of calcium into the gut, but to an absolute gain in absorption. A change in the pH of the lower intestinal tract as a result of vitamin D has been suggested as one factor. Greater acidity increases the solubility of calcium salts, such as the phosphates, and this should lead to increased absorption. Studies with isotopic Ca (Ca^{45}) have not added pro-foundly to our knowledge of absorption. Harrison and others showed increased calcium transfer through the intestinal wall in experiments using Ca^{45} with everted intestinal sacs.

Schachter arid co-workers found that in rats the calcium transfer is greater in proximal than in distal segments of the small intestine and also that calcium transfer is more readily performed by intestinal tissue of young growing rats than that of older animals. It appears that vitamin D increases the efficiency of Ca absorption only under conditions in which the intestinal Ca is poorly soluble.

Increased absorption as soon as one hour after Ca^{45} by stomach tube in the rat has been observed. Maximal absorption of Ca^{45} resulted from 10 IU of vitamin D in the rat, although a hundredfold increase in vitamin intake resulted in further elevation 'of blood Ca^{45}. This increase was apparently due to the action of vitamin D on bone salts. Bile salts are concerned in some way with calcium absorption, and vitamin D is thought by some workers to increase the activity of these molecules in enhancing calcium absorption. Certainly other .mechanisms, more subtle than these, must be in operation.

3. Vitamin D is antirachitic, Rickets is a disease of the young. The disease may involve a low blood calcium level or a low blood phosphorus level. In humans the latter type is generally seen. Either can be produced in animals by dietary means. In infancy the inorganic phosphorus level (primarily as HPO_4^- and $H_2PO_4^-$) of blood is normally 4 to 6 mg per cent. In rickets this may be decreasec to 1 or 2 mg per cent.

At the ends of bones during normal growth the osteoblasts, or bone-forming cells, appear as the cartilage cells degenerate and disappear. After capillaries grow into this site, the osteoblasts deposit bony matrix. The process is a continuous one in that new osteoblasts are always under formation. In a vitamin D deficiency the cartilage cells do not degenerate but continue to grow; consequently, capillaries and osteoblasts are not formed.

The cartilage tissue increases in size and remains uncalcified. In more severe deficiency bone mineral may be resorbed, leaving a greater area of osteoid tissue. The time and the degree of the deficiency obviously determine the extent of the abnormality. In infants and children clinical symptoms of severe deficiency are readily seen on gross examination. Enlargements of the ankle, knee, and wrist joints are noted. Other prominent features are bowed legs, delayed closure of the fontanelle, beading of the ribs at the costochondrarjunction ("rachitic rosary"), and delayed tooth eruption. The mineral content of bone decreases as the severity of the rickets progresses. The ash on a dry, fat-free basis may reach one half to one third or less of the normal value. The administration of vitamin D to rachitic animals brings about the degeneration of cartilage cells, the growth of capillaries in this area, and ossification.

How does the presence of vitamin D bring about such remarkable and rapid effects? Here again our information on actual mechanisms is almost nil. The increased absorption of calcium and phosphorus from the gut due to vitamin D is not the answer, since healing of rachitic bones in animals can take place without food or with food devoid of these minerals. Also, the intravenous administration of calcium and phosphorus salts to raise blood levels is not effective in initiating healing. The administration of parathyroid hormone or of AT 10, both of which increase blood calcium, likewise produces little or no antirachitic effect. It was demonstrated in 1928 by Hess that when rachitic bone is placed in blood serum from rachitic animals no change takes place, but if serum from a normal animal is used, calcification is initiated. Robison demonstrated the presence of an enzyme capable of splitting inorganic phosphate from organic combination (phosphatase).

Various phosphatase enzymes are now known to be in the body. In bone cartilage and especially in rachitic osteoid tissue the concentration is high. This incidentally increases the blood content of the enzyme. The level is of value in diagnosing early rickets and some other bone abnormalities. Many investigators have found rather good agreement between vitamin D deficiency and increased blood alkaline phosphatase. In a study of a large number of children, aged six months to 2.5 years, a mean value for the serum phosphatase of 9.4 Bodansky units was reported for the normals, whereas the rachitic children had in general over twice this phosphatase activity. It was also indicated that the increased enzyme activity correlated well with severity of the clinical symptoms of rickets.

Apparently the enzyme is responsible, under normal conditions, for liberating inorganic phosphate from organic combination and thus increasing at the site of ossification the ion product $Ca^{++} \times PO_4^{---}$. It has been suggested that bone is deposited in the matrix only when the solubility product of $Ca^{++} \times PO_4^{---}$ in plasma is exceeded, and perhaps that the calcium and phosphate ions must be present in a special form. At any rate the process of ossification requires phosphatase enzymes, and it now appears that a number of other enzymes are likewise intimately involved in the process.

The glycolytic process with the many enzymes involved not only may provide some as yet unidentified organic phosphate ester as the substrate for phosphatase, but may also supply the energy requirements. Rachitic bone contains glycogen and the enzymes required for the glycogenolysis process. When such bone slices are placed in the proper medium containing inorganic phosphate, calcification takes place. However, if phlorizin is also added, calcification does not proceed. This substance is known to inhibit phosphorylase, the enzyme responsible for the conversion of glycogen plus inorganic phosphate into glucose-1-phosphate.

The inference is that this block precludes the further steps leading to the required phosphate ester which might act as phosphatase enzyme substrate. Upon the addition of glucose-1-phosphate, the compound that the system is unable to produce, calcification proceeds in the presence of the inhibitor. By studying inhibitors of other enzymes in the system, the conclusion has been reached that phosphorylative glycolysis plays an important role in the calcification process.

It is likely that normal bone calcification *in vivo* likewise involves such complicated mechanisms to supply a special type of phosphate ester in order that phosphatase may liberate phosphate ions at the active calcifying site. For the normal bone formation or for calcification of rachitic bone, vitamin D is required. What part it plays is still questionable. Its activity may involve some effect on the phosphatase enzyme system, or the state of serum calcium or phosphorus or both, or the various cells concerned with laying down bone.

Phosphorylated vitamin D, but not the unaltered molecule, was shown to have a marked initial activating effect on kidney alkaline phosphatase. What part, if any, this might play in bone formation is not now apparent. The rat differs from other species in vitamin D requirement. A rachitic condition is easily induced in this animal by keeping him on a diet with an upset calcium-phosphorus balance, and bone healing will proceed in the absence of Vitamin D if the animal fails to grow or loses weight. This will be discussed further under the section on vitamin D assay. Without understanding each biochemical event in calcification, the process may be visualized as follows: As cartilage cells in the growing end of bone degenerate, the matrix is invaded by capillaries and the bone-forming cells called osteoblasts. These cells may induce the seeding of crystals of some form of calcium phosphate.

The collagen fibers of the matrix are regularly bonded with a spacing of 640 A, and this spacing appears to be critical and probably involved also in the seeding process. These workers regard alkaline phosphatase as essential in splitting organic phosphate esters to increase the Ca tims P product to a critical level. The initial precipitate is unstable and changes to hydroxyapatite. Maturation consists in crystal growth and displacement of water among other changes. A more dense bone is thus produced.

4. Harrison and Harrison demonstrated a specific function of vitamin D on kidney tubular reabsorption of phosphate. They developed rickets in dogs and then by phosphate clearance studies showed that the administration of large doses of vitamin D definitely increased phosphate reabsorption. At equilibrium the plasma phosphate .was thus elevated, and these workers postulated that such an effect is probably part of the antirachitic action of vitamin D. Such an action is definitely antirachitic since it tends to conserve phosphate and to increase the $Ca^{++} \times PO_4^{---}$ product of the plasma. It has been pointed out that the increased tubular reabsorption may be a secondary effect to the primary increase in circulating calcium.

5. Citric acid is a normal constituent of many body tissues, including bone. Rachitic rats given vitamin D show increased urinary excretion of citric acid and increased levels in blood, bone, kidney, heart, and small intestine, with no elevation in liver. Such findings indicate a rather general effect of the vitamin on citric acid metabolism. Steenbock and co-workers have continued with *in vitro* enzyme studies and demonstrated that addition of vitamin D to either a rachitogenic or a nonrachitogenic diet resulted in a depression of *in vitro* citric acid oxidation by kidney homogenates or mitochondria. In further experiments the addition of the vitamin to an *in vitro* system of kidney mitochondria reduced citrate oxidation. These findings help explain the increased tissue citrate levels in a deficiency of the vitamin.

6. It has been suggested that vitamin D increased the activity of the enzyme phytase in the rat intestine. This enzyme hydrolyzed food phytic acid (grains primarily), yielding inorganic phosphate. More phytic acid is excreted in the feces of rats and dogs in a deficiency state than when the vitamin is given. The rat intestine produces phytase, and in rachitic rats phytic acid of a high-cereal diet is completely hydrolyzed only when vitamin D is given. However, the increased enzyme activity does not liberate sufficient inorganic phosphate to account for the antirachitic action of the vitamin. Vitamin D is known to stimulate the release of calcium and strontium from isolated rat kidney mitochondria.

The vitamin D effect on release of calcium is observed only in the presence of an oxidizable substrate such as succinate and is accompanied by a release of inorganic phosphate. Thompson and DeLuca showed that rats fed a diet with added vitamin D showed a threefold increase in the incorporation of P^{32}-orthophosphate into the phospholipids of intestinal mucosa. At the same time there was no alteration of the incorporation into nonlipid organic phosphates. Such findings do not readily fit into a biochemical scheme at the present time. It is advisable at this point to digress and bring the parathyroid glands into the discussion. One of the primary actions of the parathyroid hormone is to increase the urinary excretion of phosphate and concomitantly reduce the plasma phosphate level.

It is known too that an increased plasma calcium level tends to lessen the activity of the parathyroid glands and that a low level results in increased activity. On the basis of these two facts and on an imposing accumulation of data of their own on the action of vitamin D, the parathyroids, etc., on calcium and phosphorus metabolism, Albright and Reifenstein interpreted the results of Harrison and Harrison on the basis that the increased kidney reabsorption of phosphate resulted from decreased parathyroid activity (from increased blood calcium level). Albright and co-workers showed that in a patient with idiopathic hypoparathyroidism the administration of large doses of vitamin D increased urinary calcium and phosphorus excretion more than the decrease in fecal calcium and phosphorus excretion. The extra mineral obviously came from bone, and negative balances of these two elements developed.

Albright feels that a primary action of vitamin D is to increase the urinary excretion of phosphate, but that the action is quantitatively far less than that of the parathyroid hormone. In the case of an individual with a normal parathyroid, he feels that this action is masked by the ability of the vitamin to increase the absorption of calcium from the intestine, leading to an increased serum calcium level, which inhibits parathyroid hormone production, resulting in lessened urinary phosphate excretion. Under these conditions the serum phosphate increases, whereas in the individual with a parathyroid hormone deficiency the end result on plasma phosphate is opposite after the administration of vitamin D. Increased blood calcium cannot have an effect on a nonfunctional parathyroid gland. The other primary action of the parathyroid hormone (besides increasing urinary phosphate excretion) is to increase a low blood calcium level or maintain a normal one by removal of calcium from bone.

Excess hormone causes hypercalcemia. As a result of the two established actions of the hormone, two schools of thought on its primary action have developed. One group holds that parathyroid hormone brings about increased phosphate excretion by the kidney and that changes in blood calcium and phosphate levels are secondary to this. A variety of experiments indicate increased phosphate excretion, followed by elevation of the blood calcium, upon administration of the hormone. The other school claims that parathyroid hormone brings about dissolution of bone calcium and that the renal effects are secondary.

Evidence in favour of this theory is found in the various experiments in which nephrectomized animals, unable to excrete phosphate, respond to the hormone with an increase in blood calcium. In one such experiment, nephrectomy (rats) appeared to nullify the effects of the hormone upon blood phosphate levels while the 'action as regards blood calcium was maintained. These workers were led to postulate that normally the hormone has a direct and independent effect on both calcium and phosphorus metabolism.

In a review on the mode of action of the parathyroid hormone the action of maintaining blood calcium level is stressed, and new findings do not offer an explanation of how this is accomplished. It is advisable at present to keep in mind the two well-established actions of the hormone without regard to which one is primary—they are both obviously important. Unlike vitamin D, the hormone has little effect on the absorption of either calcium or phosphorus in the gut. Overdosage with parathyroid hormone brings about bone demineralization and hypercalcemia.

Hypercalcification, on the other hand, may result from massive doses of vitamin D. The hormone has no effect on the healing of rachitic bones, but in a hormone deficiency tetany vitamin D is effective in regulating the blood calcium and phosphorus levels. Usually large doses, 50,000 to 200,000 IU of vitamin D per day, are required for this in humans. Another important drug in this respect is dihydrotachysterol (AT 10). This product has been used in human parathyroid hormone deficiency with considerable success.

It is administered orally in oil solution, a distinct advantage over parathyroid hormone, which must be given parenterally. AT 10 is very slightly antirachitic; it increases calcium absorption from the intestine, brings about resorption of calcium from bone to increase the blood level, and increases urinary excretion of phosphate. The last two actions are those of the parathyroid hormone; the first two are those of vitamin D. Quantitatively the actions of AT 10 are less than those of either the vitamin or the hormone. Albright and co-workers have studied the effects of the three substances, parathyroid hormone, AT 10, and vitamin D, important in the physiological regulation of deranged calcium and phosphorus metabolism in a variety of clinical conditions.

Table 3.5. Relative Effect of Vitamin D, Dihydrotachysterol, and Parathyroid Hormone on Intestinal Calcium Absorption and Urinary Phosphorus Excretion

	Calcium Absorption	*Urinary Phosphours Excretion*
Vitamin D	++++	++
Dihydrotachysterol	++	+++
Parathyroid hormone	+	++++

Vitamin D, especially in large doses, is somewhat effective in bringing about calcium resorption from bone; AT 10 is far more effective; and the parathyroid hormone has still greater activity. The hormone has a very short action, a matter of hours, and may bring about its maximum effect very rapidly (four to ten hours). AT 10 has a longer period of action, up to several days, and requires, a day or two to act.

The vitamin may be effective for long periods of time and is slow to exert its effect. Multiple C^{14}-labeled calciferol was prepared by Kodicek. One mg of the compound in oil was given to a rachitic rat by mouth. After 24 hours, analyses showed that only 30 per cent of the C^{14} activity remained as vitamin D. although all the C^{H} activity was recovered. The bulk of the breakdown products were not identified. The liver contained 5.7 per cent of the dose, the bones 1.4 percent, the intestines 0.6 per cent, the kidney 0.2 per cent, and the blood 1.2 per Cent. It was of interest that tissues concerned with phosphate turnover, such as kidney, bone and intestine, contained significant amounts of the labeled vitamin D.

Vitamin D Assay

Vitamin D is determined most successfully at the present by animal assay. Rats or chicks are employed, and the standardized procedures yield excellent and reproducible results. The rat assay method is used for vitamin products intended for human consumption. Chicks do not respond to vitamin D_2 as they do to D_3 on a rat unit basis. In the assay of vitamin D products for the poultry industry it is necessary to use chicks as the test animal. In this procedure the vitamin product is fed to groups of chicks at various levels with a standardized diet, and the level of bone ash after a certain number of days is the criterion used to establish the vitamin level in the test material. The rat, as indicated elsewhere, is unique as regards the development of rickets.

No extreme caution is required to deprive a ration of vitamin D for the production of rickets in this species. It is only necessary to arrange the calcium-phosphorus ratio of the diet to about 4 or 5 to 1. Of the various rachitic rations developed, the Steenbock 2965 ration is employed most generally. This is composed of wheat gluten, yellow corn, calcium carbonate, and salt. Weanling rats restricted to water and this diet for 18 to 21 days develop a rather reproducible rachitic condition. Small additions of test material are given daily for the following five days (in one procedure) or eight days to groups of the standard rachitic rats.

One group receives a known amount of vitamin daily (USP Reference cod liver oil). After a further day or two without supplement, the animals are sacrificed. The rachitic metaphysis of the distal ends of the longitudinally sectioned radius and ulna are examined for new bone growth brought about by the vitamin administered. This is readily seen on gross examination after treating the bones with $AgNO_3$ solution and then reducing the silver phosphate by light. Black areas develop wherever bone mineral exists. At the provisional zone of calcification a rather straight continuous line of new bone is seen when the proper amount of vitamin D was administered, which led to the name "line test" for such an assay.

The extent of new calcification compared with that found in the animals receiving the known amount of vitamin D is an index of the vitamin intake and thus the content of the test material. Details of the method may be found in recent volumes of the U.S. *Pharmacopoeia*. The official assay method of the *United States Pharmacopeia XVII* involves the reaction of vitamin D with antimony trichloride ($SbCl_3$). The method is that of Wilkie and co-workers. Vitamin A and cholesterol do not interfere when present in reasonable quantities. The reaction is satisfactory only with concentrated preparations of the vitamin.

Species Differences to Vitamin D, and D_3

It was early established that, vitamin D_2 and D_3 showed about 40,000 USP units of vitamin D activity per milligram by the rat assay procedure. It was also found that the chick did not respond to a certain number of rat units of D_2 as it did to the same number of units of D_3. In fact, about 100 times the number of units of D_2 as of D_3 were found to be necessary to produce equal bone ash in assay chicks. Human beings do not show such a variation in response to the two forms of vitamin D. Different response is questioned by many. Jeans and other workers, however, felt that D_3 by oral administration is about 1.5 times as active as D_2 in human beings. Other forms of the vitamin have not been studied in man.

Hypervitaminosis D

Vitamin D toxicity brings about increased absorption of Ca and P from the intestine and thus increased blood levels of these minerals. Calsification of a variety of tissues in different species has been observed. Gillman and Gilbert reported arterial lesions and kidney injury in rats. Calcification of the vascular system was prominent in D_3 toxicity in calves. In dogs toxic symptoms included a reduction of glomerular filtration and renal flow which may be secondary to renal calcification.

Obviously, human toxicity has not been studied experimentally, although Jeans and steams showed growth retardation in infants on elevated intakes. According to Kramer *et. al.*, the tolerable dose in humans is highly variable. These workers reported that in one instance 400 IU daily (recommended intake) was fatal. They also indicated that general calcinosis in humans takes at least 14 days and is a reversible process if the dose of vitamin D was not too high. The simultaneous intake of a large amount of vitamin A markedly reduces the toxic effects due to excessive vitamin D intake.

Standards and Requirements

One USP unit of vitamin D is taken as the activity of 0.025 μg of pure crystalline vitamin D_3. This is equal to the IU Other units have largely been dropped. The human requirements have not been established with accuracy. I his is not surprising, since the requirements are subject to so many variables. However, a number of clinical studies with infants and children have led to a fair understanding of the intake required to obviate symptoms and signs of suboptimal growth and bone development. The Food and Nutrition Board, National Research Council, sets the recommended intake at 400 IU per day for infants and at 400 IU daily during childhood. The allowance is set at 400 IU during pregnancy and lactation also.

Adults apparently have little or no need for the vitamin. Under certain dietary conditions,a vitamin requirement may be demonstrated. In many diseases of calcium and phosphorus metabolism the vitamin is beneficial. But under ordinary dietary conditions normal adults appear to maintain normal calcium and phosphorus metabolism without benefit of extra vitamin D.

General aspects of vitamin D are reviewed by Wagner and Folkers and by Dam and Sanderguard.

Vitamin K

In 1929 Dam reported that chicks developed a hemorrhagic condition and prolonged blood clotting time when raised on specific synthetic rations. In 1935 he proposed the term vitamin K (Koagulations-vitainin) for a factor in certain foods which protected chicks against this hemorrhagic syndrome. Primarily the affected chicks exhibited internal hemorrhage and prolonged blood-clotting time (up to several hours in some cases). Many foods were tested for vitamin K activity by Dam, and he reported that hog liver was very high, cod liver oil practically devoid of activity, and egg yolks low in activity. Among the vegetable material tested, hemp seed was found to be an excellent source.

The factor was found to be fat soluble; even large amounts of the other fat-soluble vitamins had no beneficial effect on the course of the syndrome. Thus vitamin K was differentiated from the other fat-soluble vitamins A, D, and E. Another group of investigators had been working along similar lines and reported similar findings. In 1935 Almquist and Stakstad reviewed the earlier work done in their laboratory and showed that fish meal was an excellent source of the antihcmorrhagic factor, or vitamin K.

It was especially interesting that, during slight putrefaction the fish meal improved as a source of this factor. They also showed that in alfalfa the factor was localized in the unsaponifiable fraction of the ether extract. It was logical that workers in this field initiated attempts at isolation of the active principle or principles in the products found to be the best sources. Dam, Karrer and their co-workers isolated the vitamin from alfalfa in 1939. In the same year Doisy and associates isolated the principle from both alfalfa and fish meal. It was evident that the vitamin from alfalfa was chemically different from the product obtained from fish meal. Vitamin K, was used to designate the former and vitamin K, the latter. Synthesis of vitamin K was accomplished by various workers in 1939.

Chemistry of Vitamin K

Phthiocol was the first pure chemical shown to have vitamin K activity. It is not, however, a naturally occurring vitamin, but its structure was known and thus this helped in the elucidation of the quinoid structure of the vitamin. Vitamin K, is 2-methyl-3-phytyl-l, 4-naphthoquinone. The different vitamins K_2 contain different length side chains. The original

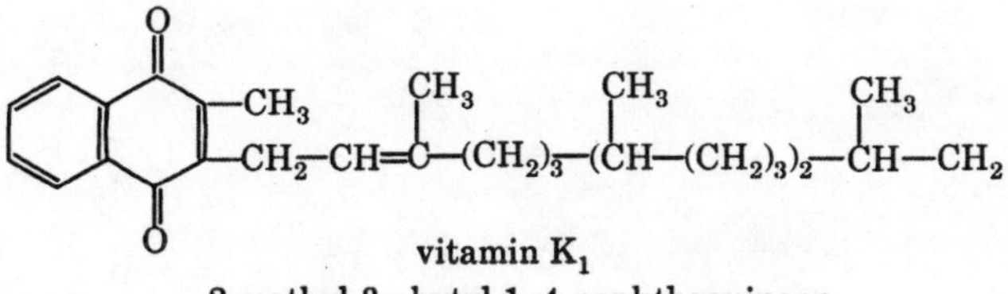

O, O, $-CH_3$, $-CH_2-CH=C(CH_3)-(CH_2)_3-(CH(CH_3)-(CH_2)_3)_2-CH(CH_3)-CH_2$

vitamin K_1
2-methyl-3-phytyl-1, 4-naphthoquinone

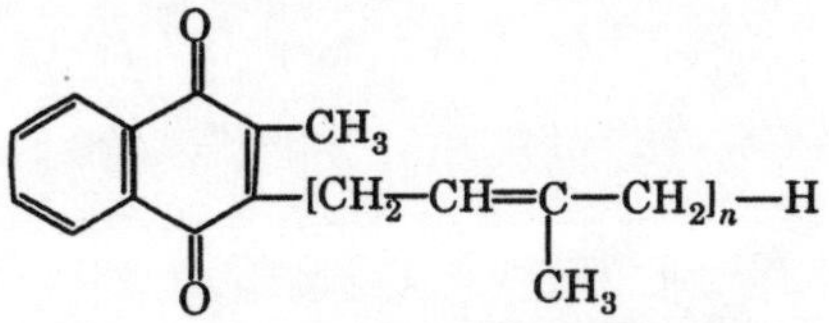

O, O, $-CH_3$, $-[CH_2-CH=C(CH_3)-CH_2]_n-H$

vitamin K_2 in which n can be 6, 7, 9

K_2 contains two farnesyl units in the side chain at position 3. This is equivalent to 6 isoprene units or 30 carbon atoms. Other K_2 molecules contain 7 isoprene units and 9 isoprene units in the side chains (35 and 45 carbon atoms). To simplify the naming of these compounds it was suggested that they be designated vitamin $K_{2(30)}$, vitamin $K_{2(35)}$, and vitamin $K_{2(45)}$. The structures are shown herewith. A number of vitamins of both the K_1 and the K_2 series have been synthesized. A simpler molecule, 2-methyl-1, 4-naphthoquinone, is known as vitamin K_3 or menadione and is readily obtainable in pure form. Some of the water-soluble derivatives are of importance.

Since the natural products as well as the synthetic menadione are fairly insoluble in water, their parenteral administration is impractical. Diphosphates, diacetates, disulfates, and a bisulfite addition compound (as salts) of menadione have been used for parenteral and oral therapy. Most, but not all, synthetic compounds appear to owe their vitamin K activity to conversion *in vivo* to 2-methyl-l, 4-naphthoquinone:

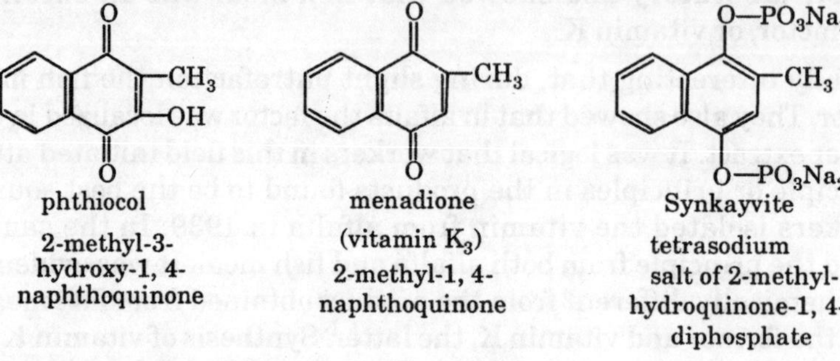

phthiocol
2-methyl-3-hydroxy-1, 4-naphthoquinone

menadione
(vitamin K_3)
2-methyl-1, 4-naphthoquinone

Synkayvite
tetrasodium salt of 2-methyl-hydroquinone-1, 4-diphosphate

CHAPTER 4 Calcium in Nutrition

Calcium is the principal cation of bone. The human body at birth contains 25 – 30 g calcium (6.25 – 7.5 mol), and at maturity, 1000 – 1500 g (25 – 37.5 mol). All of this postnatal increase must come in by way of the diet. It is not surprising, therefore, that deficiency of bony tissue was linked to a low calcium intake approx 100 yr ago when the major bone diseases were being sorted out. Appreciation of the importance of calcium for bone health has waxed and waned over the ensuing century; nevertheless, the connection between calcium and bone is intuitive and reasonable.

Moreover, evidence establishing the importance of an adequate calcium intake for skeletal health has now accumulated to the point that is incontrovertible. In the past 25 yr, other studies were reported showing an association between calcium intake and certain nonskeletal disorders, first hypertension and colon cancer, and more recently preeclampsia, renolithiasis, obesity, premenstrual syndrome (PMS), and polycystic ovary syndrome (PCO). This diversity of effects was initially puzzling, it not being clear how these seemingly unrelated disorders could have a common dietary basis.

Gradually, however, some understanding of the underlying mechanisms has emerged. The explanation ultimately goes back to paleolithic times, *i.e.,* to the conditions that prevailed during hominid evolution, to the physiological adaptations that evolved (or failed to evolve) in response to those conditions, and to ecologic changes that have occurred since the agricultural revolution, roughly 10,000 yr ago. This background is described, both physiological and ecological, and then the mechanisms by which low calcium intake produces these diverse effects are reviewed, as well as the evidence with respect to those chronic diseases for which there is today a reasonably solid evidential base. Finally, these observations are integrated and conclusions for calcium nutritional policy derived.

PHYSIOLOGICAL BACKGROUND

It is widely recognized that the calcium ion plays an essential role as an intracellular second messenger and that it mediates processes as diverse as muscle contraction, interneuronal synaptic signal transmission, glandular secretion, cell division, and blood clotting. These biochemical functions of calcium are exceedingly well protected, first by a combination of intracellular calcium stores and an immense extracellular nutrient reserve (the skeleton), and second by an elaborate endocrine control system (the parathyroid hormone-vitamin D axis and calcitonin) that maintains blood calcium concentration.

As a consequence of these protections, a low calcium intake virtually never compromises, or even threatens, the essential biochemical functions of the mineral. Instead, chronic low calcium intakes result in depletion of the reserve; a continuous homeostatic adaptive response in the attempt to compensate for low intake; and diminution of the physical chemical effects associated with high calcium levels in the intestinal lumen. It is in connection with these secondary phenomena that calcium nutritional deficiency manifestations arise.

Regulation of Extracellular Fluid Calcium Concentration

A central feature of the calcium economy of all mammals is the tight control of calcium in extracellular fluid (ECF). This comes about by adjusting both the renal calcium threshold and the flows of calcium into and out of the ECF. These processes, in turn, are regulated by parathyroid hormone (PTH), 1,25-dihydroxyvitamin D [1, 25$(OH)_2$D], and calcitonin. PTH acts to correct a fall in calcium concentration by a complex set of interacting effects. These actions include, in the probable order of the appearance of their effects, decreased renal tubular reabsorption of blood inorganic phosphate, increased resorptive effi-ciency of osteoclasts already working on bone surfaces, increased renal 1-α-hydroxylation of circulating 25(OH)vitamin D to produce the chemically most active form of vitamin D, increased renal tubular reabsorption of calcium, and activation of new bone remodeling loci. These effects interact and reinforce one another in important ways.

For example, the reduced ECF phosphate caused by the immediate fall in tubular reabsorption of phosphate is itself a potent stimulus to the synthesis of 1,25$(OH)_2$D, and it also increases the resorptive efficiency of osteoclasts already in place and working in bone, and thereby augments their release of calcium from the bony reserves. The 1, 25$(OH)_2$D directly increases intestinal absorption of both calcium ingested in food or supplement form and the endogenous calcium contained in the digestive secretions; it also is necessary for the full expression of PTH effects in bone, particularly the maturation of cells in the myelomonocytic line that produce new osteoclasts. Although a great deal more complicated in detail, these actions amount to three effects: reduced losses through the kidneys, improved utilization of dietary calcium, and withdrawal of calcium from the bony reserves.

The aggregate effect of all three is to prevent or reverse a fall in ECF [Ca^{2+}] and, at a whole body level, to conserve calcium in the face of an environmental shortage. However, the first of these takes priority, *i.e.,* the maintenance of ECF [Ca^{2+}] occurs at the expense of the skeletal reserve. A key feature of the triple end organ response to the calcium-conserving action of PTH is the balance between the three responses, and particularly the relative sensitivity of the internal (bone) and the external (gut, kidney) effector organs. In blacks, for example, the bony effects are relatively resistant to PTH (1 – 3).

In order to maintain ECF [Ca^{2+}], Blacks must secrete more PTH and 1, 25$(OH)_2$D, which produce a correspondingly greater absorptive response at the gut and a greater calcium-conserving response at the kidney. This explains why, despite lower mean calcium intakes, blacks nevertheless develop and maintain skeletons approx 10% more dense than Caucasians or Asians. A similar, although smaller, difference exists between pre- and postmenopausal women. In the presence of estrogen, bone is slightly more resistant to PTH, which explains why the calcium requirement for skeletal maintenance rises after menopause.

Inputs, Outputs, and the Requirement

On the input side, gross intestinal absorption in adults, at prevailing intakes, averages about 30%; and because substantial quantities of calcium enter the digestive tract with gastrointestinal secretions, net absorption is generally only 10 – 15%. These calcium movements are depicted schematically, which illustrates the relationship of gross and net absorption for an intake of 800 mg (20 mol) and approx 32% absorption efficiency. On the output side, many dietary components interfere with renal calcium conservation, notably sodium, net acid production, and sulfur-containing amino acids. These agencies create a floor below which urinary calcium cannot be reduced, despite the effect of PTH on increasing renal tubular calcium reabsorption.

Finally, resting dermal losses (*i.e.,* without sweating) are in the range of 1.5 mmol (60 mg)/d, and copious sweating can cause losses 5 – 10 × larger. The net result of all these excretory forces is that total obligatory calcium losses in sedentary adults on typical diets are generally in the range of 4 – 6 mmol (160 – 240 mg)/d. To stay in balance, *i.e.,* to maintain skeletal integrity, adults must absorb at least this much calcium from ingested food. At typical absorptive efficiencies, net absorption of 160 – 240 rag requires a total intake of at least 1000 – 1500 mg/d (25 – 37.5 mmol/d).

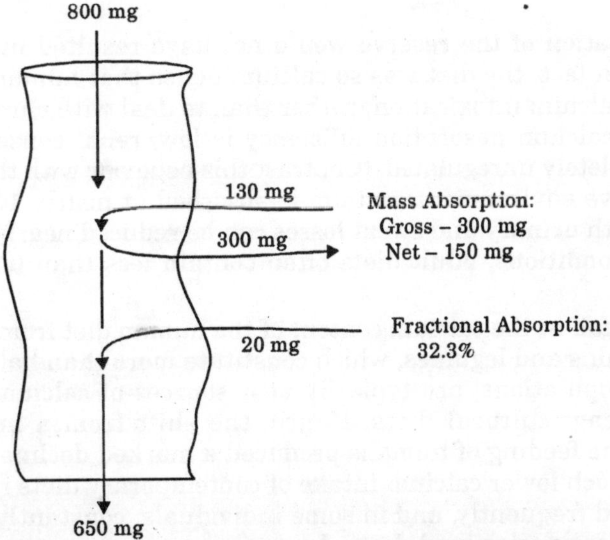

Fig. 4.1. Schematic illustration of the fluxes of calcium into and out of the intestine for a hypothetical intake of 800 mg (20 mmol) and an absorption efficiency of approx 32%. Calcium entering the gut from endogenous sources is continuous along the GI tract, but is depicted here in two components, one proximal to most of the absorbing surface and one distal, the values for which are based on measurements in adult humans.

Bccausc absoiptivc cxtraction is low, ovcn low calcium diets might seem to be adequate, since, in theory, much more calcium *couldbe* absorbed. Awareness of this potential may explain why nutritional scientists had been slow to accept the lifelong importance of a high calcium intake for contemporary humans. If the calcium was in the diet and the body was not fully accessing it, then—it was argued—the body did not really need it.

That conclusion proved to be wrong, as was definitively shown on publication of several randomized, controlled trials that demonstrated that the body would indeed use extra calcium and improve calcium balance if the mineral were provided by a high enough intake. Failure to retain calcium at low intakes reflected not absence of need but the fact that human absorption and conservation efficiencies for calcium are simply not up to the challenge of a low intake, particularly when the bony reserves are so accessible. As noted in the following section, there was no evolutionary need for hominids to develop the type of absorptive and excretory conservation that today's diets demand, when then-available foods provided a surplus of calcium. In brief: although our diets are modern, our physiologies are paleolithic.

PALEOLITHIC CONDITIONS

Calcium is the fifth most abundant element in the environment in which life evolved, and was present in high concentrations in the foods consumed by evolving hominids and hunter-gatherer humans. During the evolution of human physiology, calcium intake would have averaged approx 0.8 mmol (32 mg)/kg body weight/d. When adequate food was available, the gut would have served as a nearly continuous source of calcium. Despite low absorption efficiency, such a diet provided sufficient calcium to offset obligatory losses and to adjust to day today variability in intake and excretion. The mechanisms for dealing with a fall in ECF [Ca^{2+}], described earlier, would have been called into play only infrequently, *i.e.,* at times of food shortage or for long intervals between meals. Thus, PTH secretion is basically an emergency measure.

Intermittent utilization of the reserve would not have resulted in lasting effects under primitive conditions. In fact, the diet was so calcium dense that human physiology has been optimized to prevent calcium intoxication, rather than to deal with chronic shortages. This is the reason intestinal calcium absorption efficiency is low, renal conservation is weak, and dermal losses are completely unregulated. (Contrast this behavior with that of sodium, a scarce mineral in the primitive environment: sodium is absorbed at nearly 100% efficiency and, in trained individuals, both urinary and sweat losses can be reduced nearly to zero.) By contrast, under contemporary conditions, adult diets often contain less than 0.2 mmol (8 mg) Ca/kg body weight/d.

The reason for the fall in the calcium content of the human diet from paleolithic to historic times is that cereal grains and legumes, which constitute more than half the total food intake of agriculture-based populations, are typically poor sources of calcium. Neither food group was prominent in preagricultural diets. Hence, the shift from a hunter-gatherer to an agricultural base for the feeding of humans produced a marked decline in calcium density of the food supply. The much lower calcium intake of contemporary diets requires that adaptive mechanisms be invoked frequently, and in some individuals, constantly. The ability to adapt to low intake or high loss varies, and depends upon genetics, life stage, and environmental factors. As already noted, blacks adapt with high efficiency. Similarly, most young individuals can increase absorptive efficiency sufficiently to permit building a skeleton during growth. However, even in the young, adaptation is often not fully optimal, *i.e.,* bone mass probably does not reach its full genetic potential on prevailing calcium intakes.

Moreover, this ability to get by on low intakes declines with age. Various policymaking bodies have recently recognized this decline by recommending higher calcium intakes for the elderly. But beyond simply recognizing a special need in the elderly, there is a sense in which the requirement in the elderly uncovers the true requirement for all ages, that is, it reveals the intake that protects the skeleton without requiring constant adaptation. It is the spectrum of responses to this constant adaptation that constitutes the basis for the role of low calcium intakes in nonskeletal disorders.

MECHANISMS OF CALCIUM DEFICIENCY DISEASE

In regulating ECF [Ca^{2+}], and specifically in adapting to low calcium intake, effects occur not just for bone health but for nonskeletal systems such as blood pressure and body fat regulation. Adaptation, although obviously a necessary capacity, may nevertheless exert undesirable effects when constantly invoked. In addition to sustaining ECF [Ca^{2+}] by withdrawing calcium from the reserve, PTH elicits responses in extraskeletal tissues that may be inappropriate. Medical science is familiar with this phenomenon in the case of stress—with its high adrenergic hormone levels and high secretion of adrenal glucocorticoids.

Less well understood, but gradually becoming clearer, are the counterpart effects following from constant high secretion of PTH with its cascade of mechanisms. First is an increase in the parathyroid cell mass itself. Associated with this phenomenon is an age-specific increase in incidence of hyperparathyroidism among postmenopausal women. Whether the two phenomena are causally related is not settled, but it is true for other organ systems that constant stimulation promotes neoplasia, and it would be surprising if this relationship were not true for the parathyroid glands as well. All of the disorders currently linked to calcium relate either to decrease in the size of the calcium reserve, to decrease in residual calcium in the intestinal contents as they reach the lower bowel, or to nonskeletal responses to the hormones mediating adaptation to low calcium intake.

The principal disorders linked to low calcium intake are classified on this mechanistic basis in table 4.1. Osteoporosis is the disorder that results when the size of the calcium reserve (the skeleton) is depleted for nutritional reasons. (As described briefly below, osteoporosis has other causes as well.) The hormonal responses regulating ECF [Ca^{2+}] succeed in maintaining that critical value, but they do so by depleting the nutrient reserve (bone mass). The risk of both colon cancer and kidney stones rises as the calcium content of the diet residue falls, both for the same basic reason: failure to complex potentially harmful chemicals in the food residue.

The best available explanation for the remaining disorders lies in the fact that, in addition to its classical effects regulating ECF [Ca^{2+}], PTH and/or 1, 25$(OH)_2$D elevate cytosolic calcium ion concentrations in many tissues, thereby altering their basal level of functional activity. Hyperparathyroidism, listed as a consequence of the adaptive mechanism, is placed in parentheses because the evidence linking this disorder to low calcium intake is less clear than for the other disorders.

Table 4.1. Classification of Disorders Related to Low Calcium Intake.

	As a result of	
Decreased Ca nutrient reserve	*Decreased food residue Ca in chyme*	*Adaptive mechanisms maintaining ECF [Ca^{2+}]*
Osteoporosis	Colon cancer	Hypertension
	Renolithiasis	Preeclampsia
		Premenstrual syndrome
		Obesity
		Polycystic Ovary Syndrome
		(Hyperparathyroidism)

The first two mechanisms are straightforward, related directly to reduction in calcium mass—either the total body mass difference between calcium absorption and calcium excretion or to the mass of calcium in the intestine left over from the diet. Both are inescapable aspects of low intake. The mechanisms for the other disorders are more subtle and are set forth schematically. In addition to the effect of PTH on cytosolic calcium concentration, expression of these disorders almost certainly requires other defects. The constant high blood PTH level and the consequent high blood 1, 25$(OH)_2$D level stress the system beyond the reserve capacity of sensitive individuals. In this sense, such response to chronically high PTH levels is analogous to the hemolytic anemia that develops, for example, on exposure to certain drugs or to fava beans in individuals with glucose-6-phosphate dehydrogenase deficiency.

In the absence of a stressor agent, little or no clinical evidence of anemia is present. Presumably, the disorders that follow on a hypersensitivity to PTH and/or 1, 25$(OH)_2$D manifest themselves in individuals who lack sufficient redundancy in the cell control mechanisms to compensate for the unsignalled elevation in cytosolic calcium ion concentration.

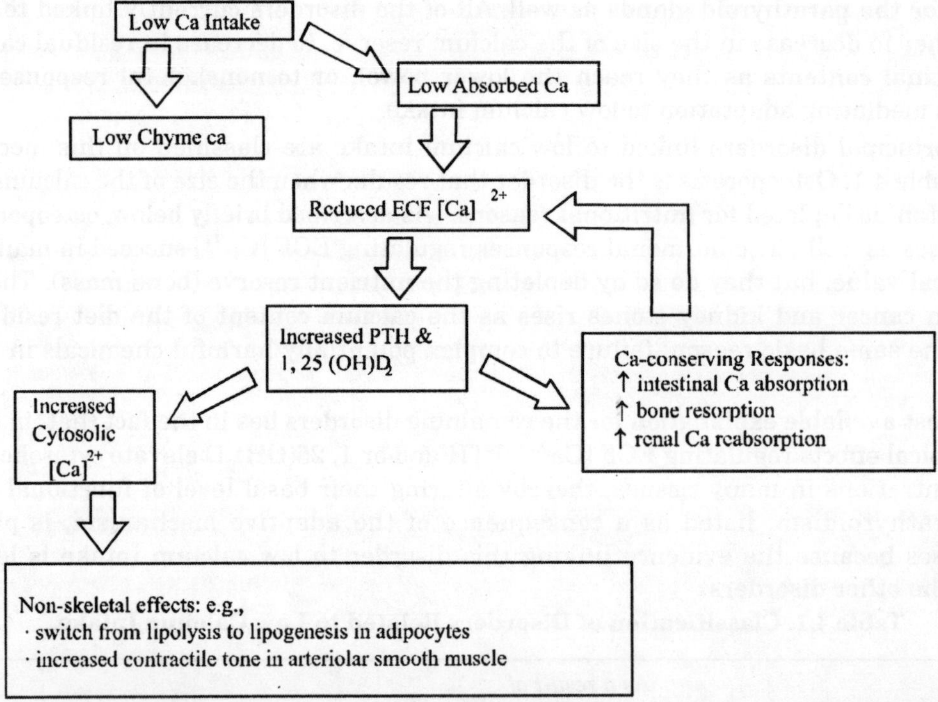

Fig. 4.2. Schematic depiction of the consequences of low levels of ingested and/or absorbed calcium. The primary hormonal response (*i.e.*, increased PTH secretion) initiates not only the well described calcium-conserving responses that are a part of the negative feedback loop regulating ECF [Ca^{2+}], but also elevates cytosolic [Ca^{2+}] in certain tissues, thereby falsely signaling responses in cells that are not a part of the calcium homeostatic control loop.

At the same time, it must also be emphasized that all of the disorders in all three categories are, in their own right, multifactorial. For example, there are many ways to have a reduced skeletal mass in addition to inadequate nutrient intake, and there are many ways to have elevated blood pressure in addition to high endogenous PTH levels. Thus, low calcium intake is only one factor in the genesis of these disorders, and if one could optimize calcium intake in the entire population, it is certain that none of the diseases concerned would be totally eradicated. Nevertheless, altering calcium intake will reduce the total disease burden for all of these disorders. Furthermore, doing so is a cost-effective intervention within society's grasp, and one that therefore demands attention.

CALCIUM DEFICIENCY DISORDERS

Reduction in the Size of the Nutrient Reserve

As noted earlier, bone is the body's calcium reserve, and mechanisms designed to protect ECF [Ca^{2+}] tear down bone to scavenge its calcium when excretory and dermal losses exceed absorbed dietary intake. In providing the calcium needed to maintain critical body fluid concentrations, the reserve is functioning precisely as it should. But sooner or later there has to be payback, or the reserve becomes depleted, with an inescapable weakening of skeletal structures. During growth, on any but the most severely restricted of intakes, some bony accumulation does occur, but the result is usually failure to achieve the full genetic potential for bone mass. Later in life, the result is failure to maintain the mass achieved. As also noted earlier, both osteoporotic fractures and low bone mass have many causes other than low calcium intake.

Nevertheless, under prevailing conditions in the industrialized nations at high latitudes, the effect size for calcium is large. Calcium supplemented trials, even of short duration, have resulted in 30+% reductions in fractures in the elderly. In addition to the effect size's being large, the evidence for calcium's role is itself very strong. There have been roughly 50 published reports of investigator-controlled increases in calcium intake with skeletal endpoints, most of them randomized controlled trials and most of them published since 1990. All but two demonstrated either greater bone mass gain during growth, reduced bone loss with age, and/or reduced osteoporotic fractures.

The sole exceptions among these studies were a supplementation trial in men in which the calcium intake of the control group was itself already high (nearly 1200 mg/d), and a study confined to early postmenopausal women in whom bone loss is known to be predominantly due to estrogen deficiency. Complementing this primary evidence are roughly 80 observational studies testing the association of calcium intake with bone mass, bone loss, or fracture. It has been shown elsewhere that such observational studies are inherently weak, not only for the generally recognized reason that uncontrolled or unrecognized factors may produce or obscure associations between the variables of interest, but because the principal variable in this case, lifetime calcium intake, cannot be directly measured and must be estimated by dietary recall methods.

The errors of such estimates are immense and have been abundantly documented. Their effect is to bias all such investigations toward the null. Nevertheless, more than three-fourths of these observational studies reported a significant calcium benefit. Given the insensitivity of the method, the fact that most of these reports are positive emphasizes the strength of the association; at the same time, it provides reassurance that the effects achievable in the artificial context of a clinical trial can be observed in real world settings as well.

Calcium is a unique nutrient in several respects. It is the only nutrient for which the reserve has required an important function in its own right. We use the reserve for structural support, *i.e.,* we literally walk on our calcium nutrient reserve. Calcium is unique also in that our bodies cannot store a surplus, unlike, for example, energy or the fat-soluble vitamins. Calcium is stored not as such but as bone tissue, and regulation of bone mass is cell mediated, with the responsible bone cellular apparatus controlled through a feedback loop regulated by mechanical forces.

In brief, given an adequate calcium intake, we have only as much bone as we need for the loads we currently sustain. These features are the basis for the designation of calcium as a threshold nutrient with respect to skeletal status, a term that means that calcium retention rises as intake rises, up to some threshold value that provides optimal bone strength; then, above that level, increased calcium intake produces no further retention and is simply excreted. This threshold intake is the lowest intake at which retention is maximal, *i.e.,* it is the minimum daily requirement (MDR) for skeletal health. The MDR varies with age, and is currently estimated to be about 20 – 25 mraol (800 – 1000 mg)/d during childhood, 30 – 40 mmol (1200 – 1600 mg)/d during adolescence, approx 25 mmol (1000 mg)/d during the mature adult years, and 35 – 40 mmol (1400 – 1600 mg)/d in the elderly.

As previously noted, the rise in the requirement in old age reflects a corresponding decline in ability to adapt, *i.e.,* to respond to low intakes with improved absorption and retention. At the same time, as already noted, osteoporosis is a distinctly multifactorial disorder, both in the sense that many factors contribute to bone loss and fracture risk, and also in the sense that yet other factors influence the body's handling of calcium. Of most importance in the latter respect are dietary factors such as caffeine, which reduces calcium absorption very slightly, and protein and sodium, both of which increase urinary calcium loss. The urinary effects are the larger, but for all three nutrients, the impact on the calcium economy disappears at high calcium intakes (40 mmol/d or above). As already noted, such intakes, although high by today's standards, are typical of the paleolithic intake. The behaviour of the calcium homeostatic system in response to a challenge such as sodium or protein-induced hypercalciuria illustrates beautifully how the fine tuning of the calcium economy presumes a high intake. A provoked increase in urinary calcium loss, if not offset, leads to a fall in ECF [Ca^{2+}], which evokes an increase in PTH and in 1, 25$(OH)_2$D levels. These hormones act on bone as well as on the gut, and it is the combined response of both effector organs that offsets the drop in ECF [Ca^{2+}].

However, the balance between the bone and dietary components of this response depends entirely on the calcium content of the diet. For example, the increase in 1, 25 OH$)_2$D evoked by a 1 mmol/d additional loss of calcium leads to an increase in absorption efficiency of 1.5 – 2.0 absorption percentage points. At an intake of 50 mmol/d, that absorptive increase is entirely adequate to compensate for the additional urinary loss. On the other hand, at intakes such as those at the bottom quartile in NHANES-III (National Health and Nutrition Examination Survey), the absorptive increase offsets less than one-eighth of the urinary calcium loss. The rest must come from bone. Of the factors that influence bone more directly (rather than by way of the calcium economy), one can list smoking, alcohol abuse, hormonal status, body weight, and exercise.

Smoking and alcohol abuse exert slow, cumulative effects by uncertain mechanisms that result in reduced bone mass and increased fracture risk. Low estrogen status and hyperthyroidism produce similar net effects, although probably by very different mechanisms. Bone mass rises directly with body weight, again by uncertain mechanisms.

Exercise, particularly impact loading, is osteotrophic and is important both for building optimal bone mass during growth and for maintaining it during maturity and senescence. In a sense, bone health is like a three-legged stool; one leg is nutrition, another is hormones, and the third is lifestyle. All three legs must be strong if the stool (or the skeleton) is to support us. Strengthening of one leg will not compensate for weakness of another.

Reduction of Residual Calcium in the Chyme

An inevitable consequence of reduced calcium intake is a reduction of calcium in the food residue delivered to the ileum, cecum, and colon. This is both because less calcium is ingested and because fractional absorption rises as intake drops. The quantitative features of the relationship between diet and fecal calcium are illustrated, which is based on over 500 calcium balance studies performed by the author in middle-aged women. With the primitive diet, net calcium absorption would have been approx 10%, and fecal calcium would thus have been nearly as high as dietary intake.

The upshot of low intakes is that there is less calcium available to form complexes with other food residue substances that may be harmful in some individuals. The two conditions for which this effect is best established are colon cancer and renolithiasis. The potentially harmful chyme substances are free fatty acids and bile acids for colon cancer, and oxalic acid for renolithiasis. There is no good chemical theory for a mixture as complex as the chyme at any location through the gastrointestinal tract, let alone in its distal segments. Nevertheless, a few observations concerning the quantities of reactants required to Complex possibly noxious substances in the chyme may be useful. Fat absorption is highly efficient; nevertheless from 1 – 5% of ingested fat normally escapes absorption in the small bowel and is delivered to the cecum as free fatty acids.

For a typical North American diet with 35% of calories from fat, that translates to 3 – 1.5 mmol of unabsorbed fatty acid, plus a smaller quantity of unreabsorbed bile acid. If there were no other ligands for the calcium in the chyme, this quantity of fatty acid would require, for its full neutralization, the presence in the mixture of 1.5 – 15 mmol (60 – 600 mg) of calcium [depending on the mixture of $Ca(FA)^+$ and $Ca(FA)_2$]. As the relationship, it would require a diet containing up to 680 mg calcium to achieve that level. This is slightly above the median for U.S. women. But fatty acids are not the only ligand competing for calcium. Phosphate, oxalate, phytate, and various organic acids produced by bacterial fermentation of colonic contents are also present in varying abundance.

Unabsorbed phosphate from a typical diet alone amounts to 10 – 15 mmol, and would be expected to bind with an equivalent quantity of calcium (forming $CaHPO_4$). Thus, it can be estimated that chyme calcium levels of more than 30 mmol/d are required to neutralize the potentially noxious products of digestion left in the food residue. As the relationship of indicates, that level in the chyme requires an intake of at least 34 mmol (1360 mg) Ca/d. This value should be taken as a lower estimate, as the efficiency of ligand binding and the respective binding constants in a medium such as chyme are unknown.

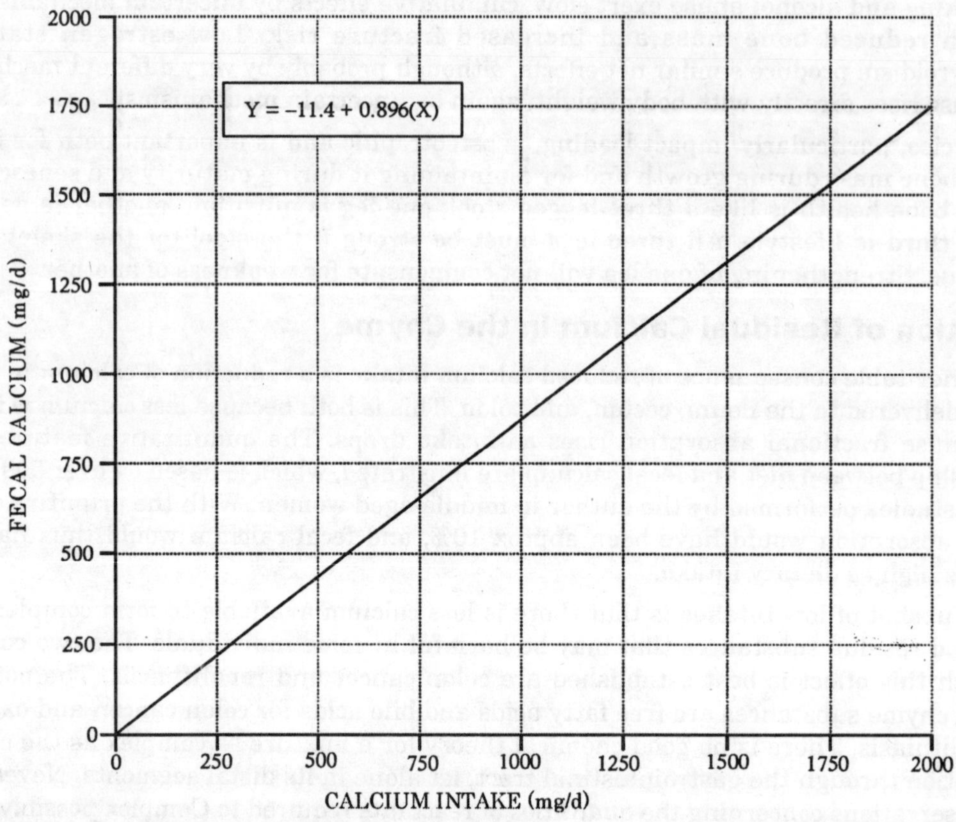

Fig. 4.3. The regression line (and its equation) for fecal calcium on calcium intake from over 500 balance studies in healthy middle-aged women.

Colon Cancer

As just described, unabsorbed dietary calcium forms complexes with unabsorbed fatty acids and bile acids that would otherwise serve as mucosal irritants (and hence cancer promoters). Such complexation effectively renders these irritants harmless, thereby breaking one of the chains of potential colon cancer development. Early epidemiological studies noted a strong inverse relationship between calcium intake and colon cancer risk, with several studies indicating that the risk was minimal at intakes in the range of 45 mmol (1800 mg)/d or higher. As with observational studies in most disorders, the evidence implicating calcium intake is mixed, and even when positive, the results could not by themselves be used to establish a causal connection.

However, animal carcinogenesis studies followed the observational data, showing that unabsorbed fatty acids and bile acids serve as cancer promoters in systems using potent carcinogens, and that a high calcium intake complexed (and hence effectively neutralized) both types of promoter. More importantly, calcium reduced the numbers of cancers that developed. Similarly, studies in humans with a familial cancer tendency showed lower levels of cellular atypia and reduced mitotic indices in those whose calcium intakes were high. At least two randomized trials of calcium supplementation have been recently reported.

Neither used actual cancer as the endpoint, but in one, adenoma recurrence was significantly reduced in calcium-supplemented individuals and in the other, various indices of mucosal hyperplasia were significantly reduced. Thus, although the bulk of the evidence points to protection, final proof is still not available. This may be because sample size requirements for a cancer endpoint are large, and no controlled trials to date have involved the requisite number of subject-years of observation. In summary, the epidemiological data, the animal models, the chemistry of the chyme, and the still limited clinical trials are all concordant, pointing to a protective effect of high calcium intakes for colon cancer risk.

Renolithiasis

Renal stones develop primarily in individuals who, for whatever reason, lack a solution stabilizer in their urine. In such individuals, solute concentration, calcium loads, and oxalate load are some of the risk factors for oxaiate stone formation (the most common variety in North America today). Normally, 75 – 85% of the renal oxalate burden comes from endogenous (*i.e.*, metabolic) sources, and the remainder is dietary in origin, coming mainly from vegetable sources and from bacterial fermentation of food residues in the distal bowel. Chyme calcium complexes this oxalate in the intestine before it can be absorbed, thereby reducing the renal oxalate burden.

Renal oxalate load is a more powerful risk factor for stone formation than is the renal calcium load, which explains the seemingly paradoxical effect of calcium intake on calcium stone risk. It should be noted that this is not a new observation. Therapeutic nutritionists have long used very high calcium intakes (up to several grams per day) to treat patients with short bowel syndromes who develop the syndrome of intestinal hyperoxalosis, the principal manifestation of which is severe renal calcinosis and renolithiasis. There have, to date, been no clinical trials of calcium supplementation in stone formers, but two prospective studies have reported an inverse relationship between stone incidence and calcium intake.

HYPERSENSITIVITY TO THE ADAPTIVE RESPONSE TO LOW CALCIUM INTAKE

As already noted, cytosolic [Ca^{2+}] functions as a ubiquitous second messenger-mediating cell response to a wide variety of specific control signals. High basal levels presumably trigger inappropriate (*i.e.*, unsignaled) cell responses or, equivalently, lower the response threshold for appropriate stimuli. This effect of PTH is found, for example, in arteriolar smooth muscle, in platelets, and in fat tissue. The first is associated with increased vascular tone, and the last with enhanced lipogenesis. Presumably, the pathogenesis of the premenstrual and polycystic ovary syndromes have an analogous basis.

Hypertensive Disorders

The relationship of the hypertensive disorders to low calcium intake is also discussed Weinberger. Here, we need only mention that, although data from observational studies are mixed, the controlled trials have almost always shown a blood pressure lowering effect of calcium supplementation, and meta-analyses of calcium intervention trials in essential hypertension, preeclampsia, and pregnancy-induced hypertension have unfailingly found a lowering of risk in individuals given augmented calcium intake (38 – 41) even when the data have been assembled so as to minimize the calcium effect size. At a population level, the effect of high calcium intake is probably small (a blood pressure lowering on the order of 2 – 5 mm Hg), probably because hypertension is a heterogeneous disorder, with only a fraction of cases exhibiting sensitivity to PTH, and with the calcium effect necessarily confined to only these patients.

More recently, the Dietary Approaches to Stop Hypertension (DASH) study exhibited a much larger effect, using food sources of calcium, combined with a high fruit and vegetable diet. The full DASH diet effect (which is probably more than the sum of its parts) was estimated to be capable of producing a 27% reduction in stroke and a 15% reduction in coronary artery disease at a population level—the largest effect reported to date. The impressive size of the effect found in this study has been attributed not solely to the calcium content of the diet, but to the higher potassium intake as well, and to the shift to an alkaline ash diet—or both. It seems likely that the relative resistance of the skeletal remodeling apparatus to PTH in blacks, coupled with their typically low calcium intakes, is a part of the explanation for the high prevalence of hypertensive disease in this racial group.

As already noted, blacks exhibit high serum levels of PTH and 1, 25$(OH)_2$D, both of which, of course, decrease when calcium intake is increased. It may also be that the larger overall response to the high calcium diet in DASH (relative to other calcium supplementation trials) was due precisely to the high proportion of African-Americans in the DASH study cohort (62%). In addition to the results of calcium intervention trials, there are several corollary lines of evidence implicating dysregulation of the calcium economy in the genesis of the hypertensive disorders. McCarron and associates have shown both a renal calcium leak and lower bone mineral density in hypertensives and Brickman and colleagues have shown higher circulating PTH and 1, 25$(OH)_2$D levels and higher platelet cytosolic [Ca^{2+}] in hypertensives. Such observational studies do not establish the causal direction in the association, but low bone mineral density and high PTH levels are inescapable reflections of negative calcium balance, whether due to inadequate intake or excessive loss.

Hence the human data are highly consistent at several levels. Also, it is interesting to note that the salt-sensitive rat models of hypertension require low calcium diets (with consequent elevation of PTH secretion and of 1, 25$(OH)_2$D levels) for their expression and, for any given calcium intake level, the spontaneously hypertensive rat (SHR) exhibits lower serum [Ca^{2+}] and higher PTH than normotensive animals. Unfortunately, widespread acceptance of the role of calcium in the hypertensive disorders has been impeded by the anti-salt advocates as if sodium sensitivity and calcium dependency could not each be involved in different sub-sets of the population.

Premenstrual Syndrome

Longstanding primary hyperparathyroidism has been classically associated with mood responses, sometimes characterized as "difficult to deal with" and "cussedness." Thys-Jacobs and coworkers, reflecting on the symptom complex known as premenstrual syndrome (PMS), noted a similarity to certain of the symptoms of hyperparathyroidism and hypothesized that there might be a connection between PMS and the PTH-calcium economy. In a small, randomized crossover trial, they showed significant reduction of PMS symptoms on daily supplementation with 1000 mg calcium. Because of the inherently subjective character of PMS symptoms, a multicenter trial was organized so as to circumvent potential objections of nonreproducibility. This trial, treating 466 women at 12 US medical centers, showed conclusively that supplementation with 30 mmol (1200 mg) calcium reduced PMS symptom severity by as much as 50% or more. A small number of other studies have yielded concordant collateral data. Thus, Lee and Kanis showed that women with established vertebral osteoporosis were more likely to have had severe PMS during their reproductive years than age-matched control subjects. This association between the two conditions is unsurprising, as low calcium intake may be a predisposing factor for each.

Similarly, Thys-Jacobs and coworkers have shown lower values for bone mass at the spine and hip in menstruating women with PMS than in controls, as well as significantly higher PTH values at the time of the luteinizing hormone peak in women with PMS. As with the hypertensive disorders, there is consistency across all the lines of evidence. Nevertheless, how the pieces of the puzzle fit together remains unclear. Even whether PTH is the trigger is uncertain.

However, reduced PTH production is one of the most rapidly occurring changes produced by increasing calcium intake, and elevated PTH can be considered a plausible candidate for the pathogenesis of PMS until proved otherwise. Still, since not every woman with low calcium intake and/or a high PTH gets PMS, and since relief of symptoms with calcium is often not complete, there must be other contributing factors. However, this is true of this entire group of disorders associated with the adaptive response to a low calcium intake.

Obesity

The linkage of obesity to low calcium intake has only recently become apparent. At least four sets of clinical observations have shown that children and adolescents with high milk intakes weigh less and have less body fat than nonmilk drinkers. Moreover, Zemel and associates have shown, using the NHANES-III database, that risk of obesity exhibits a highly significant stepwise, inverse correlation with dietary calcium intake. In other work, the same authors have demonstrated that low calcium intake increases cytosolic [Ca^{2+}] in the adipocyte (whether through increased PTH or 1, 25$(OH)_2$D is unclear).

The result of this cell biologic effect is to switch adipose cell metabolism from lipolysis to lipogenesis. This change means a higher average fat content per fat cell in the presence of high PTH levels and provides a plausible mechanism to explain the clinical and epidemiological observations. In transgenic mice expressing the *ago-uti* gene, low calcium diets provoke obesity, whereas high calcium diets suppress the obesity and elevate core body temperature. Rather than an untoward response (as with blood pressure), this effect may well reflect a primitive adaptation to food shortage. With the primitive diet naturally rich in calcium, low levels of absorbed calcium would signal low food intake and possibly, therefore, impending starvation.

Under such circumstances, a shift from fat breakdown to fat synthesis and a reduction in thermogenesis would help conserve critical energy reserves. It is still unclear how large the effect may be at a population level. Nor are there randomized clinical trial data showing the extent to which high calcium intakes will reduce weight gain or augment weight loss. So the ultimate significance of the observations cited remains uncertain. At very least, this linkage may simply be a fortuitous instance illustrating how PTH can affect the function of cells and tissues not usually considered a part of the calcium homeostatic regulatory system.

Polycystic Ovary Syndrome

Also still at an early stage of understanding is the relationship of calcium intake to polycystic ovary syndrome (PCO). PCO is one of the most common causes of infertility today and is generally considered to be a condition of abnormal oocyte differentiation of uncertain etiology. The calcium-PTH-vitamin D axis has not previously been evaluated in PCO. Recently, Thys-Jacobs and colleagues reported a series of 13 patients with PCO who had high serum PTH levels and very low serum 25(OH)D.

Although this represented a new finding, interesting in its own right, what is more impressive is the fact that treatment with calcium and vitamin D reversed not only the abnormal indices of the calcium economy, but produced a remission in the syndrome, with two of the patients becoming pregnant. There are, as yet, no randomized trials of calcium in this disorder, and so the results, however provocative, must be considered tentative. Nor is there yet an experimental basis to explain the effect. The various meiotic and mitotic divisions through which the oocyte goes during its maturation are known to be triggered by rises in cytosolic [Ca^{2+}], so there is at least a plausible area in which to look for an effect of high circulating levels of PTH. But whether such an effect can explain the findings in this syndrome is still only speculative.

Nevertheless, the dramatic effect on the disease process produced by lowering serum PTH and raising serum 25(OH)D implicates these substances in the pathogenesis of the disorder (and with them, of course, calcium intake), even if the details of the mechanism must, for now, remain unclear.

INTEGRATION

The primitive diet, the one to which human physiology has adapted over the millennia of evolution, was calcium dense, with a calcium content of 70 – 80 mg/100 kcal (0.42 – 0.48 mmol/kJ). At energy expenditure levels that prevailed until the very recent advent of private automobiles and other labour-saving devices, such a diet would have provided individuals of contemporary body size with total calcium intakes in the range of 50 – 75 mmol (2000 – 3000 mg)/d. Additionally, evolution provided mechanisms to tide us over transient periods of dietary calcium deficiency. Modern diets, based on the calcium-poor foods that served as the basis of the agricultural revolution, provide total calcium intakes far short of those to which our physiologies are adapted.

For this reason, the same modern diets cause us to call into constant play the compensatory mechanisms that, under primitive conditions, allowed us to function during what would usually have been transient periods of food shortage. Except for some decline in the size of the calcium reserve, a substantial fraction of contemporary humans appears to be able to do this perfectly well, without sustaining any apparently untoward consequences. But many cannot cope so well. For some, the decreased calcium reserve translates to increased bony fragility. Still others, better able to protect their skeletons, develop hypertension. Others, with a potential for colon carcinogenesis, increase that risk by virtue of absence of a natural protective factor in the food residue from their diets. For the various reasons reviewed in this chapter, skeletal maintenance must be judged as a necessary, but insufficient, functional endpoint for calcium nutrition. Instead, policymakers need to calculate in terms of *total* health, recognizing that different racial, ethnic, age, and gender groups may have different sensitivities to, or abilities to adjust to, low calcium intakes.

As already noted, the skeletal endpoint may not be the right functional indicator of calcium nutritional status in African-Americans, who are manifestly able to amass and maintain strong bones on low calcium intakes. In them, the hypertensive disorders would seem to provide a more relevant indicator, but current data do not allow very precise estimation of the intake needed to minimize the calcium deficiency component of these nonskeletal disorders. As noted earlier, the rise in the calcium requirement for skeletal health in the elderly reflects mainly a decline in ability to adapt to a low intake. In that sense, the skeletal requirement in the elderly may reflect the true requirement at all ages, *i.e.*, the intakes for which no compensatory adaptive response is required.

McKane and associates showed substantial involution of the parathyroid glands in healthy elderly women given 60 mmol (2400 mg) calcium for 3 yr. This seemingly high intake did not suppress PTH to subnormal levels; instead, the usually high PTH levels of the elderly were reduced to healthy young adult normal levels. The same study also showed restoration of bone remodeling activity to young adult normal levels. PTH and bone remodeling rise with age, not because these components of the calcium economy fail to adapt, but because the external end organs of the control loop—calcium absorption and renal calcium retention—become less and less responsive with age. Available data suggest that a calcium intake in the range of 37.5 mmol (1500 mg)/d to 60 mmol (2400 mg)/d may be optimal for all body systems. Lest such an intake be thought "heroic" or "pharmacologic," it is useful to recall that diets with such calcium contents exhibit a calcium nutrient density less than, or in the range of, those of our hominid ancestors, and substantially below those of diets we feed our laboratory animals, including chimpanzees and other primates.

CHAPTER

5 Sodium in Nutrition

The most recent report of the Sixth Joint National Committee on Prevention, Detection, Evaluation, and Treatment of High Blood Pressure (JNC VI) has recommended a trial of lifestyle intervention as initial therapy in individuals with high-normal (130 – 139 mm Hg systolic/85 – 89 diastolic) or stage I (140 – 159/90 – 99) blood pressure levels without end-organ disease, concomitant cardiovascular disease or diabetes mellitus. This chapter examines the evidence in support of dietary alterations and their effect on blood pressure. The constituents that are considered include calories (body weight), salt (sodium chloride), potassium, calcium, and combinations of these minerals, other dietary components and alcohol. Rather than attempting an encyclopedic review of the studies of all these dietary elements, this chapter succintly summarizes what is currently known about the factors outlined.

BODY WEIGHT

Body weight has long been linked to blood pressure levels. More recent findings indicate that the distribution of body fat may be a more important determinant of blood pressure elevation and the risk for cardiovascular disease. An increase in visceral abdominal fat (the central or "apple" form) in contrast to the lower body adiposity pattern (the "pear" shape) is linked not only to blood pressure elevation but also to insulin resistance, dyslipidemia and an increased risk for cardiovascular events. These associations have been based largely on epidemiological evidence. However, there are now several intervention trials in which it has been demonstrated that weight loss, often as little as 5 kg rather than a reduction to "ideal" body weight, is associated with a decrease in blood pressure and an improvement in insulin sensitivity.

Studies on both humans and experimental animals suggest that the sympathetic nervous system is involved in the pathophysiology of the weight-blood pressure-insulin resistance relationship, but therapeutic interventions based on these findings are not yet available to confirm this connection.

Another mediator of the body weight-blood pressure relationship appears to be the kidney. In experimental animals, obesity has been associated with alterations in renal blood flow and glomerular filtration rate or intraglomerular pressure. In humans, urinary microalbumin excretion was increased among obese subjects, supporting an abnormality in renal function in humans as well. Moreover, microalbuminuria has been linked to an increased risk of cardiovascular events in hypertensive individuals.

Table 5.1. Review of Nutrition Related Factors that Are Known to Influence Blood Pressure

Factors that increase blood pressure	*Factors that decrease blood pressure*
Increased sodium intake (Especially for salt-sensitive subjects)	Increased potassium intake
Increased body weight	Increased fruit and vegetable intake
Increased alcohol intake	Increased calcium intake (Especially for salt-sensitive subjects)

SALT (SODIUM CHLORIDE)

The relationship between dietary salt consumption and elevated blood pressure is well known. The prevalence of hypertension and its consequences is linearly related to dietary salt intake in societies throughout the world. Hypertension and its sequelae are virtually absent in societies in which habitual salt intake is <50 – 100 mmol/d. However, there are other differences between these groups and those that habitually consume larger amounts of salt. The "low-salt" societ-ies tend to be isolated, genetically homogeneous, physically fit, and consume increased amounts of potassium and calcium in the form of fresh fruits and vegetables. Increased salt intake is associated with societal "acculturation." This implies a crowded and sedentary lifestyle as well as many other behavioral factors that may affect blood pressure and cardiovascular risk.

In addition, the age-related increase in blood pressure is observed only in societies in which salt intake is high. Many of the elderly individuals in low-salt cultures have blood pressure levels that are no higher than those of young adults. Despite this convincing evidence, controversy still exists concerning the importance of salt in human blood pressure in general, in hypertension, and as a treatment modality. Without considering the various reasons for this controversy, suffice it to say that the magnitude of the effect of salt intake on blood pressure is diluted by the fact that there is substantial heterogeneity in the blood pressure responses of humans to alterations in salt intake. Numerous studies have demonstrated that salt-sensitive and salt-resistant individuals can be identified within both the hypertensive and normotensive populations.

Salt-sensitive subjects will demonstrate a decrease in blood pressure with dietary sodium reduction, usually to the level of 100 mmol/d (2.4 g/d). The human need for sodium is about 10 mmol/d (230 mg/d). Thus, the threshold for blood pressure responsiveness to a reduction in salt intake is many times higher than the physiological requirements. It is often difficult to differentiate between salt-sensitive and salt-resistant subjects without sophisticated research techniques. However, a trial of modest dietary salt restriction or diuretic administration should identify those most likely to benefit from this dietary intervention. Moreover, there have been no adverse reports when a modest reduction in salt intake such as 80 – 100 mmol/d have been followed. Certain population groups have been reported to be more likely to be salt sensitive than others. Hypertensive individuals are more salt sensitive than those with normal blood pressure. Among hypertensive subjects, salt sensitivity of blood pressure is more frequent among African-Americans (75%) than Caucasians (50%) and increases with increasing age. The latter finding is also observed in the normotensive population, with the finding that significant salt sensitivity of blood pressure is not seen until the age decade of 60 yr or more.

Individuals with reduced renin responses to sodium and volume depletion, the so-called low-renin subjects, are more likely to be salt sensitive than those with brisk renin responses. In addition to a possible permissive effect of sluggish renin responses to salt sensitivity, a variety of substances have been reported to be involved in the pathophysiology of salt sensitivity of blood pressure.

An extensive scientific critique of the many studies that have been conducted in this area is beyond the scope of this chapter; however, note that the sympathetic nervous system, endothelin, insulin sensitivity, atrial natriuretic factor, alterations in renal hemodynamics, and leptin have all been implicated in the pathophysiology of salt, sensitivity. It remains to be determined which of these many factors are primary events and which are simply compensatory responses or epiphenomena. It has been shown that salt sensitivity requires the administration of sodium as the chloride salt and that other forms of sodium do not have the same pressor effect. However, this is a moot point because over 95% of the sodium found in foods is in the chloride form.

Moreover, most of the salt found in food is added in the preparation, processing, and preservation of food and only 15% is added as the discretionary form (as table salt). Thus, it is important for the food preparer as well as the patient to become familiar with identifying the salt content of foods at the grocery store and restaurant as well as in cooking. Another important recent finding related to salt and blood pressure is the observation that long-term follow-up of salt-sensitive normotensive subjects over a period of 10 yr or more demonstrated an eightfold greater rate of blood pressure increase compared with those who were initially salt-resistant. This finding supports the epidemiological observations relating the age-associated rise in blood pressure to increased salt intake.

POTASSIUM

As mentioned earlier, in societies in which there is a low prevalence of hypertension and its complications as well as little age-related increase in blood pressure, people tend to consume increased amounts of potassium (and calcium, as discussed in the next section) and follow a reduced salt diet. Fewer studies have examined the relationship between potassium and blood pressure than those involving sodium. However, the findings regarding potassium tend to be consistent. In general, the effect of potassium is smaller than that of sodium based on interventional trials. Again, heterogeneity in responsiveness of blood pressure to alterations in potassium intake or balance has been demonstrated. Among potassium-replete normotensive subjects, typically those consuming 60 mmol/d of potassium or more, little effect on blood pressure can be demonstrated with additional potassium administration.

However, in hypertensive populations, particularly those comprising substantial numbers of individuals in whom dietary potassium intake is traditionally deficient (the elderly, African-Americans) or those in whom diuretic-induced potassium loss occurs, potassium supplementation has been shown to lower blood pressure. It has also been observed that potassium is more likely to lower blood pressure in hypertensive individuals consuming a high-salt intake, further suggesting a link between sodium and potassium intake in their effects on blood pressure. The amount of potassium intake required for optimal reduction in blood pressure in those who are sensitive to this mineral is not clear. Most studies indicate that dietary potassium deficiency begins at levels of intake 50 mmol/d and is clearly observed at intakes of 30 mmol/d or less.

Dietary sources of potassium are largely fresh fruits and vegetables. In environments where these are scarce, *e.g.*, because of cold climate or high cost, potassium deficiency is more likely. Among a group of normotensive nurses in whom dietaiy intakes of potassium, calcium, and magnesium were deficient, only potassium supplementation reduced blood pressure. Recent studies using diets involving multiple mineral manipulations, such as the Dietary Approaches to Stop Hypertension (DASH) Trial are discussed in the Combination Diets section.

CALCIUM

Epidemiological surveys have suggested a relationship between reduced dietary calcium intake and hypertension. Several studies have demonstrated a small and inconsistent effect of calcium supplementation to lower blood pressure. Again, this appears to be largely owing to the heterogeneity in human responses to calcium supplementation. Subgroup analyses of some of the larger studies suggest that those in whom dietary calcium intake is often reduced (African-Americans, the elderly) are more likely to demonstrate a reduction in blood pressure with calcium supplementation than other groups in whom intake is higher. Since both subgroups are traditionally salt sensitive, we conducted a study of calcium supplementation in a group of normal and hypertensive subjects who had been previously categorized with respect to salt sensitivity of blood pressure. We found no significant effect of calcium supplementation on blood pressure for the entire group.

However, when the subjects were separated on the basis of salt-sensitivity status, we found a significant decrease in blood pressure when the salt-sensitive subjects received calcium supplements and a significant increase in blood pressure when calcium was given to the salt-resistant subjects. These findings suggested a reciprocal relationship between the effects of calcium and sodium on blood pressure that was confirmed by the results of the DASH Trial.

OTHER DIETARY CONSTITUENTS

There is, at present, no convincing evidence to link alterations in magnesium intake with blood pressure, and thus the JNC VI report did not advocate an increase in this mineral for the purpose of lowering blood pressure. Caffeine may raise blood pressure acutely in caffeine naive individuals but does not appear to be a factor in the chronic elevation of blood pressure. Moreover, there is no evidence that withdrawal of caffeine in habitual consumers produces a decrease in blood presssure. While some studies have suggested a beneficial effect of large amounts of omega-3 fatty acids in reducing blood pressure, intolerance of these doses makes this an impractical approach for most individuals.

COMBINATION DIETS

A variety of studies have examined the effect of combined dietary approaches on blood pressure as nonpharmacological treatment of hypertension or for the prevention of hypertension in those at increased risk (high-normal blood pressure). These combined studies have been fraught with problems resulting from recidivism, inadequate achievement of dietary goals, or relatively short duration. In general, it can be stated that weight loss appears to be the most effective single intervention as long as the weight loss can be maintained. There does not appear to be an additive benefit when potassium supplementation is combined with modest dietary salt restriction beyond that seen with salt restriction alone. However, in the DASH trial, when a specific diet incorporating modest salt restriction with an increase in fresh fruits and vegetables and low-fat dairy products (presumably increasing potassium, calcium, and magnesium intake) was followed, a significant reduction in blood pressure was observed over the 8-wk study period.

This benefit appeared to be greatest among African-Americans and those with higher initial blood pressure levels. Another study exami ning multiple dietary changes was a subgroup of the Nurses Health Study II, which compared the effects of supplemental potassium, calcium, magnesium, or all three minerals to placebo in normotensive nurses in whom dietary deficiencies of these minerals were documented. As previously mentioned, potassium supplementation alone, but not combination supplementation, lowered blood pressure.

ALCOHOL

Alcohol consumption has been shown to have a biphasic effect on blood pressure. Small amounts of alcohol appear to lower blood pressure, presumably secondary to a vasodilator effect, but as alcohol consumption increases, blood pressure rises. The dose-response characteristics vary from individual to individual and may be based on factors such as body surface area, gender, and race. The racial differences may be explicable, in part, by virtue of genetic differences ip alcohol metabolism. The mechanism for the alcohol-induced increase in blood pressure appears to be related to activation of, or increased responsiveness to, the sympathetic nervous system. This is manifested by an increase in cardiac output when more than one ounce of alcohol is consumed. Thus, a prudent recommendation to hypertensive subjects is to limit their daily alcohol consumption to no more than 2 oz (60 mL) of 100-proof spirits (or 2.5 oz of 80-proof whiskey), 24 oz (720 mL) of beer or 10 oz (300 mL) of wine. For those hypertensive individuals in whom habitual alcohol consumption exceeds these levels, a reduction in intake may lower blood pressure or make it easier to control.

SUMMARY

A variety of dietary and lifestyle factors can influence blood pressure, these factors are reviewed. The ideal recommendation for individuals who are hypertensive or are at increased risk for its development are to maintain a body weight as close to ideal as possible; to consume a diet modest in salt content and enriched with fresh fruits, vegetables, and low-fat dairy products, and to consume no more than the recommended optimal amounts of alcohol.

CHAPTER 6

Dietary Antioxidants

This chapter discusses evidence linking antioxidant nutrients with the inhibition of low-density lipoprotein (LDL) oxidation. The possibility that increasing antioxidant intake could protect against coronary heart disease (CHD) is reviewed. The chapter also provides information regarding what ideal antioxidant intakes should be. Hypercholesterolemia is universally accepted as a major risk factor for athero-sclerosis. However, at any given concentration of plasma cholesterol, there is still great variability in the occurrence of cardiovascular events. One of the major breakthroughs in atherogenesis research has been the realization that oxidative modification of LDL may be a critically important step in the development of the atherosclerotic plaque.

The formation of foam cells from monocyte-derived macrophages in early atherosclerotic lesions is not caused by native LDL but only following modification of LDL by various chemical reactions such as oxidation. The cholesterol-laden foam cell is a characteristic feature of the atherosclerotic lesion. The rapid uptake of oxidatively modified LDL occurs through scavenger receptors, which are not down regulated by cholesterol accumulation. Recognition of LDL by the scavenger receptor depends on alteration of key lysine residues of apolipoprotein B, which can be brought about by aldehydes produced during the spontaneous decomposition of lipid hydroperoxides. Some reports have suggested the presence of oxidatively modified LDL in plasma, but most oxidation is believed to occur in the arterial wall. There, LDL may be in a microenvironment where the antioxidants, which normally prevent lipid peroxidation, can become depleted.

All the cells of the vessel wall—endothelial cells, smooth muscle cells, macrophages, and lymphocytes—can modify LDL in vitro. LDL oxidation is believed to be caused by highly reactive free radicals but the nature and source of these have yet to be fully defined. Several mechanisms are likely to be involved, including transition metal ion-mediated generation of hydroxyl radicals, production of reactive oxygen species by enzymes such as myeloperoxidase and lipoxygenase, and direct modification by reactive nitrogen species. Oxidized LDL may also be atherogenic by mechanisms other than its rapid uptake into macrophages the scavenger receptor.

Oxidized forms of LDL are chemotactic for circulating macrophages and smooth muscle cells and facilitate monocyte adhesion to the endothelium and entry into the subendothelial space. Oxidized LDL is also cytotoxic toward arterial endothelial cells and inhibits the release of nitric oxide and the resulting endothelium-dependent vasodilation. There is therefore a potential role for oxidized LDL in altering vasomotor responses, perhaps contributing to vasospasm in diseased vessels.

In addition, oxidized LDL is immunogenic; autoantibodies against various epitopes of oxidized LDL have been found in human serum, and immunoglobulin (IgG) specific for epitopes of oxidized LDL can be found in lesions. Oxidized LDL may be able to induce arterial wall cells to produce chemotactic factors, adhesion molecules, cytokine, and growth factors that have an important role in the development of the plaque.

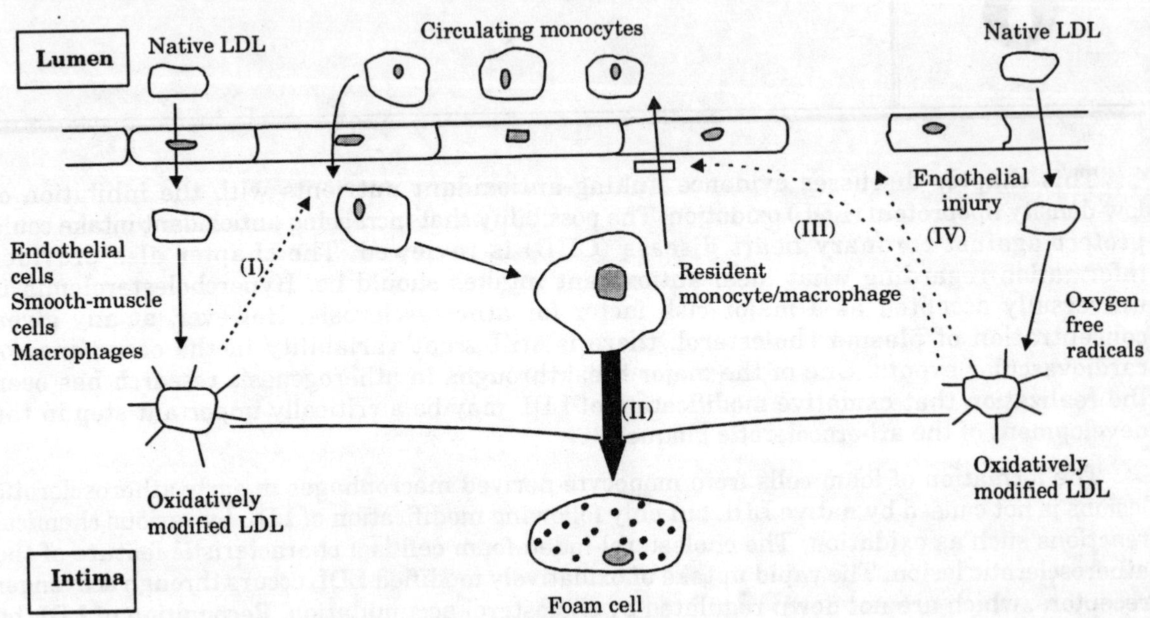

Fig. 6.1. Mechanisms by which oxidation of LDL may contribute to atherogenesis. Oxidized LDL is chemotactic for circulating monocytes (I), which are phenotypically modified and become macrophages. Oxidized LDL is recognized by the scavenger receptor on the macrophage and becomes internalized rapidly. As more lipid is ingested by the macrophage, a foam cell is formed (II). This eventually bursts and a fatty streak, the first their ability to leave the intima (III). Oxidized LDL is cytotoxic to endothelial cells, leading directly phase of an atherosclerotic lesion, results. Oxidized LDL inhibits the motility of resident macrophages and therefore to endothelial cell damage (IV). Oxidation can occur via the effects of reactive oxygen species or due to the oxidation of the cell's own lipids.

Evidence for LDL oxidation in vivo is now well established. In immunocyto-chemical studies, antibodies against oxidized LDL stain atherosclerotic lesions but not normal arterial tissue. LDL extracted from animal and human lesions has been shown to be oxidized and is rapidly taken up by macrophage scavenger receptor. In young myocardial infarction (MI) survivors, an association has been demonstrated between increased susceptibility of LDL to oxidation and the degree of coronary atherosclerosis, while the presence of ceroid, a product of lipid peroxidation, has been shown in advanced atherosclerotic plaques. A recent study however, has suggested that atherosclerotic plaques contain very little oxidized LDL compared to the amounts of activated complement and enzymatically altered LDL an observation that requires further study.

DIETARY ANTIOXIDANTS AND LDL OXIDATION

The role of dietary factors in protecting against the change from native to oxidized LDL has received considerable attention. The antioxidant vitamins are derived from fresh fruits and vegetables, and from vegetable oil and polyunsaturated margarine to which vitamin E is usually added as an antioxidant; they cannot be synthesized from simple precursors. Thus, dietary intake, absorption, metabolism, and storage determine concentrations of vitamins in plasma and body tissues. An overview of epidemiological research suggests that individuals with the highest intake of antioxidant vitamins, whether through diet or supplements, tend to experience 20 – 40% lower risks of CHD than those with the lowest intake or blood levels.

Vitamin E is the major lipid-soluble antioxidant present in LDL, preventing the formation of lipid hydroperoxides from polyunsaturated fatty acids. Vitamin C can scavenge free radicals in the aqueous phase and may also regenerate vitamin E. β-Carotene, a vitamin A precursor, does not have a confirmed antioxidant mechanism although.it and other carotenoids are contained within LDL. There is some evidence that carotenoids may protect LDL against oxidation more efficiently at low pO_2 levels, which could have relevance to the levels of protection provided in the arterial wall in vivo. When LDL is exposed to oxidative stress in vitro, lipid peroxidation can only proceed after the sequential loss of its antioxidants in the order ubiquinol-10, α-tocopherol, γ-tocopherol, lycopene, and β-carotene.

Accordingly, LDL supplemented with vitamin E in vitro is much harder to oxidize in vitro. Similarly, the α-tocopherol content of the LDL has been shown to be the most important determinant of susceptibility to oxidation in an in vitro model. In addition to antioxidant content of LDL, several other factors can influence the susceptibility to oxidation. Polyunsaturated fatty acids appear to be the most vulnerable moiety following the application of oxidative stress. The fatty acid composition of the diet is, therefore, an important factor determining the susceptibility of LDL to oxidation, with monounsaturated fatty acids protecting LDL against oxidation. Small dense LDL particles are also easier to oxidize than more buoyant particles. The binding of LDL particles to proteoglycans, glycation, and the presence of preformed lipid peroxides are also important factors.

INTERVENTION STUDIES IN ANIMALS

Studies of antioxidant supplementation in laboratory animals have provided evidence of the importance of oxidized LDL in vivo. Many such studies have been performed, and they generally provide support for the antioxidant hypothesis. For example, in a study in atherosclerotic rabbits, uptake of LDL into foam cells was approx 4 times greater for oxidized LDL than for native LDL. Adding of vitamin E to the system resulted in a 30 – 55% decrease in accumulated radio-labeled LDL in plaques and foam cells. The effects of a diet high in saturated fat with and without fruits and vegetables and antioxidant vitamins (C, E, and β-carotene) on oxidative stress and development of atherosclerosis has also been evaluated in rabbits. Blood lipid peroxide levels decreased significantly in rabbits supplemented with fruits and vegetables or antioxidant vitamins. In contrast, blood lipid peroxide levels increased significantly in the unsupplemented groups and precipitated coronary thrombosis.

INTERVENTION STUDIES IN HUMANS WITH BIOCHEMICAL ENDPOINTS

A number of studies have evaluated the effects of vitamin E on copper-catalyzed LDL oxidation in healthy volunteers. In one study, men supplemented with 268, 537, or 805 mg vitamin E per day for 8 wk showed a decreased susceptibility of LDL to oxidation. There was no significant effect of daily supplementation with 40 or 1.34 mg. In a study of the effects of low-dose vitamin E supplementation (100 mg/d for 1 wk, then 200 mg/d for 3 wk), there was a significant increase in lag time before the onset of LDL oxidation and a significant decrease in the propagation rate. Princen and associates have evaluated the minimal supplementary dose of vitamin E necessary to protect LDL against oxidation in vitro in healthy young adults.

Resistance of LDL to oxidation increased in a dose-dependent manner with resistance time differing significantly from baseline even after ingestion of only 17 mg/d of vitamin E. However, the progression of lipid peroxidation in LDL was only reduced after intake of 268 and 536 mg/d. In a study of 200 mg α-tocopherol supplementation over a 2-mo period in smoking men, lag time was increased in LDL + VLDL after both copper induction and hemin/hydrogen peroxide induction. Plasma α-tocopherol, VLDL + LDL α-tocopherol, and LDL total antioxidant capacity were all increased in the intervention group. Vitamin E supplementation appears to promote a clear reduction in the susceptibility of LDL to oxidation. The case for (3-carotene and vitamin C, however, is less clear cut. The effect of high-dose vitamin C supplementation (1000 mg/d) on LDL oxidation was evaluated in a study of 19 smokers.

The vitamin C-supplemented group had a significant reduction in the susceptibility of LDL to oxidation after 4 wk. Two months of vitamin C supplementation with freshly squeezed orange juice (estimated 500 mg/d) also produced a significant increase in lag time in 36 healthy males consuming a diet high in saturated fatty acids. Jialal and associates found that supplementation with β-carotene (1-2 μmol) inhibited the oxidative modification of LDL in healthy subjects. By contrast, Lin and colleagues reported that a natural-food diet containing 0.93 (μmol β-carotene per day was insufficient to alter either plasma or LDL (3-carotene or carbonyl groups in LDL.

A higher dose of 6.2 μmol β-carotene per day did increase both plasma and LDL β-carotene and reduce LDL carbonyl production. When the effectiveness of β-carotene, vitamin C, and vitamin E supplements were compared by Reaven and associates, susceptibility of LDL to oxidation did not change during β-carotene supplementation (60 mg/d) but decreased 30-40% with the addition of a vitamin E supplement (1600 mg/d). Addition of vitamin C supplementation (2 g/d) did not further reduce the susceptibility to oxidation. In another study, a combined supplement—β-carotene (30 mg/d), vitamin C (1 g/d), and vitamin E (530 mg/d)—was given to men for 3 mo. It produced a twofold prolongation of the lag phase of LDL oxidation and a 40% reduction in the oxidation rate.

The effects of combined antioxidant supplementation were not significantly different from the effects of vitamin E supplementation alone. Finally, we have recently shown that a combination of low-dose antioxidant vitamins (150 mg ascorbic acid, 67 mg α-tocopherol, 9 mg β-carotene daily) over a period of 8 wk significantly prolonged the lag time to oxidation. The effects of antioxidant supplementation on LDL oxidation may depend on smoking status. In a group of smokers and nonsmokers, resistance of LDL to oxidation increased significantly and the rate of LDL oxidation decreased significantly after vitamin E supplementation (671 mg/d for 7 d). There was some indication of a small increase in the resistance of LDL to oxidation in smokers after supplementation with β-carotene (20 mg/d for 2 wk, then 40 mg/d for 12 wk). In summary, therefore, experimental evidence suggests that antioxidant vitamins can reduce the susceptibility of LDL to oxidation in vitro.

Vitamin E would appear to be the most effective antioxidant; both β-carotene and vitamin C have produced extensions in lag time to oxidation only in a minority of studies, although it remains possible that they may have a beneficial effect in individuals with poor baseline status.

EPIDEMIOLOGICAL STUDIES LINKING ANTIOXIDANTS AND LDL OXIDATION

Vitamin E

Two large longitudinal studies in the United States examined the association between vitamin E intake and risk of CHD. In a group of 39, 910 male health professionals, those who took vitamin E supplements in doses of at least 100IU per day for over 2 yr had a 37% lower relative risk of CHD compared to men who did not take vitamin E supplements, after adjustment for age, coronary risk factors, and intake of vitamin C and β-carotene. In the Nurses' Health Study of 87, 245 female nurses, women who took vitamin E supplements for more than 2 yr had a 41 % lower relative risk of major coronary disease. This effect persisted after adjustment for age, smoking, obesity, exercise, blood pressure, plasma cholesterol, and use of postmenopausal estrogen replacement, aspirin, vitamin C, and β-carotene.

It must be noted that this effect was limited to vitamin E supplement use.

High vitamin E intakes from dietary sources were not associated with a significant decrease in risk, although even the highest dietary vitamin E intakes were far lower than intakes among supplement users. Support from case-control studies based on biological samples is sparse, although Gey and coworkers found that plasma levels of vitamin E in men aged 40 – 49 yr correlated strongly and inversely with the age-specific mortality from CHD in 16 European regions. A population case-control study also evaluated the relation between undiagnosed angina pectoris and plasma antioxidant levels in men aged 35 – 54 yr. Plasma levels of vitamin C, vitamin E, and carotene manifested a significant inverse correlation with undiagnosed angina.

The inverse association between vitamin E levels and angina remained signifi-cant after adjustment for smoking habits, age, blood pressure, relative weight, and blood lipid levels. However, these results have been offset by several negative studies. There was no association between plasma vitamin E and prevalence of CHD in a cross-sectional survey of 1132 Finnish men. Similarly, most nested case-control studies found no relationship between plasma vitamin E levels and subsequent coronary mortality or risk of MI. The reason for these disparate results is unknown, but may include changes in diet following disease diagnosis, poor classification of controls, and lack of variation in plasma levels within populations not using supplements.

Vitamin C

The evidence linking the water-soluble vitamin C with cardiovascular disease is less strong than that for vitamin E. In the Physicians' Follow-Up Study, a high intake of vitamin C was not associated with a lower risk of CHD in men, whereas in women from the Nurses Health Survey, an initial effect was attenuated after adjustment for multivitamin use. Only one prospective study, which involved 11,348 adults, demonstrated an inverse relationship between vitamin C intake and cardiovascular mortality. This effect was due largely to the use of vitamin C in supplements and may have been a reflection of other antioxidant vitamins in multivitamin preparations. A link between intake and carotid artery wall thickness has also been suggested.

Plasma levels were not correlated with coronary mortality rates among four European populations or with prevalent coronary disease in Finland. In the Basle Prospective Study, low levels of vitamin C alone did not increase the risk of CHD, although the risk of disease at low levels of both vitamin C and β-carotene was greater than that for β-carotene alone. However, a prospective population study of 1605 healthy men aged 42 – 60 in Finland has recently shown that men who had vitamin C deficiency had a relative risk of MI of 2.5 after adjusting for the main risk factors for MI.

β-Carotene

There is some indication that increased dietary intake of β-carotene is associated with reduced risk of CHD although again the evidence is less convincing than that for vitamin E. In the prospective Nurses Health Survey, consumption of vitamin A and β-carotene in food and supplements weakly predicted the incidence of CHD; Gaziano and Hennekens reported a 22% risk reduction for women in the highest quintile of β-carotene compared with those in the lowest. No adjustment was made for the potentially confounding effect of other antioxidant vitamins in multivitamin preparations. However, a small prospective study on 1271 elderly people also demonstrated an inverse relationship between β-carotene intake in fruit, and vegetables and subsequent cardiovascular death. Similar findings have been shown for serum carotenoid level and CHD risk, and carotenoid intake and carotid artery plaque thickness.

Several studies indicate that dietary and circulating levels of β-carotene affect smokers more than nonsmokers. In the Health Professionals Follow-Up Study, high β-carotene intake was associated with reduced CHD risk in current smokers and exsmokers (70% risk reduction) but not never-smokers, after adjustment for cardiovascular risk factors and vitamin C and E intake. It was suggested that a high dietary intake of β-carotene is especially important in smokers who have both an increased demand for antioxidants (to combat smoking-induced free radicals) and a correspondingly lower circulating level for a given dietary intake. However, (3-carotene supplements in smokers may have harmful effects as discussed in the next section.

Intervention Studies in Humans with Clinical Endpoints

Observational studies, of course, only infer a cause-and-effect relationship. Even in the most well-designed observational studies, the amount of uncontrolled and uncontrollable confounding could easily be as large as the small to moderate reductions in risk that are most plausible. Individuals who select higher intakes of antioxidants may also adopt other dietary or nondietary lifestyle characteristics that account for the apparent benefits of antioxidants seen in observational studies. Prospective studies with a clinical endpoint are required, and only randomized trials of sufficient sample size, dose, and duration of treatment and follow-up can provide reliable data. The limitations of intervention trials are that they can often only be interpreted for the particular study population, and for the antioxidant dose provided during the trial.

Numerous intervention studies designed to test the hypothesis that increased antioxidant intake will protect against atherosclerosis are currently in progress. In general, these studies are using supplements of antioxidant vitamins or other antioxidants, and results so far have not been encouraging. The Alpha-Tocopherol, Beta-Carotene Cancer Prevention Trial (ATBC), conducted among 29, 133 male heavy smokers in Finland, found no reduction in CHD morbidity or mortality during 5-8 yr of treatment with vitamin E (50 mg/d) or (3-carotene (20 mg/d). Those assigned vitamin E had no significant decrease in CHD deaths, but a 50% excess of deaths from cerebral hemorrhage, whereas those assigned to β-carotene experienced an 11% increase in CHD deaths. In a further analysis, a subgroup of the original subjects with a previous MI were considered.

The endpoint of this substudy was the first major coronary event after randomization. The proportion of major coronary events was not decreased with either α-tocopherol or β-carotene supplements. In fact, β-carotene conferred an excess of fatal CHD (75% increase in risk). There was a beneficial effect of vitamin E on nonfatal MI with a risk reduction of 38%. The CARET (Beta-Carotene and Retinol Efficacy Trial), designed to test the effects of a combined supplement of 30 mg (3-carotene and 25,000 IU retinol daily among 18, 314 cigarette smokers and individuals with occupational asbestos exposure, was ended early when researchers detected an elevated risk of death from lung cancer in those receiving β-carotene and, again, no beneficial effect on cardiovascular disease (C VD) was found. For C VD mortality, there was a nonsignificant 26% increase in the treated group ($p = 0.06$). The Physicians Health Study followed more than 22,000 US male doctors treated with 50 mg β-carotene or placebo every other day for an average of 1.2 yr.

The trial appears to have been conducted meticulously and its results would seem to seriously question any beneficial effect with such supplementation on CVD in well-nourished populations. There were no significant effects on indi-vidual outcomes, or on a combined endpoint of nonfatal MI, nonfatal stroke, and cardiovascular death, for which the relative risk was 1.0. There was also no evidence of harm (or benefit) among the 11% of participants who were current smokers at baseline. Greenberg and associates studied the effect of β-carotene supplementation (50 mg/d) in 1720 male and female subjects for a median period of 4.3 yr with a median follow-up of 8.2 yr. Subjects whose plasma levels of β-carotene were in the highest quartile at the beginning of the study had the lowest risk of death from all causes compared with those in the lowest quartile. Supplementation, however, had no effect on either all-cause or cardiovascular mortality. Thus for β-carotene supplementation, it would appear that there are no overall benefits among individuals with a good nutritional status who are at low or average risk of developing CHD. The situation may be different, however, for those with a previous history of such disease.

Intervention Studies in Humans—Secondary Prevention

Hodis and coworkers have shown a reduction in CHD progression (as measured angiographically) in men given 100 IU vitamin E daily, although no benefit was found for vitamin C. Singh and colleagues found that a combi-nation of vitamins A, C, E, and β-carotene administered within a few hours after acute MI and continued for 28 d led to significantly fewer cardiac events and a lower prevalence of angina pectoris in the supplemented group. The Cambridge Heart Antioxidant Study (CHAOS), a trial of vitamin E supplementation on 2002 patients with angiographic evidence of coronary disease, was carried out with a mean treatment duration of 1.4 yr. This short-term supplementation with α-toco-pherol (268 or 537 mg/d) reduced CHD morbidity in patients (77% decreased risk of subsequent nonfatal MI). No benefit was found, however, in terms of cardiovascular mortality, with a nonsignificant excess among vitamin E-allocated participants.

The recently published GISSI-P trial investigated the independent and combined effects of n-3 PUFA and vitamin E on morbidity and mortality after MI. Although n-3 PUFA reduced the primary combined endpoint of death, nonfatal MI, and stroke, vitamin E had no benefit. When each supplement was compared with no treatment in a four-way analysis, however, there was a nonsignificant reduction of 11 % in events by vitamin E (the study was designed so that only a reduction of 20% in events would be statistically significant).

Summary and Evaluation of Human Intervention Studies

Thus, for vitamin E in Western populations, the only available trial data in primary prevention are from the ATBC trial which shows a negative effect. In secondary prevention, the accumulating trial data for vitamin E are more encouraging, particularly in terms of nonfatal endpoints, but far from conclusive in demonstrating net benefits. Each of these studies is open to specific criticism—the CHAOS study was too small to measure mortality, hemodynamic data were limited to only 706 of the 2002 patients, the amount of vitamin E was changed during the trial, and the patients were only followed up for 17 mo on average. Subjects in the ATBC study received a relatively low dose (50 mg) of vitamin E; there are doubts about the length of treatment; and it is uncertain whether the potential vitamin E-mediated protective effect could have any impact in such high-risk subjects (middle-aged heavy smokers).

This last uncertainty also applies to the CARET study where those treated were already at elevated risk for lung cancer due to exposure to asbestos or cigarette smoking; therefore, preclinical cancerous change may already have been present. How should we interpret the discordance between data from cohort studies and the results so far available from clinical trials? In general, it may be that the duration of clinical trials is too short to show a benefit, and that antioxidant intake over many years is required to prevent atherosclerosis. Thought needs to be given to trial design; with dose, duration of treatment and follow-up period, initial antioxidant levels and dietary intake, and extent and distribution of existing atherosclerosis being taken into consideration.

Animal models have nearly always tested the effects of antioxidants on the early atherosclerotic lesions. Whether or not antioxidants have inhibitory effects on the later stages remains to be seen. In addition, the complex mixture of antioxidant micronutrients found in a diet high in fruit and vegetables may be more effective than large doses of one or a small number of antioxidant vitamins. It could be that several of these compounds work together but have no effect individually, or that other dietary components (*e.g.*, trace elements) may be effectors of carotenoid action. The significant results linking antioxidant intake with CHD risk observed in cohort studies may be due to confounding by other lifestyle behaviours.

Slattery and associates examined dietary antioxidants and plasma lipids in the Coronary Artery Risk Development in Young Adults (CARDIA) study and found that a higher intake of antioxidants was associated with other lifestyle factors such as physical activity and not smoking. Plasma concentrations of antioxidants are linked with social class, being higher in more affluent groups. Although these variables can be individually controlled for in analyses, it may be that a complex lifelong behaviour pattern needs to be studied before conclusions regarding antioxi-dants and CHD can be made. For example, passive smoking has recently been shown to have an atherogenic effect on LDL, yet exposure to smoke is a difficult lifestyle variable to control for in cohort analyses and is rarely measured.

Further Clinical Trials of Antioxidant Supplementation

There are several clinical trials of antioxidant supplementation underway at present that are designed to clarify the effects of antioxidant supplementation on CVD. In each of these trials, doses of vitamin E greater than 200 mg/d are being used, which should be sufficient to increase serum levels at least two- to three-fold. The Women's Health Study is a primary prevention trial investigating the effects of vitamin E, β-carotene, and aspirin on CVD and cancer in 40,000 women aged 50 yr and over. In France, the Supplementation vitamins, minerals and antioxidant (SU.VI. MAX) Trial is testing a combination of antioxidant vitamins including vitamin E, vitamin C, and β-carotene in 15,000 healthy men and women.

The Heart Protection Study in Oxford is investigating the effects of vitamin E, vitamin C, and β-carotene in 18,000 subjects with above average risk of MI, and a secondary prevention trial using the same three vitamins in 8000 women has been established in the United States (Women's Antioxidant Cardiovascular Disease Trial, WACDR). The Heart Outcomes Protection Study is also assessing vitamin E in 9000 persons with previous MI, stroke, or peripheral vascular disease and diabetic patients.

WHOLE FOODS AND LDL OXIDATION

Although intervention trials are important in evaluating possible beneficial effects of antioxidants against development or progression of CHD, they have limitations and should be considered as only one component in the totality of available research evidence. It may be that a lifetime of intake is required to show a protective effect, or that a mixture of natural antioxidants found in fruits and vegetables provide the necessary protective mixture. A number of studies of people who eat a diet rich in fruit and vegetables, and therefore rich in antioxidant nutrients, have tried to test the hypothesis that fruits and vegetables lower the risk of CHD. In general, observational studies of vegetarians and those with diets rich in fruits and vegetables support the hypothesis that such a diet might lower the risk of CHD.

Vegetarians generally have high intakes of cereals, nuts, vegetable oils, vegetables, and fruit. However, vegetarians differ from the rest of the population in a number of important ways: they tend to smoke less, have a lower BMI and alcohol intake, and come predominantly from higher social classes, all of which are known to confer a health advantage. Few studies have looked at the effects of whole foods on biochemical end-points. A recent study asked subjects with normal lipid concentrations who ate three or fewer servings of fruit and vegetables daily to consume eight servings per day.

Plasma concentrations of vitamin C, retinol, α-tocopherol, α- and β-carotene, lipids, and lipoproteins were assessed before and after an 8-wk intervention period. The plasma vitamin C, α-carotene, and β-carotene concentrations increased, whereas concentrations of retinol, α-tocopherol, lipids and lipoproteins remained unchanged, despite some increase in dietary vitamin E and a small reduction in saturated fat intake. An interesting addition to the results would have been the inclusion of data on the susceptibility of LDL to oxidation. The authors concluded that more specific dietary advice to modify fat intake may be necessary to reduce the risk of CVD. By contrast, Singh and colleagues found over a 12-wk period that fruit and vegetable administration to subjects at high risk of CHD lowered total and LDL cholesterol and triglyceride levels, and increased HDL-cholesterol.

Another study by Wise and coworkers using dehydrated fruit and vegetable extracts over a period of 28 d in 15 healthy adults aged 18 – 53 yr produced increases of 50 – 2000-fold in plasma carotenoid and tocopherol levels. During the same intervention period, plasma lipid peroxides decreased fourfold, with much of this lowering taking place during the first week. Other antioxidant micronutrients may be important in increasing the resistance of LDL to oxidation. Flavonoids are plant-derived compounds that inhibit in vitro copper-catalyzed LDL oxidation (100 – 103). The inhibition of oxidation of human LDL by consumption of red wine or tea has been attributed to the presence of antioxidants such as flavonoids and other polyphenols in red wine and catechin in tea.

Isoflavonoids such as genistein also reportedly increase LDL resistance to oxidation, although this has not been confirmed, whereas aged garlic extract has recently been shown to have antioxidant properties. Ubiquinol-10 is another effective lipid-soluble antioxidant that inhibits LDL oxidation due to aqueous or lipid-phase peroxyl radicals. The effect of antioxidant supplementation may be sex-specific with 17 β-estradiol at physiological levels increasing the resistance of LDL to oxidation in some studies. Further studies, especially in humans, are required to validate the role of these antioxidants in inhibiting LDL oxidation.

OTHER ANTI-ATHEROGENIC EFFECTS OF ANTIOXIDANTS

The difficulty in linking inhibition of LDL oxidation with inhibition of atherosclerosis may stem from the dynamic nature of CHD, CHD involves not only the development of an atherosclerotic plaque, but also plaque rupture, vasoconstriction, and local thrombosis, resulting in partial or total arterial obstruction. Some antioxidants may limit the clinical expression of atherosclerosis by stabilizing the plaque rather than by affecting its size. Scavenger receptor activity has been shown to be downregulated in macrophages after incubation with α-tocopherol. Several other mechanisms can contribute to the vascular effects of α-tocopherol: it has been shown to reduce monocyte adhesion and transmigration into the intima, it can inhibit the proliferation of smooth muscle cells, and it prevents the cytotoxic effects of oxidized LDL by reducing endothelial cell damage. Oxidized LDL also impairs the release of nitric oxide from normal arteries, which usually prevents inappropriate adhesion of leukocytes and platelets, and vasospasm.

Thus, the presence of oxidized LDL might contribute to the platelet adhesion and vasospasm that are involved in the pathogenesis of acute coronary syndromes. LDL derived from patients treated with Probucol (a synthetic antioxidant) and oxidized in vitro does not impair the action of nitric oxide to the same degree as LDL derived from normal subjects. Antioxidant protection of LDL by Probucol and improved endothelial function has also been found in patients treated for hypercholesterolemia. These alternative vascular effects of antioxidants and decreased LDL oxidation may explain the positive findings for antioxidants in the prevention of secondary CHD or the clinical expression of established CHD, rather than in primary prevention or the initial development of atherosclerotic lesions.

LEVELS OF ANTIOXIDANT INTAKE FROM FOOD SOURCES

In the absence of conclusive evidence linking antioxidant intake with a reduced risk of CHD, dietary recommendations are difficult. Vitamin C and β-carotene are available from fruits and vegetables, and vitamin E from vegetable oils. Results from southern European countries consuming the classical Mediterranean diet show that high plasma levels of these antioxidants can be achieved. This diet is characterized by a preference for fresh products and frequent consumption of fruits, vegetables, legumes, and oils with a high vitamin E content. In contrast, major parts of populations in the United States or in northern parts of Europe do not consume optimal amounts of antioxidant nutrients. The availability of lower-priced convenience foods in the United States acts against the consumption of freshly prepared foods. Thus, only 40% of Americans con-sume five servings of fruit and vegetables daily, as recommended by the US national food guide, the Food Guide Pyramid. Only a quarter consumed fruits or vegetables rich in vitamin C or the carotenoids, and on any given day 54% ate no fruit at all.

A recent reanalysis of the Second National Health and Nutrition Examination Survey (NHANES II) data showed that vitamin supplements are the major contributors of the principal antioxidant micronutrients in the US diet (28% of vitamin C and 46% of vitamin E). Estimates of the requirement for micronutrients are based on the minimum quantity necessary to prevent a deficiency. The translation of minimum requirement to a dietary recommendation for population groups necessitates that allowance be made for a number of variables, including periods of low intake, increased utilization, individual variability, and bioavailability.

In the United States, the Recommended Daily Allowances (RDAs) incorporate "margins of safety" intended to be sufficiently generous to encompass the variability in the minimum requirement among people, and bioavailability from different food sources. To date, recommendations on micronutrient intake have therefore been primarily intended to prevent clinically overt deficiencies such as scurvy. If, however, observational and experimental evidence continues to show that the prevention of slow multistage processes such as CVD and cancer might require a higher intake of some essential antioxidants, then a recommended intake should be devised referring to amounts considered sufficient for the avoidance of these disease states.

Gey suggests that the present recommendations will require either an upgrading or an additional term, *e.g.*, a recommended optimum intake (ROI) that will vary with gender and age, with special requirements for smokers, pregnancy, and the elderly. The ROI could be defined as sufficient (culture-and/or regionspecific) intake to achieve blood levels associated with the observed minimum relative risk of disease. For example, Carr and Frei have suggested a doubling of the current RDA for vitamin C for optimum reduction in chronic disease risk. A ROI would simply quantify specific dietary constituents of conceivably crucial importance within the still desirable "five servings of fruit and vegetables daily."

CONCLUSION

There is strong evidence to support a link between antioxidant intake and a protective effect on the susceptibility of LDL to oxidation, but the relevance of this to CHD, and the possibility of other important bioactive micronutrients and phytochemicals in vegetables and fruit, require further research.

CHAPTER 7 Homocysteine—An Amino Acid

This chapter reviews how diet affects the occurrence and severity of hyperhomocysteinemia and coronary heart disease (CHD). The chapter will also provide an update on clinical studies and review the relative importance of this issue for nutritional considerations in health care. Homocysteine is a sulfur-containing amino acid that is an intermediary product in methionine metabolism. In 1969, based on studies of postmortem findings in patients with very high homocysteine levels due to rare genetic defects, McCully proposed that homocysteine may promote the development of vascular lesions. More recent investigations have focused on the possibility that moderate elevations may also be associated with increased risk of vascular disease. To date, more than 80 clinical and epidemiological studies, including prospective studies, have shown that an elevated total homocysteine level is a common cardiovascular risk factor in the general population. This has been shown to be the case for coronary artery disease (CAD), cerebrovascular disease, and peripheral vascular disease.

METABOLISM OF HOMOCYSTEINE

Intracellular homocysteine can be remethylated to methionine—the transmethylation pathway, converted to cystathionine—the transsulfuration pathway, or exported from the cells (see Fig. 7.1). Pathway 1 is catalyzed by the enzyme methionine synthase, which requires cobalamin (vitamin B_{12}) as a cofactor and folate (in the form methyltetrahydrofolate) as cosubstrate. The remethylation pathway is favoured during relative methionine deficiency, and this recycling and conservation of homocysteine ensures adequate methionine maintenance. During pathway 2, the vitamin B_6-dependent enzyme cystathionine β-synthase catalyzes the irreversible condensation of homocysteine with serine to form cystathionine, which is then broken down to cysteine and α-ketobutyrate. In the presence of excess methionine, the transsulfuration pathway is favoured by the upregulation of cystathionine β-synthase and downregulation of the remethylation pathway.

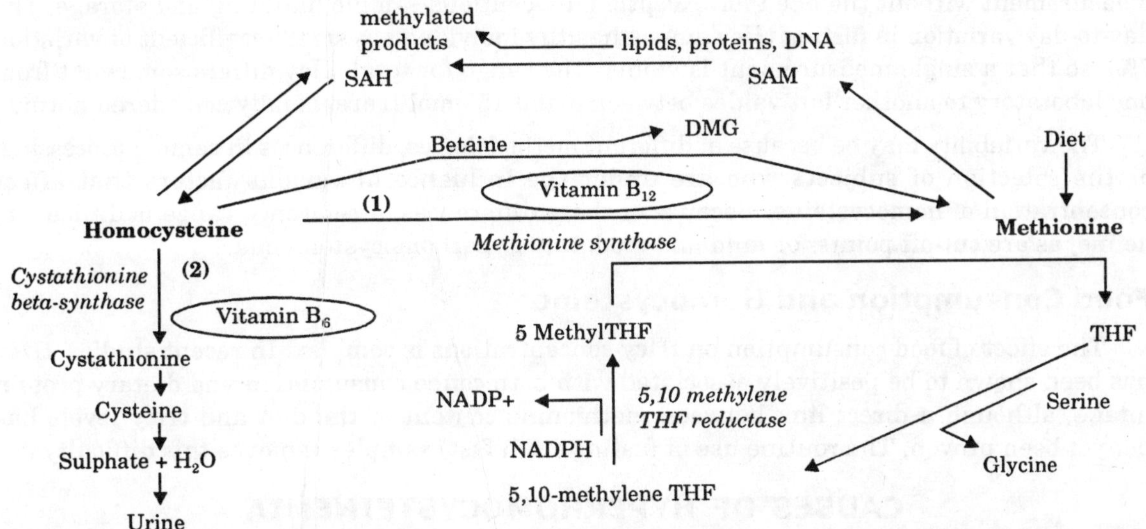

Fig. 7.1. Methionine cycle: metabolic cycle of homocysteine metabolism. THF = tetrahydrofolate; DMG = dimethylglycine; SAH = S-adenosyl homocysteine; SAM = S-adenosyl methionine; NADP+ = nicotinamide adenine dinucleotide phosphate (oxidized form); NADPH = nicotinamide adenine dinucleotide phosphate (reduced form).

Release of homocysteine into the extracellular medium represents the third route of homocysteine removal from the cell. Such export is enhanced when production of homocysteine is increased, and reduced when production is inhibited. Thus, the amount of homocysteine in the extracellular media, such as plasma and urine, reflects the balance between intracellular production and utilization.

Redox Status of Homocysteine

Homocysteine in plasma exists in different forms, including the major protein-bound fraction (approx 65%), free oxidized fraction (approx 30%), and reduced homocysteine, which is present in only trace amounts (1.5 – 4%). Because homocysteine in blood is rapidly oxidized, and is associated with plasma proteins, assessment of its redox status and protein binding requires immediate derivatization of the reduced homocysteine and separation of the free and bound forms. It is thought that the redox status of homocysteine may differ between patient groups and be linked with its atherogenic effects. Reduced homocysteine acts as a pro-oxidant in vitro and may be the atherogenic agent.

In patients with early-onset peripheral vascular disease, levels of reduced, oxidized, and protein-bound homocysteine are reportedly elevated whereas in patients with hyperhomocysteinamia because of cobalamin deficiency and in patients infected with human immunodeficiency virus reduced homocysteine concentrations are markedly above normal.

MEASUREMENT OF HOMOCYSTEINE

The sum of all homocysteine species in plasma (free plus protein bound) is referred to as total homocysteine (tHcy or homocyst(e)ine). As a marked redistribution between free and bound homocysteine takes place after preparation of plasma, this gives an accurate measurement without the need for exceptionally cautious sample handling and storage. The day-to-day variation in fasting tHcy among healthy individuals is small (coefficient of variation 7%), so that a single measurement is viable. The range for total tHcy differs somewhat from one laboratory to another but values between 5 and 15 μmol/L are usually considered normal.

The variability may be because of different methodologies, differences in sample processing, or the selection of subjects who are under the influence of various factors that affect concentration of homocysteine. Because of these differences, a reference range is difficult to define, as are cut-off points for mild and moderate hyperhomocysteinemia.

Food Consumption and Homocysteine

The effect of food consumption on tHcy concentrations is complex. In recent studies, tHcy has been shown to be positively associated with both coffee consumption and dietary protein intake, although a direct link between methionine content of the diet and tHcy levels has not yet been proven. The routine use of fasting (12 h fast) samples removes this difficulty.

CAUSES OF HYPERHOMOCYSTEINEMIA

A number of enzymes, essential cofactors, and the availability of the important cosubstrate methyltetrahydrofolate regulate plasma homocysteine concentrations. Predictably, therefore, the causes of hyperhomocysteinemia are multifactorial.

Environmental Factors

Age and Gender

Plasma tHcy increases with age in both genders, for reasons that have not been elucidated. Decreases in cofactor levels or coexisting renal impairment often seen in older patients may be responsible, and age-dependent reductions in cystathionine β-synthase activity may also play a part. In general, men have higher plasma levels than women. After menopause, fasting tHcy seems to increase although this has not been confirmed, and hormone replacement therapy can lower elevated tHcy levels in postmenopausal women. Although gender differences may be explained by the effect of sex hormones on homocysteine metabolism, they may be related to higher creatinine values or the greater muscle mass of men than women. Homocysteine is decreased by up to 50% during pregnancy, returning to normal 2 – 4 d postpartum. The authors suggest several reasons for this decrease, including the hemodilution known to occur in pregnancy, or an increased demand for methionine by the fetus, leading to increased remethylation of homocysteine.

Ethnic Group

Despite a high prevalence of CHD risk factors such as hypertension, obesity, and smoking, CHD incidence rate is much lower among westernized black Africans compared with the white population. A group of 27 black men, aged 18 – 25 yr, had tHcy 46% lower than similarly aged white men. By contrast, in a study examining tHcy and B-vitamin levels in American black pre-menopausal women, who have higher rates of CAD than white women, black women had higher tHcy and lower folate concentrations, largely because of lifestyle factors. Another study examined homocysteine and CAD in the Hong Kong Chinese population. Although the prevalence of hyperhomocysteinemia was similar to that in white subjects, elevated tHcy was not an independent risk factor, being associated with smoking. Serum vitamin B_{12} did not differ between patients and control subjects. The observation of higher serum folate in those with elevated tHcy does not seem compatible with what is known about tHcy metabolism.

Coexistent Disease

Elevated homocysteine levels are found in a number of disease states. Impaired renal function is associated with hyperhomocysteinemia. There is a positive correlation between fasting plasma tHcy and serum creatinine although the mechanism is unclear. Markedly elevated homocysteine levels have also been seen in acute lymphoblastic leukemia various carcinomas (including breast, ovary and, pancreas, severe psoriasis, and diabetes mellitus.

HYPERHOMOCYSTEINEMIA AND CORONARY HEART DISEASE

Numerous studies have indicated that mild hyperhomocysteinemia is an independent risk factor for CHD. In the Physicians' Health Study, a total of 14, 91.6 US male physicians, aged 40 – 84 yr, were followed up for 6 yr. Men with homocysteine levels above the 95th percentile (based on control distribution) had a threefold increased risk of myocardial infarction compared with those within the bottom 90%. The findings were also statistically compatible with a graded risk increase across the distribution, a suggestion made by Perry and coworkers in a prospective study of stroke in middle-aged British men. Similar findings have been reported for myocardial infarction, carotid artery thickening, and angiographically defined coronary artery stenosis.

In addition, Selhub and associates demonstrated a gradual increase in the prevalence of carotid artery stenosis with increasing levels of homocysteine. A meta-analysis by Boushey and colleagues showed an increase in risk of CAD of about 70% for each 5 μmol/L rise in fasting homocysteine. They concluded that a total of 10% of CAD risk appeared to be attributable to homocysteine. Two recent studies have confirmed the relevance of homocysteine as a risk factor for vascular disease. The first was a multicenter, case-control study where the risk of atherosclerotic vascular disease (cardiac, cerebral, and peripheral) was examined, as well as the association of plasma tHcy with conventional risk factors.

The subjects comprised 750 cases and 800 controls, both male and female, below 60 yr of age. The relative risk (RR) for vascular disease in the top fifth of the fasting tHcy distribution, when compared with the bottom four-fifths, was 2.2 (95% confidence interval [CI] 1.6 – 2.9), and a dose-response effect was noted. This level of risk is similar to that observed for hypercholesterolemia and smoking. The second study examined the prognostic value of homocysteine in patients with established CAD. A total of 587 patients with angiographically confirmed CAD were followed-up for a median period of 4.6 yr. Homocysteine levels were strongly associated with levels of folate and vitamin B_{12}, history of myocardial infarction, the left ventricular ejection fraction, and serum creatinine.

A strong, graded relationship was found between plasma tHcy and overall mortality. After 4 yr, 3.8% of patients with tHcy below 9 μmol/L had died compared with 24.7% of those with tHcy greater than 15 μmol/L. The association was not altered significantly when adjusted for other, possibly con-founding, factors such as age, sex, serum creatinine, left ventricular ejection fraction, and history of myocardial infarction. However, not all studies examining the effect of homocysteine levels on the incidence of cardiovascular disease (CVD) have shown a positive association, highlighting the need for further investigation. An updated analysis of the Physicians' Health Study data yielded a relative risk for elevated tHcy of only 1.3 (95% CI 0.5 – 3.1).

An analysis of the Multiple Risk Factor Intervention Trial (MRFIT) cohort showed no effect of tHcy after adjustment for other variables (RR = 0.94; 95% CI 0.56 – 1.56). An analysis of tHcy and CHD in the Caerphilly cohort showed tHcy to be only weakly predictive of CHD events. Finally, the ARIC study showed no association between the incidence of CHD and tHcy, although there was a possibility that vitamin B_6 offered independent protection, a finding also suggested by others.

MECHANISMS BY WHICH HOMOCYSTEINE MAY LEAD TO VASCULAR DISEASE

Several mechanisms may be involved in the genesis of vascular disease by homocysteine, including effects on connective tissue, smooth-muscle cells, platelets, endothelial cells, blood lipids, coagulation factors, and nitric oxide—summarized Table 7.1. The relative importance of each of these mechanisms is not fully understood.

Table 7.1. Possible Vascular Damaging Mechanisms of Homocysteine

Endothelial cell injury
- Impaired nitric oxide production
- Overproduction of reactive oxygen species
- Increased von Willebrand factor and thrombomodulin
- Increased tissue factor production
- Decreased antithrombin III production
- Increased smooth muscle cell production
- Increased monocyte adhesion to the vessel wall

Coagulation pathways
- Impaired platelet survival time
- Increased production of thromboxane A2 by platelets and decreased production of prostacyclin
- Increased activation of factors V, X, and XII
- Inhibition of antithrombin III and factor C production
- Increased fibrinogen levels
- Enhanced binding of lipoprotein (a) to fibrin

Oxidative stress
- Overproduction of reactive oxygen species
- Decreased plasma antioxidant activities
- Increased lipid peroxidation

Homocysteine is toxic to endothelial cells in vitro, and in vivo. Hyperhomo-cysteinemia is associated with impaired endothelium-dependent vasodilation and impaired endogenous tissue-type plasminogen activator activity. Homocysteine promotes increased platelet aggregation as a consequence of increased synthesis of thromboxane A_2 and decreased synthesis of prostaglandin. Hyperhomocysteinemia is associated with abnormalities of the clotting cascade. Homocysteine promotes the binding of lipoprotein (*a*) to fibrin and the growth of smooth muscle cells, and tHcy levels correlate with levels of fibrinogen, an independent risk factor for CVD. It must be noted, however, that many of these effects are not specific to homocysteine; a variety of free thiol-group amino acids, particularly cysteine, show similar tendencies. Much of the research into mechanisms has been carried out at millimolar concentrations of homocysteine, which are 100 – 1000-fold higher than those observed in vivo. In addition, the complex redox reac-tions involving the various homocysteine forms and their relation to other thiols in vivo are difficult to represent accurately in vitro where a single homocysteine species is used.

The complexities of elucidating an atherogenic mechanism are illustrated by the fact that oxidative modification of LDL by tHcy has been demonstrated in vitro and in animal models, but has not been observed in hyperhomocysteinemic patients. Similarly, supplementation with B-group vitamins, postulated to reduce tHcy and thus inhibit lipid peroxidation, had no effect on the susceptibility of LDL to oxidation. Homocysteine's effect on endothelial dysfunction, however, has been confirmed in a clinical setting. The blocking of the effects of hyperhomocysteinemia on endothelial dysfunction by pretreatment with antioxidant vitamins still suggests the involvement of an oxidative mechanism.

Similarly, in a cross-sectional study, plasma tHcy was associated strongly ($r = 0.40$, $p < 0.001$) with plasma F-2-isoprostane levels, a marker for in vivo lipid peroxidation. The observation by Glueck and associates of a higher risk of myocardial infarction in hyperlipidemic patients with hyperhomocysteinemia and low HDL also suggests the possibility of an interaction between these risk factors.

EFFECT OF VITAMIN SUPPLEMENTATION AND DIET ON HOMOCYSTEINE

Cross-sectional and experimental evidence suggests that mild hyperhomo-cysteinemia may be related to subclinical deficiencies of folate, vitamin B_6, and vitamin B_{12}—all cofactors or cosubstrates in homocysteine metabolism (84 – 93). In a study of vitamin and tHcy levels in an elderly population, Selhub and coauthors (21) found a strong inverse correlation between tHcy and plasma folate, and weaker inverse correlations between tHcy and both cobalamin and vitamin B_6 (pyridoxal-5-phosphate). The authors concluded that elevated tHcy levels could, in great part, be due to poor vitamin status, and these results have been confirmed in many other studies. Many studies have assessed the effects of vitamin supplementation on plasma tHcy. Levels in folate-deficient patients can be reduced by oral folate supplementation.

In one study, the elevated tHcy in patients with renal failure decreased after only 2 wk of folate therapy. Maximal effects may be seen after 4 to 6 wk of therapy. The lowest effective dose for folate supplementation has not yet been determined. Doses of 5 mg or 10 mg alone or 1 mg in conjunction with vitamins B_{12} and B_6 may be effective. Ubbink and colleagues confirmed by intervention that a daily supplement of folate (1 mg), B_6 (10 mg) and B_{12} (0.4 mg) could normalize elevated tHcy (> 16.3 μmol/L, $n = 44$) within 6 wk. Ubbink and colleagues then looked at the effect of supplementation with the individual vitamins in 100 men, aged between 20 and 73 yr, with tHcy greater than 16.3 μmol/L, over 6 wk. Folate supplementation (0.65 mg/d) reduced plasma tHcy by 42% whereas a daily vitamin $B_{,2}$ supplement (0.4 mg/d) lowered it by 15%. The vitamin B_6 supplement (10 mg/d) had no significant effect. The combination of the three vitamins reduced circulating tHcy by 50% that was not significantly different from the reduction achieved by folate supplementation alone.

Brattstrom and coworkers noted significantly lower tHcy in middle-aged and elderly subjects taking multivitamins containing doses of folic acid ranging from only 200 – 400 µg. Homocysteine values increase if vitamin therapy is discontinued. Ward and colleagues carried out an uncontrolled study examining the effect of low-dose folate supplementation in healthy male subjects. Folate supple-ments were administered daily at doses increasing from 100 µg for 6 wk to 200 µg for 6 wk and then up to 400 pig for 14 wk. A dose of 200 µg/d appeared to be the optimum tHcy-lowering dose as there was no apparent benefit of increasing the dose to 400 µg/d. The subjects had reported a mean dietary folate intake of 281 ± 60 µg/d. Their total folate intake, with the additional 200 pig supplement, therefore corresponds well with the observation by Selhub and colleagues that total intakes over 400 µg/d are associated with desirable tHcy in a healthy elderly US population.

When the analysis in the former study was carried out by tertiles of baseline tHcy, the lowest tertile group showed no increase in red-cell folate over the 6-mo supplementation period, suggesting that their baseline folate status was optimal to begin with. Ward and associates suggests that, although there does not appear to be a threshold in the relationship between elevated tHcy and CVD risk, there may be a threshold of plasma tHcy in terms of ability to respond to folate. A further randomized placebo-controlled study in healthy women aged 18 – 40 yr confirmed that both 250 µg and 500 µg folate/d for 4 wk decreased tHcy. Eight weeks after the end of folate supplementation, tHcy had not returned to baseline. A meta-analysis of 12 studies using B-group vitamins to lower tHcy has recently been carried out.

The magnitude of tHcy reduction was related to the pretreatment tHcy and folate levels. Folate reduced tHcy by 25% (95% CI 23 – 28%), and the effects were similar at daily doses ranging from 0.5 – 5 mg. Vitamin B_{12} yielded a further tHcy reduction of 5 – 6%, but vitamin B_6 had no significant effect. None of these trials, assessed the effect on tHcy after methionine loading, however, which is determined by the transsulfuration pathway where vitamin B_6 is a cofactor. The effectiveness of vitamins B_{12} and B_6 in homocysteine-lowering therapies needs further study.

Alien and associates have shown that folate supplementation will not correct hyperhomocy steinemia that is primarily the result of vitamin B_{12} deficiency, whereas Robinson and colleagues found that vitamin B_6 was inversely related to risk of CVD independently of homocysteine, although this may be due to its ability to modulate the homocysteine peak after the oral methionine load test. Administration of vitamin B_6 alone does not lower fasting tHcy. Dose-optimizing studies for these vitamins are also required. Woodside and colleagues have tested the hypothesis that simultaneous administration of antioxidant vitamins may potentiate the effect of B-group vitamins on elevated tHcy. An interaction between antioxidants and folate is conceivable because the latter is susceptible to inactivation by freeradical mediated oxidation, which can be prevented both in vitro and in vivo by vitamin C.

Although B-group vitamins lowered tHcy by roughly 30%, the addition of antioxidant vitamins to the supplement had no effect on this. It is clear from the foregoing evidence above that B-group vitamin supplementation effectively lowers plasma homocysteine. The question therefore arises whether a dietary change in B-vitamin intake or food fortification can achieve similar results. To date, the lowest effective tHcy-lowering supplement was 200 μg folate daily.

Selhub and coworkers found that a total folate intake of 400 μg/d was necessary to prevent elevation of tHcy, and dietary intake approximating this value was only attained by 30 – 40% of participants in the Framingham study. Boushey and coworkers suggested that if the US population were to eat two to three more servings of fruit and vegetables per day, this would lead to a reduction of 4% in CVD deaths per year in the United States through tHcy reduction. Food fortification at 350 μg of folate/100 g of grain products would lead to a reduction in CVD deaths of 8%. Cuskelly and associates looked at the effects of increasing dietary folate on red-cell folate, with a view to preventing neural tube defects.

Red-cell folate concentration increased significantly over 3 mo in patients taking folate supplements or food fortified with folate (both given so as to increase consumption by 400 μg/d). There was no increase in those given food naturally rich in folate (again to provide an estimated increase in intake of 400 μg/d) or in those offered dietary advice only. The likely explanation lies in the higher bioavailability of folate supplements compared to food folates. Further investigation is needed using larger numbers of subjects to assess the relative effectiveness of a high-folate diet, food fortification with folate, and folate supplementation on homocysteine. In 1996, the Food and Drug Administration issued a regulation requiring all enriched grain products to be fortified with folate (140 μg/100 g, providing approximately an additional 70 – 120 fig folate/d). The main aim of this public health action was to reduce the risk of

Table 7.2. Ongoing Vitamin Supplementation Studies to Lower Homocysteine and Prevent Cardiovascular Disease

CHD prevention	*Number of subjects*
NORVIT (Norwegian study of homocysteine lowering with B Vitamins in myocardial infarction; University of Tromso, Norway)	3000
PACIFIC (Prevention with a Combined Inhibitor and Folate in Coronary heart disease; University of Sydney, Australia)	10,000
SEARCH (Study of the Effectiveness of Additional Reductions in Cholesterol and Homocysteine; University of Oxford, England)	12,000
WACS (Women's Antioxidant and Cardiovascular disease Study; Harvard Medical School, USA)	8000
CHAOS-2 (Cambridge Heart Antioxidant Study; University of Cambridge, England)	4000
BERGEN (Bergen Vitamin Study; University of Bergen, Norway)	2000
Storke prevention	3600
VISP (Vitamin Intervention for Stroke Prevention; Wake Forest University, USA)	
HOPE-2 (Health Outcome and Prevention Evaluation number 2; Canada)	5000

neural tube defects in newborns. A survey of folate and tHcy in the Framingham Offspring Study cohort before and after fortification showed an increase in folate and decrease in tHcy, and also a reduction in the prevalence of low folate and elevated tHcy, respectively. In this middle-aged and elderly population, fortification has significantly improved folate status and may therefore contribute to reduced CVD prevalence. This is in contrast to the findings of Malinow and colleagues who, in a crossover trial in 75 subjects, tested the effects of fortified breakfast cereal on plasma tHcy and folate.

These researchers found that cereal providing approximately the intake resulting from the PDA enrichment policy only decreased tHcy by 3.7%, a difference that was not statistically significant. Jacques and coworkers suggested that several features of this study may have limited its applicability to the general population. The length of the treatment was only 5 wek, which may have not been sufficient time to achieve a new steady-state concentration, while the study was performed in CAD patients who may require higher folate intake to lower tHcy.

CLINICAL TRIALS

There is a wealth of epidemiological evidence linking homocysteine with the development of vascular disease, including large prospective cohort studies with confirmation by meta-analysis, and clear evidence that B-group vitamins effectively lower homocysteine. There is presentlyno clinical trial evidence to show that lowering homocysteine by nutritional or pharmacological intervention will prevent cardiovascular events, however, such evidence is clearly crucial in providing final confirmatio of the homocysteine hypothesis, and a number of large clinical trials will report in the next few years.

CONCLUSION

In conclusion, there is strong support for tHcy as an independent risk factor for atherosclerosis. Dose-finding studies are required to elucidate the optimum combination of vitamins that will lower tHcy in the general population, whereas further work must be carried out on tHcy's atherogenic mechanism. Intervention studies with clinical endpoints are also required to assess the effect of strategies designed to lower plasma homocysteine, and initial studies suggesting a role for tHcy in other diseases must also be developed.

CHAPTER

8

Coenzymes: Nature's Special Reagents

Most of the reactions are catalyzed by enzymes that contain only those functional groups found in the side chains of the constituent amino acids. *Coenzymes* are nonprotein molecules that function as essential parts of enzymes. Coenzymes often serve as "special reagents" needed for reactions that would be difficult or impossible using only simple acid-base catalysis. In many instances, they also serve as *carriers*, alternating catalysts that accept and donate chemical groups, hydrogen atoms, or electrons. Coenzymes will be considered here in three groups:

1. Compounds of high group transfer potential such as ATP and GTP that function in energy coupling within cells. Because it is cleaved and then dissociates from the enzyme to which it is bound, ATP may be regarded as a substrate rather than a coenzyme. However, as a phosphorylated form of ADP it may also be viewed as a carrier of high-energy phospho groups.
2. Compounds, often derivatives of *vitamins* that, while in the active site of the enzyme, alter the structure of a substrate in a way that permits it to react more readily. Coenzyme A, pyridoxal phosphate, thiamin diphosphate, and vitamin B_{12} coenzymes fall into this group.
3. Oxidative coenzymes with structures of precisely determined oxidation-reduction potential. Examples are NAD^+, NADP+, FAD, and lipoic acid. They serve as carriers of hydrogen atoms or of electrons. Some of these coenzymes, such as NAD^+ and $NADP^+$, can usually dissociate rapidly and reversibly from the enzymes with which they function. Others, including FAD, are much more tightly bound and rarely if ever dissociate from the protein catalyst. Heme groups are covalently linked to proteins such as cytochrome *c* and cannot be dissociated without destroying the enzyme. Very tightly bound coenzyme groups are often called *prosthetic groups*, but there is no sharp line that divides prosthetic groups from the loosely bound coenzymes. For example, NAD^+ is bound weakly to some proteins but tightly to others.

A. ATP AND THE NUCLEOTIDE "HANDLES"

The role of ATP in "driving" biosynthetic reactions has been considered, where attention was focused on the polyphosphate group which undergoes cleavage. What about the adenosine end? Here is a shapely structure borrowed from the nucleic acids. What is it doing as a carrier of phospho groups? At least part of the answer seems to be that the adenosine monophosphate (AMP) portion of the molecule is a "handle" which can be "grasped" by catalytic proteins. For some enzymes, such as acetyl-CoA synthetase, the handle is important because the intermediate acyl adenylate must remain bound to the protein. Without the large adenosine group, there would be little for the protein to hold onto.

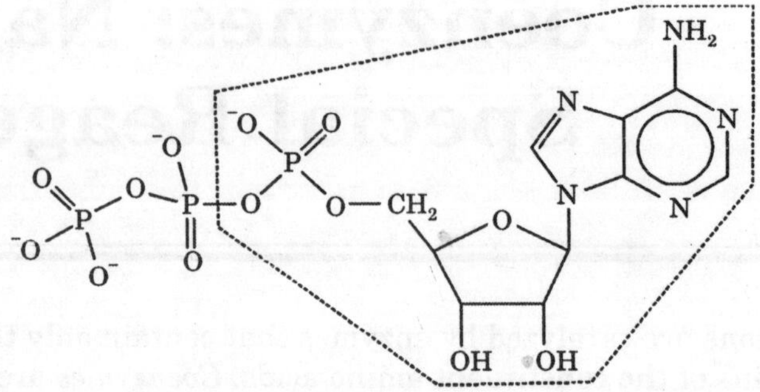

The AMP "handle" in ATP

AMP is only one of several handles to which nature attaches phospho groups to form di- and triphosphate derivatives. Like AMP, the other handles are nucleotides, the monomer units of nucleic acids. Thus, one enzyme requiring a polyphosphate as an energy source selects ATP, another CTP, or GTP. The nucleotide handles not only carry polyphosphate groups but also are present in other coenzymes, such as CoA, NAD+, NADP+, and FAD. In addition, they serve as carriers for small organic molecules. For example, *uridine diphosphate glucose* (UDP-Glc) Chapter 10, is a carrier of active glucosyl groups important in sugar metabolism and *cytidine diphosphate choline* is an intermediate in synthesis of phospholipids. Recalling that acetyl adenylate (acetyl-AMP) is an intermediate in synthesis of acetyl-CoA, and com-paring the biosynthesis of sugars, phospholipids, and acetyl-CoA, we see that in each case the enzyme involved requires a different nucleotide handle.

The handle may provide a means of recognition which can help an enzyme to pick the right bit of raw material out of the sea of molecules surrounding it. The shapes of the four purine and pyrimidine bases forming the most common nucleotide handles. The distinctive differences both in shape and in hydrogen bond patterns are obvious. In binding to proteins, the hydrogen bond-forming groups in the purine and pyrimidine bases sometimes interact with precisely positioned groups in the protein. However, in some enzymes the "handle" is not precisely bound. This seems to be the case for adenine, which makes surprisingly few hydrogen bonds to proteins. Hydroxyl groups of the ribose or deoxyribose ring also often form hydrogen bonds and the negatively charged oxygen atoms of the 5′-phosphate may interact with positively charged protein side chains.

B. Coenzyme A and Phosphopantetheine

The existence of a special coenzyme required in biological acetylation was recognized by Fritz Lipmann in 1945. The joining of acetyl groups to other molecules is a commonplace reaction within living cells, one example being the formation of the neuro-transmitter *acetylcholine*. In the laboratory acetylation is carried out with reactive compounds such as acetic anhydride or acetyl chloride.

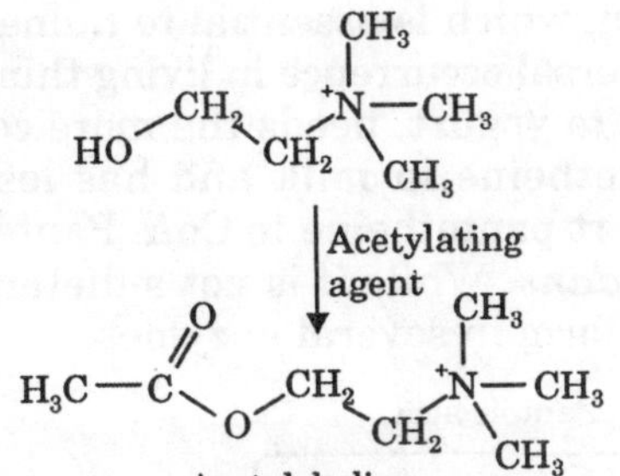

$HO-CH_2-CH_2-N^+(CH_3)_3$

Acetylating agent

$H_3C-C(=O)-O-CH_2-CH_2-N^+(CH_3)_3$

Acetylcholine

Lipmann wondered what nature used in their place. His approach in seeking the biological "active acetate" is one that has been used successfully in solving many biochemical problems. He first set up a test system to examine the ability of extracts prepared from fresh liver tissue to catalyze the acetylation of sulfanilamide. A specific color test was available for quantitative determination of very small amounts of the product. The rate of acetylation of sulfanilamide under standard conditions was taken as a measure of the activity of the biochemical acetylation system. Lipmann soon discovered that the reaction required ATP and that the ATP was cleaved to ADP concurrently with the formation of acetyl-sulfanilamide. He also found that dialysis or ultrafiltration rendered the liver extract almost inactive in acetylation.

Apparently some essential material passed out through the semipermeable dialysis membrane. When the dialysate or ultrafiltrate was concentrated and added back, acetylation activity was restored. The unknown material was not destroyed by boiling, and Lipmann postulated that it was a new coenzyme which he called *coenzyme A* (CoA). Now the test system was used to estimate the amount of the coenzyme in a given volume of dialysate or in any other sample. When small amounts of CoA were supplied to the test system, only partial restoration of the acetylation activity was observed and the amount of restoration was proportional to the amount of CoA. With test system in hand to monitor various fractionation methods, Lipmann soon isolated the new coen-zyme in pure form from yeast and liver.

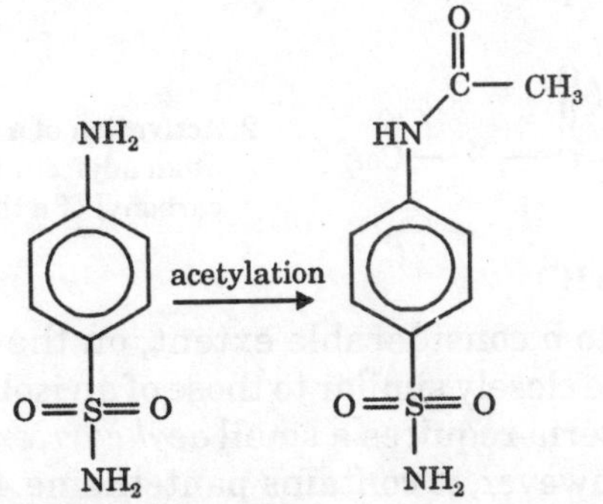

$NH_2-C_6H_4-SO_2-NH_2 \xrightarrow{\text{acetylation}} CH_3-C(=O)-HN-C_6H_4-SO_2-NH_2$

Sulfanilamide

Coenzyme A Fig. 8.1 is a surprisingly complex molecule. The handle is AMP with an extra phospho group on its 3′-hydroxyl. The phosphate of the 5′-carbon is linked in anhydride (pyrophosphate) linkage to another phosphoric acid, which is in turn esterified with *pantoic acid*. Pantoic acid is linked to β-*alanine* and the latter to β-*mercaptoethylamine* through amide linkages, the reactive SH group being attached to a long (1.9-nm) semi-flexible chain. Coenzyme A can be cleaved by hydrolysis to *pantetheine, pantetheine 4′-phosphate, and pantothenic acid.* These three compounds are all *growth factors.*

Pantothenic acid is a vitamin, which is essential to human life. Its name is derived from a Greek root that reflects its universal occurrence in living things. The bacterium *Lactobacillus bulgaricus,* which converts milk to yogurt, needs the more complex pantetheine for growth. It finds a ready supply of pantetheine in milk and has lost its ability to synthesize this compound. However, it can convert pantetheine to CoA. Pantetheine 4′-phosphate is required for growth of *Acetobacter suboxydans.* While it is not a dietary essential for most organisms, it is found in a covalently bound form in several enzymes.

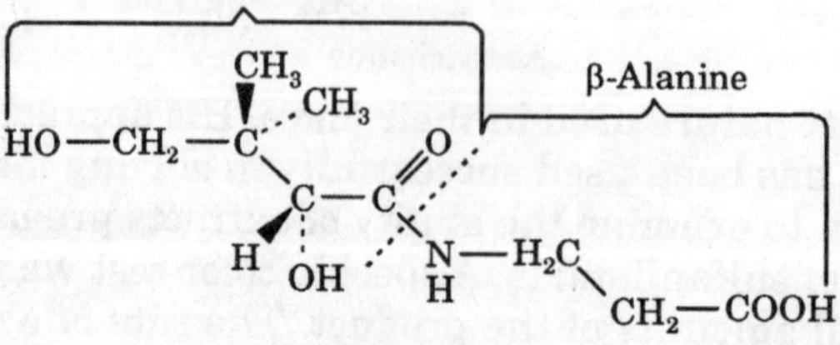

Pantoic acid β-Alanine

Pantothenic acid

While CoA was discovered as the "acetylation coenzyme," it has a far more general function. It is required, in the form of acetyl-CoA, to catalyze the synthesis of citrate in the citric acid cycle. It is essential to the β oxidation of fatty acids and carries propionyl and other acyl groups in a great variety of other metabolic reactions. About 4% of all known enzymes require CoA or one of its esters as a substrate.

Coenzyme A has two distinctly different biochemical functions, which have already been considered briefly, can be summarized as follows:

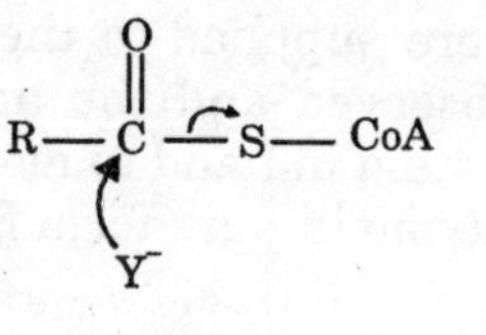

1. Activation of acyl group, R—C(=O)— , toward transfer by nucleophilic displacement

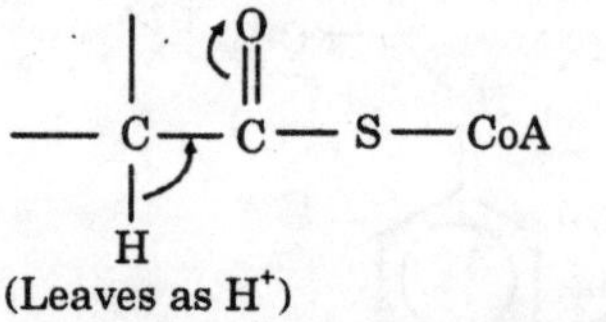

2. Activation of a hydrogen atom adjacent to the carbonyl of a thioester

(Leaves as H^+)

These functions depend, to a considerable extent, on the fact that the properties of the carbonyl group of a thioester are closely similar to those of an isolated carbonyl group in a ketone. Synthesis of fatty acids in bacteria requires a small *acyl carrier protein (ACP)* whose functions are similar to those of CoA. However, it contains pantetheine 4′-phosphate covalently bonded through phosphodiester linkage to a serine side chain. In *E. coli* this is at position 36 in the 77-residue protein. Here the nucleotide handle of CoA has been replaced with a larger and more complex protein which can interact in specific ways with the multiprotein fatty acid synthase complex described.

In higher organisms ACP is usually not a separate protein but a domain in a large synthase molecule. Bound phosphopantetheine is also found in enzymes involved in synthesis of peptide antibiotics and polyketides. It is also present in subunits of cytochrome oxidase and of ATP synthase of *Nenrospora* but appears to play only a structural role, being needed for proper

assembly of these multimeric proteins. In a citrate-cleaving enzyme, and in a bacterial malonate decar-boxylase, phosphopantethine is attached to a serine side chain as 2′-(5′-phosphoribosyl)-3′-dephospho-CoA.

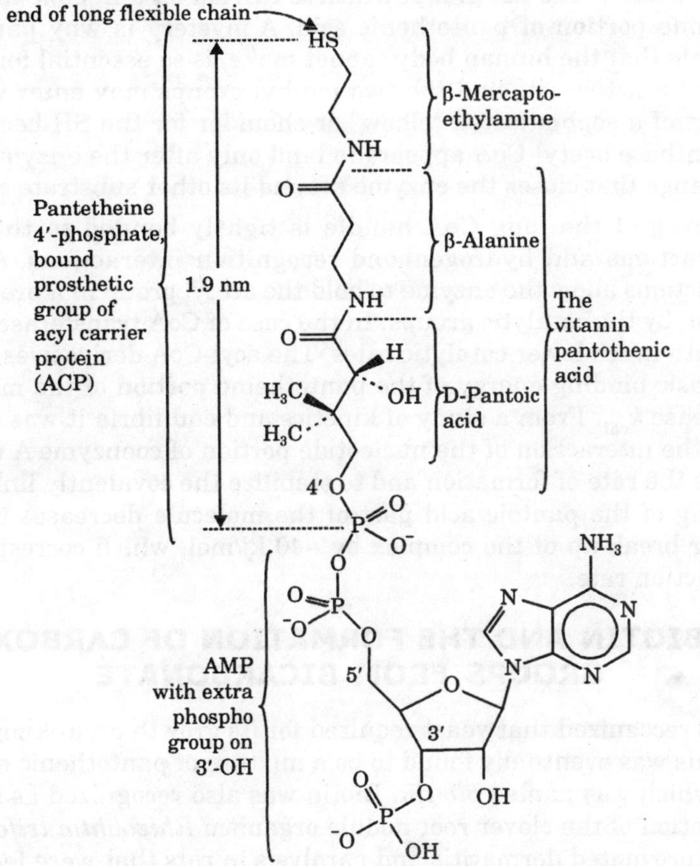
Reactive—SH group at end of long flexible chain
HS
β-Mercapto-ethylamine
NH
O
Pantetheine 4′-phosphate, bound prosthetic group of acyl carrier protein (ACP)
β-Alanine
1.9 nm
NH
The vitamin pantothenic acid
O
H
H3C
OH
D-Pantoic acid
H3C
4′
O
P
O
O⁻
NH3
O
O
P
O
N
N
N
N
AMP with extra phospho group on 3′-OH
5′
O
3′
O
O
OH
P
O
OH

Fig. 8.1. Coenzyme A, an acyl-activating coenzyme containing the vitamin pantothenic acid.

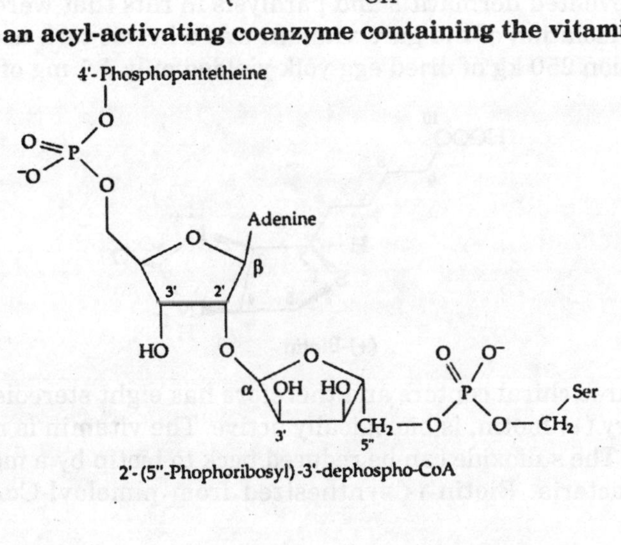
4′-Phosphopantetheine
O
O
P
O
O
Adenine
O
β
3′
2′
HO
O
O
α
OH
HO
O
O⁻
P
Ser
3′
CH2
O
O
CH2
5″

2′-(5″-Phophoribosyl)-3′-dephospho-CoA

Attachment of phosphopantetheine to proteins is catalyzed by a phosphotransferase that utilizes CoA as the donor. A phosphodiesterase removes the phospho-pantetheine, providing a turnover cycle. A variety of synthetic analogs have been made. The reactive center of CoA and phosphopantetheine is the SH group, which is carried on a flexible arm that consists in part of the β-alanine portion of pantothenic acid. A mystery is why pantoic acid, a small odd-shaped molecule that the human body cannot make, is so essential for life. The hydroxyl group is a potential reactive site and the two methyl groups may enter into formation of a "trialkyl lock", part of a sophisticated "elbow" or shoulder for the SH-bearing arm. When it binds to citrate synthase acetyl-CoA appears to bind only after the enzyme has undergone a conformational change that closes the enzyme around its other substrate oxaloacetate.

The adenine ring of the long CoA handle is tightly bonded to the protein through hydrophobic interactions and hydrogenbond recognition interactions. Additional specific hydrophobic interactions allow the enzyme to hold the acetyl group in a precise position where it can be acted upon by the catalytic groups. In the case of CoA transferase smaller thiols can replace CoA but with much lower catalytic rates. The acyl-CoA derivatives are weakly bound, the expected intrinsic binding energy of the pantetheine portion of the molecule apparently being used to increase k_{cat}. From a study of kinetics and equilibria it was concluded that the binding energy of the interaction of the nucleotide portion of coenzyme A with the enzyme is utilized to increase the rate of formation and to stabilize the covalently linked E-CoA. On the other hand, binding of the pantoic acid part of the molecule decreases the stability of the transition state for break-up of the complex by ~40 kj/mol, which corresponds to a 10^7-fold increase in the reaction rate.

C. BIOTIN AND THE FORMATION OF CARBOXYL GROUPS FROM BICARBONATE

By 1901 it was recognized that yeast required for its growth an unknown material which was called *bios*. This was eventually found to be a mixture of pantothenic acid, inositol, and a third component which was named *bio tin*. Biotin was also recognized as a factor promoting growth and respiration of the clover root nodule organism *Rhizobhtm trifolii* and as vitamin H, a material that prevented dermatitis and paralysis in rats that were fed large amounts of un-cooked egg white. Isolation of the pure vitamin was a heroic task accomplished by Kogl in 1935. In one preparation 250 kg of dried egg yolk yielded only 1.1 mg of crystalline biotin.

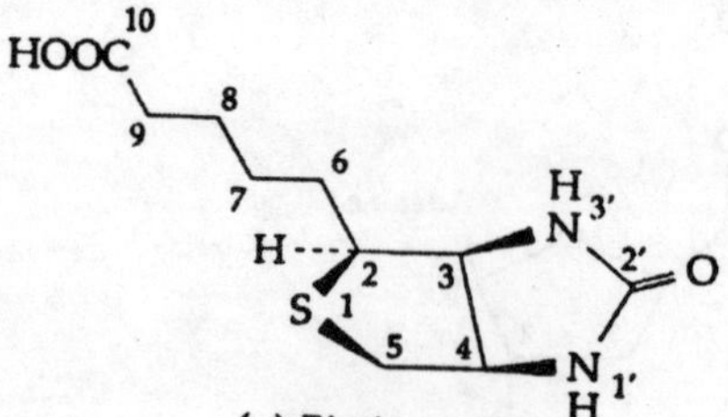

HOOC 10 9 8 7 6 H 2 3 4 5 S 1 H N 3' 2' O N 1' H

(+)-Biotin

Biotin contains three chiral centers and therefore has eight stereoisomers. Of these, only one, the dextrorotatory (+)-biotin, is biologically active. The vitamin is readily oxidized to the sulfoxide and sulfone. The sulfoxide can be reduced back to biotin by a molybdenum-containing reductase in some bacteria. Biotin is synthesized from pimeloyl-CoA. Four enzymes are required. Two of them, a synthase that catalyzes step *a* (banner) and a transaminase that catalyzes step *b,* contain the coenzyme pyridoxal phosphate (PLP). The final step is insertion of a sulfur atom from an iron-sulfur center.

Biotin-Containing Enzymes

Within cells the biotin is covalently bonded to proteins, its double ring being attached to a 1.6-nm flexible arm. The first clue to this fact was obtained from isolation of a biotin-containing material *biocytin,* ε-N-biotinyl-L-lysine, from autolysates (self-digests) of rapidly growing yeast. It was subsequently shown that the lysine residue of the biocytin was originally present in proteins at the active sites of biotin-containing enzymes usually within the sequence AMKM or VMKM. Other conserved features also mark the attachment site.

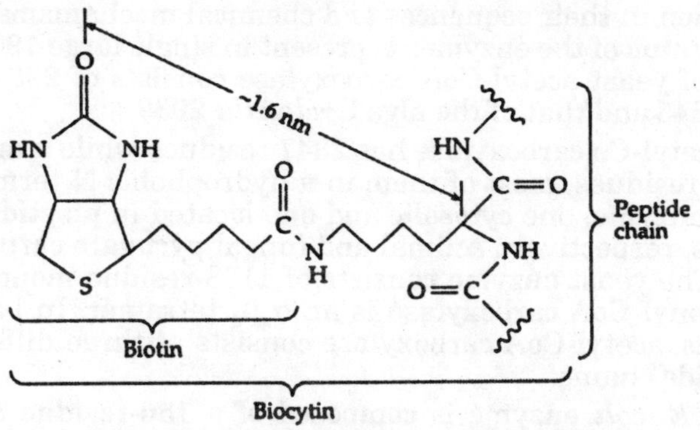

O
HN NH
~1.6 nm
HN
C=O
C
Peptide chain
N H
NH
S
O=C
Biotin
Biocytin

Biotin acts as a *carboxyl group carrier* in a series of carboxylation reactions, a function originally suggested by the fact that aspartate partially replaces biotin in promoting the growth of the yeast *Torula cremonis.* Aspartate was known to arise by transamination from oxaloacetate, which in turn could be formed by carboxylation of pyruvate. Subsequent studies snowed that biotin was needed for an enzymatic ATP-dependent reaction of pyruvate with bicarbonate ion to form oxaloacetate. This is a β carboxylation coupled to the hydrolysis of ATP.

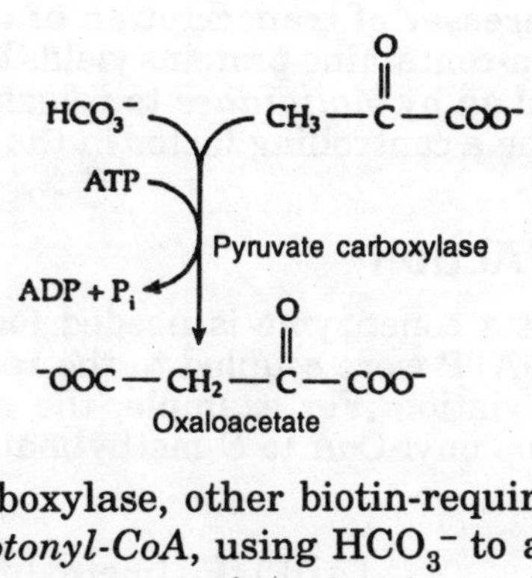

HCO_3^- + $CH_3-C(=O)-COO^-$
ATP → ADP + P_i
Pyruvate carboxylase
↓
$^-OOC-CH_2-C(=O)-COO^-$
Oxaloacetate

In addition to pyruvate carboxylase, other biotin-requiring enzymes act on *acetyl-CoA, propionyl-CoA, and β-methylcrotonyl-CoA,* using HCO_3^- to add carboxyl groups at the sites indicated by the arrows in the accompanying structures. Because of the presence of the C = C double bond conjugated with the carbonyl group, the carboxylation of β-methylcrotonyl-CoA is electronically analogous to β-carboxylation.

$$\longrightarrow H_3C-\overset{O}{\overset{\|}{C}}-S-CoA \quad \longrightarrow H_2C-\overset{CH_3\ O}{C-\overset{\|}{C}}-S-CoA$$

Acetyl-CoA   Propionyl-CoA

$$\longrightarrow H_3C-CH=\underset{CH_3}{\underset{|}{C}}-\overset{O}{\overset{\|}{C}}-S-CoA$$

β-Methylcrotonyl-CoA

Human cells, as well as those of other higher eukaryotes, carry genes for all four of these enzymes. Acetyl-CoA carboxylase is a cytosolic enzyme needed for synthesis of fatty acids but the other three enzymes enter mitochondria when they function. The biotin-dependent carboxylases, which are listed, have a variety of molecular sizes and subunits but show much evolutionary conservation in their sequences and chemical mechanisms. In higher eukaryotes all of the catalytic apparatus of the enzymes is present in single large 190- to 200-kDa subunits. The 251-kDa subunit of yeast acetyl-CoA carboxylase consists of 2337 amino acid residues. That of rats contains 2345 and that of the alga *Cyclotella* 2089.

Human cytosolic acetyl-Co carboxylase has 2347 residues while a mitochondria-associated form has an extra 136 residues, most of them in a hydrophobic N-terminal extension. Plants have two forms of the enzyme, one cytosolic and one located in plastids. In wheat, they have 2260 and 2311 residues, respectively. Animal and fungal pyruvate carboxylases are also large ~500-kDa tetramers. The yeast enzyme consists of 1178-residue monomers. In contrast, the 560-kDa human propionyl-CoA carboxylase is an $\alpha_4\beta_4$ tetramer. In bacteria and in at least some plant chloroplasts, acetyl-CoA carboxylase consists of three different kinds of subunit and four different peptide chains.

The much studied *E. coli* enzyme is composed of a 156-residue *biotin carboxyl carrier protein*, a 449-residue *biotin carboxylase*, whose three-dimensional structure in known, and a *carboxyltransferase* subunit consisting of 304 (α)- and 319 (β)- residue chains. These all associate as a dimer of the three subunits (eight peptide chains).

Biotin becomes attached to the proper ε-amino groups at the active centers of biotin enzymes by the action of *biotin holoenzyme synthetase* (biotinyl protein ligase), which utilizes ATP to form an intermediate biotinyl-AMP. Hereditary deficiency of this enzyme has been observed in a few children and has been treated by administration of extra biotin.

The *E. coli* biotin holoenzyme synthetase, whose three-dimensional structure is known, has a dual function. It is also a represser of transcription of the biotin biosynthetic operon. Intracellular degradation of biotin-containing proteins yields biotincontaining oligopeptides as well as biocytin. These are acted on by *biotinidase* to release free biotin. The action of this enzyme in recycling biotin may be a controlling factor in the rate of formation of new biotin-dependent enzymes.

The Mechanism of Biotin Action

It may seem surprising that a conenzyme is needed for these carboxylation reactions. However, unless the cleavage of ATP were coupled to the reactions, the equilibria would lie far in the direction of decarboxylation. For example, the measured apparent equilibrium constant *K′* for conversion of propionyl-CoA to *S* methylmalonyl-CoA at pH 8.1 and 28°C is given by Eq. 8.4.

$$K' = \frac{[\text{ADP}]\,[\text{P}_i]\,[\text{methylmalonyl}-\text{CoA}]}{[\text{ATP}][\text{HCO}_3^-]\,[\text{propionyl-CoA}]} = 5.27$$

$$\Delta G' = -4.36 \text{ kJ mol}^{-1} \qquad \text{...(8.4)}$$

Table 8.1. Enzymes Containing Bound Biotin

1. Catalyzing beta carboxylation using HCO_3^- with coupled cleavage of ATP to ADP + P_i	
	Acetyl-CoA carboxylase
	Propionyl-CoA carboxylase
	Pyruvate carboxylase
	β-Methylcrotonyl-CoA carboxylase (δ carboxylation)
2. Carboxyl group transfer without cleavage of ATP	
	Carboxyltransferase of *Propionobacterium*
3. Biotin-dependent Na^+ pumps	
	Oxaloacetate decarboxylase
	Methylmalonyl-CoA decarboxylase
	Glutaconyl-CoA decarboxylase
4. Other	
	Malonate decarboxylase
	Urea carboxylase

The function of biotin is to mediate the coupling of ATP cleavage to the carboxylation, making the overall reaction exergonic. This is accomplished by a two-stage process in which a *carboxybiotin* intermediate is formed. There is *one known biotin-containing enzyme that does not utilize ATP.* Propionic acid bacteria contain a *carboxyltransferase* which transfers a carboxyl group reversibly from methylmalonyl-CoA to pyruvate to form oxaloacetate and propionyl-CoA. This huge enzyme consists of a central hexameric core of large 12S subunits to which six 5S dimeric subunits and twelve 123-residue bio tiny la ted peptides are attached. No ATP is needed because free HCO_3^- is not a substrate. However, biotin serves as the carboxyl group carrier in this enzyme too.

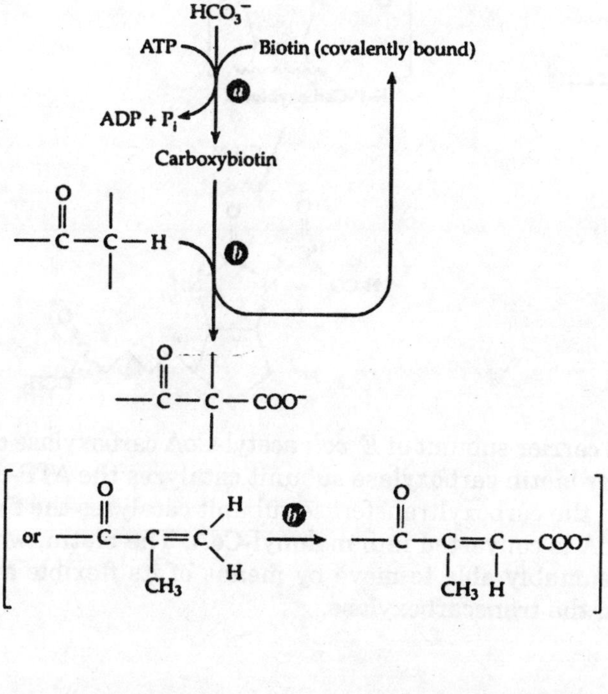

Carboxybiotin

The structure of biotin suggested that bicarbonate might be incorporated reversibly into its position 2'. However, this proved not to be true and it remained for F. Lynen and associates to obtain a clue from a "model reaction." They showed that purified β-methylcrotonyl-CoA carboxylase promoted the carboxylation of free biotin with bicarbonate ($H^{14}CO_3^-$) and ATP. While the carboxylated biotin was labile, treatment with diazomethane gave a stable dimethyl ester of *N-1'-carboxybiotin.* The covalently bound biotin at active sites of enzymes was also successfully labeled with $^{14}CO_2$ Treatment of the labeled enzymes with diazomethane followed by hydrolysis with trypsin and pepsin gave authentic N-1'-carboxybiocytin. It was now clear that the cleavage of ATP is required to couple the CO_2 from HCO_3^- to the biotin to form carboxybiotin. The enzyme must then transfer the carboxyl group from carboxybiotin to the substrate that is to be carboxylated. Enzymatic transfer of a carboxyl group from chemically synthesized carboxybiotin onto specific substrates confirmed the proposed mechanism.

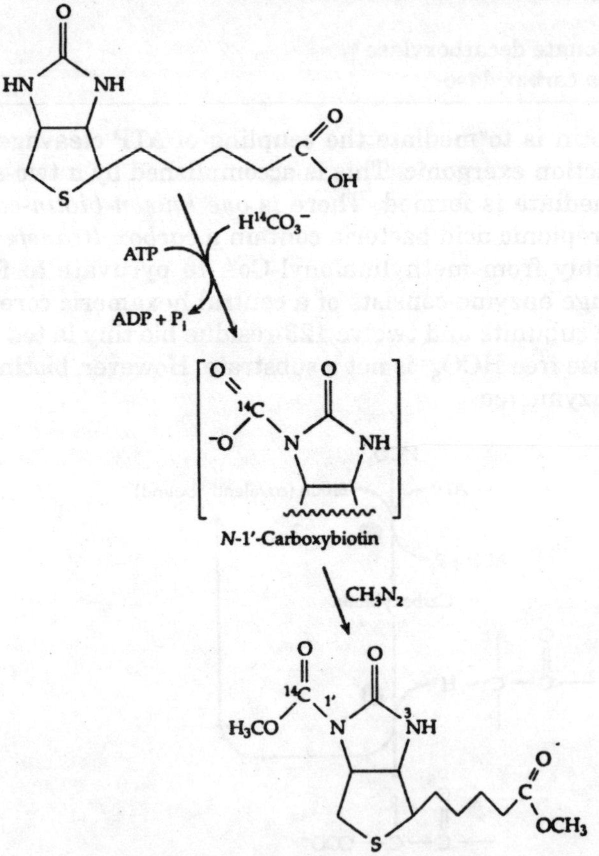

O
HN NH
O
C
OH
S
$H^{14}CO_3^-$
ATP
ADP + P_i
O O
^{14}C
^-O N NH
N-1'-Carboxybiotin
CH_3N_2
O O
^{14}C 1'
H_3CO N 3 NH
O
C
OCH_3
S

The biotin carboxyl carrier subunit of *E. coli* acetyl-CoA carboxylase contains the covalently bound biotin. The larger biotin carboxylase subunit catalyzes the ATP-dependent attachment of CO_2 to the biotin and the carboxyltransferase subunit catalyzes the final transcarboxylation step by which acetylCoA is converted into malonyl-CoA. The biotin, which is attached to the carrier protein, is presumably able to move by means of its flexible arm from a site on the carboxylase to a site on the transcarboxylase.

Carboxyphosphate

During the initial carboxylation step ^{18}O from labeled bicarbonate enters the P_i that is split from ATP. This suggested transient formation of *carboxyphosphate* by nucleophilic attack of HCO_3^- on ATP. The carboxyl group of this reactive mixed anhydride could then be transferred to biotin. This mechanism is supported by the fact that biotin carboxylase catalyzes the transfer of a phospho group to ADP from carbamoyl phosphate, an analog of carboxyphosphate in a reaction that is analogous to the reverse of that, and also by a slow bicarbonate-dependent ATPase activity that does not depend upon biotin.

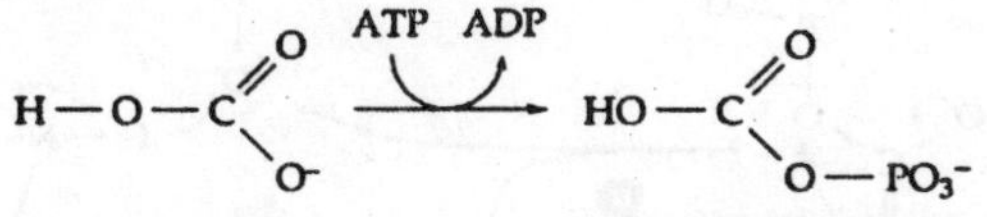

ATP ADP
H—O—C(=O)—O⁻ → HO—C(=O)—O—PO₃⁻
Carboxyphosphate

The simplest mechanism for transfer of the carboxyl group of carboxyphosphate to biotin would appear to be nucleophilic displacement of the phosphate leaving group by Nl′ of biotin. The enzyme could presumably first catalyze removal of the N1 hydrogen to form a ureido

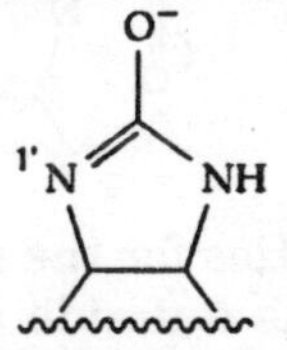

Ureido anion

anion. Another reasonable possibility would be for the terminal phospho group of ATP to be transferred to biotin to form an O-phosphate which could react with bicarbonate as *b*, Cleavage of the enol phosphate by attack of HCO_3^- would simultaneously create a nucleophilic center at N1 and carboxyphosphate ready to react with N1. However, this could not easily explain the ATPase activity in the absence of biotin.

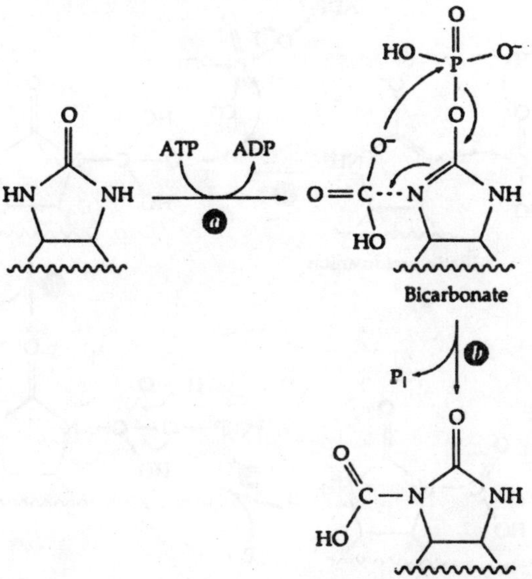

ATP ADP (a)
Bicarbonate
(b) P_i
Carboxybiotin

Either of the foregoing mechanisms requires that the ureido anion of biotin attack the rather unreactive carbon atom of carboxyphosphate. Another alternative, which is analogous to that suggested for PEP carboxylase is for carboxyphosphate to eliminate inorganic phosphate to give the more electrophilic CO_2. The very basic inorganic phosphate trianion PO_2 that is eliminated could remove the proton from N1 of biotin to create the biotin ureido anion which could then add to CO_2.

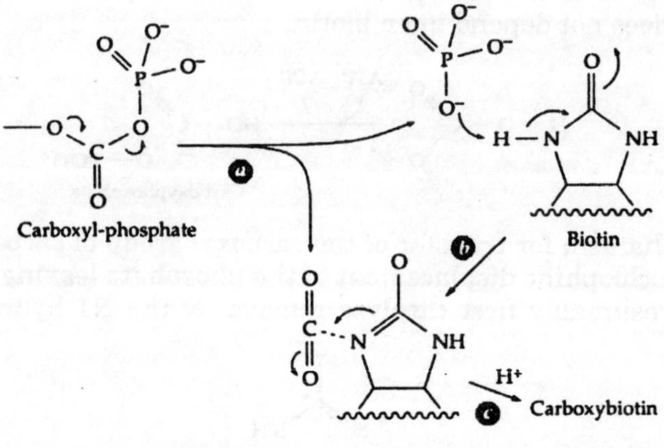

Carboxyl-phosphate

Biotin

Carboxybiotin

Have we checked all of the possibilities for the mechanism of biotin carboxylation? Kruger and associates suggested that biotin, as a ureido anion, might add to bicarbonate to form a highly unstable intermediate which, however, could be phosphorylated by ATP. This intermediate could undergo elimination of inorganic phosphate driving the reaction to completion. The observed transfer of isotope from ^{18}O-containing bicarbonate into ADP would be observed.

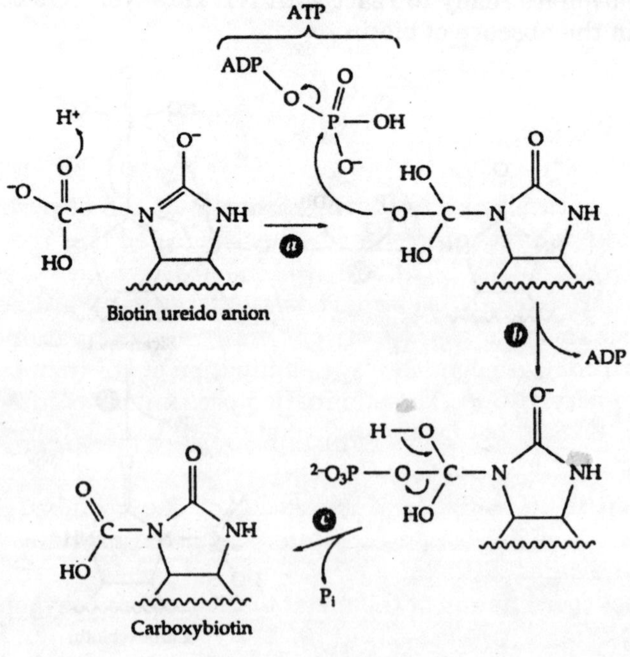

Biotin ureido anion

Carboxybiotin

The β-carboxylation step

Once formed, the carboxybiotin "head group" could swing to the carboxyltransferase site where transfer of the carboxyl group into the final product takes place. This might occur either by nucleophilic attack of an enolate anion on the carbonyl carbon or on CO_2 generated by reversal of the reactions.

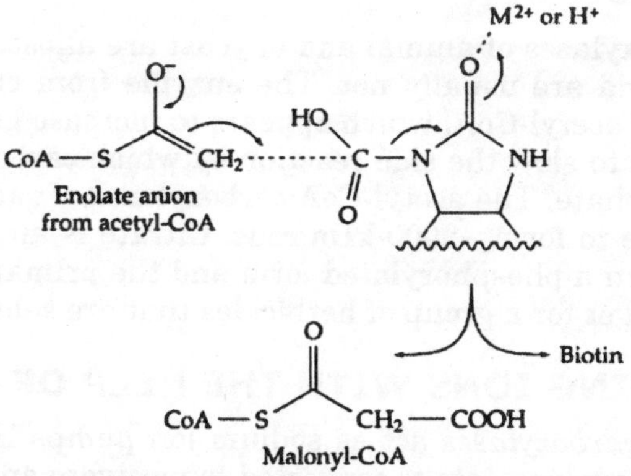

M^{2+} or H^+

CoA—S—C(O⁻)=CH₂ ······· C(HO)(=O)—N ... NH

Enolate anion from acetyl-CoA

Biotin

CoA—S—C(=O)—CH₂—COOH

Malonyl-CoA

When pyruvate with a chiral methyl group is carboxylated by pyruvate carboxylase the configuration at C-3 is retained. The carboxyl enters from the 2-*si* side, the same side from which the proton (marked H*) was removed to form the enolate anion. Comparable stereochemistry has been established for other biotin-dependent enzymes.

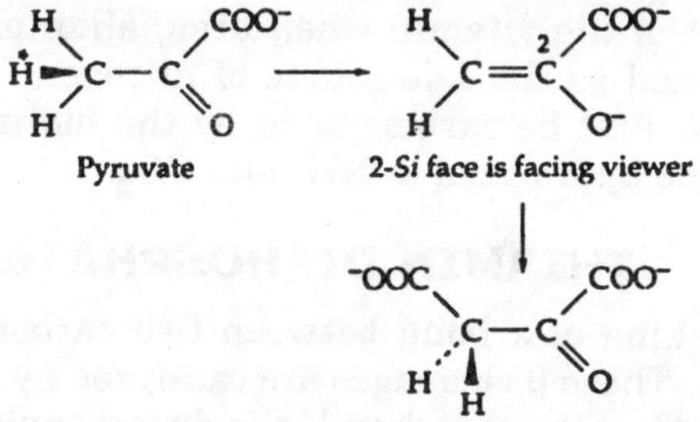

Pyruvate

2-*Si* face is facing viewer

These enzymes do not catalyze any proton exchange at C-3 of pyruvate or at C-2 of an acyl-CoA unless the biotin is first carboxylated. This suggested that removal of the proton to the biotin oxygen and carboxylation might be synchronous. However, ^{13}C and ^{2}H kinetic isotope effects and studies of ^{3}H exchange support the existence of a discrete enolate anion intermediate. This mechanism is also consistent with the observation that propionyl-CoA carboxylase and transcarboxylase both catalyze elimination of HF from β-fluoropropionyl-CoA to form the unsaturated acrylyl-CoA. The elimination presumably occurs via an enolate anion intermediate as in Eq. 8.28. A bound divalent metal ion, usually Mn^{2+}, is required in the transcarboxylation step.

A possible function is to assist in enolization of the carboxyl acceptor. However, measurement of the effect of the bound Mn^{2+} on ^{13}C relaxation times in the substrate for pyruvate carboxylase indicated a distance of –0.7 nm between the carbonyl carbon and the Mn^{2+}, too great for direct coordination of the metal to the carbonyl oxygen. Another possibility is that the metal binds to the carbonyl of biotin as indicated (in Eq. 8.11). Pyruvate carboxylase utilizes two divalent metal ions and at least one monovalent cation.

What is the role of the sulfur atom in biotin? Perhaps it interacts with CO_2, helping to hold it in a correct orientation for reaction. Perhaps it helps to keep the ureido ring of biotin planar, or perhaps it has no special function.

Control Mechanisms

Most pyruvate carboxylases of animal and of yeast are allosterically activated by acetyl-CoA, but those of bacteria are usually not. The enzyme from chicken liver has almost no activity in the absence of acetyl-CoA, which appears to increase greatly the rate of formation of carboxyphosphate and to slow the side reaction by which carboxyphosphate is hydrolyzed to bicarbonate and phosphate. The acetyl-CoA carboxylases of rat or chicken liver aggregate in the presence of citrate to form ~8000-kDa rods. Citrate is an allosteric activator for this enzyme but it acts only on a phosphorylated form and the primary control mechanism. This enzyme in plants is a target for a group of herbicides that are selectively toxic to grasses.

PUMPING IONS WITH THE HELP OF BIOTIN

Biotin-dependent *decarboxylases* act as sodium ion pumps in *Klebsiella* and in various anaerobes. For example, oxaloacetate is converted to pyruvate and bound carboxybiotin. The latter is decarboxylated to CO_2 at the same time that two Na^+ ions are transported from the inside to the outside of the cell. The function of this pump, like that of the Na^+, K^+-ATPase is to provide an electrochemical gradient that drives the transport of other ions and molecules through the membrane. Similar ion pumps are operated by decarboxylation of methylmalonyl-CoA and glutaconyl-CoA. Yeast (*Saccharomyces cerevisiae*) cannot make biotin and requires an unusually large amount of the vitamin when urea, allantoin, allantoic acid, and certain other compounds are supplied as the sole source of nitrogen for growth. The reason is that in this organism urea must first be carboxylated by the biotin-containing *urea carboxylase* (see Eq. 8.25) before it can be hydrolyzed to NH_3 and CO_2.

THIAMIN DIPHOSPHATE

In considered the breaking of a bond between two carbon atoms, one of which is also bonded to a carbonyl group. These β cleavages are catalyzed by simple acidic and basic groups of the protein side chains. On the other hand, the decarboxylation of 2-oxo acids (Eq. 8.13) and the cleavage and formation of α-hydroxyketones (Eq. 8.14) depend upon thiamin diphosphate (TDP). These reactions represent a second important method of making and breaking carbon-carbon bonds which we will designate *a condensation and a cleavage*. The common feature of all thiamin-catalyzed reactions is that the bond broken (or formed) is *immediately adjacent to the carbonyl group,* not one carbon removed, as in β cleavage reactions. No simple acid-base catalyzed mechanisms can be written; hence the need for a coenzyme.

$$\mathrm{R{-}\overset{\overset{\displaystyle O}{\|}}{C}{-}COO^- + H^+ \longrightarrow R{-}\overset{\overset{\displaystyle O}{\|}}{C}H + CO_2} \quad \text{...(8.13)}$$

$$\mathrm{R{-}\underset{\underset{\displaystyle H}{|}}{\overset{\overset{\displaystyle HO}{|}}{C}}{-}\overset{\overset{\displaystyle O}{\|}}{C}{-}R' \longrightarrow R{-}\overset{\overset{\displaystyle O}{\|}}{C}{-}H + H{-}\overset{\overset{\displaystyle O}{\|}}{C}{-}R'} \quad \text{...(8.14)}$$

Chemical Properties of Thiamin

The weakly basic portion of thiamin or of its coen-zyme forms is protonated at low pH, largely on N-1 of the pyrimidine ring. The pX_a value is ~4.9. In basic solution, thiamin reacts in two steps with an opening of the thiazole ring Eq. 8.15 to give the anion of a thiol form which may be crystallized as the sodium salt. This reaction, like the competing reaction and which leads to a yellow unstable form of the thiamin anion, is an example of a cooperative two-proton dissociation with linked structural changes. A very low concentration of the

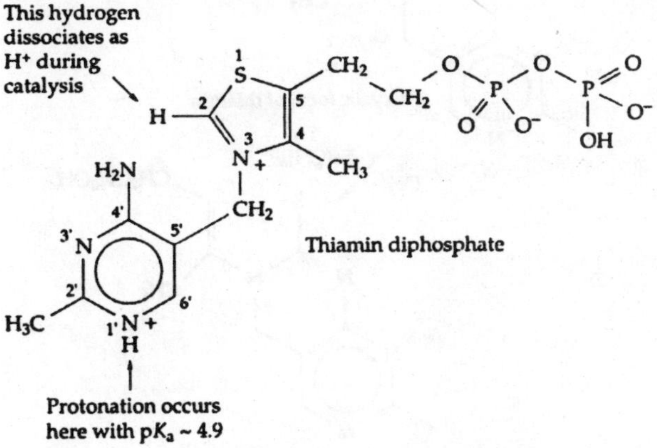

This hydrogen dissociates as H^+ during catalysis

Protonation occurs here with $pK_a \sim 4.9$

Thiamin diphosphate

intermediate "pseudobase" is present during the titra-tion. This property, which is unusual among small molecules, was instrumental in leading Williams *et al.* to the correct structure for the vitamin. A still unanswered question is, What biological significance is associated with these reactions? Perhaps the thiol form depicted in eq. 8.15 or the "yellow form" becomes attached to active sites of some proteins through disulfide linkages.

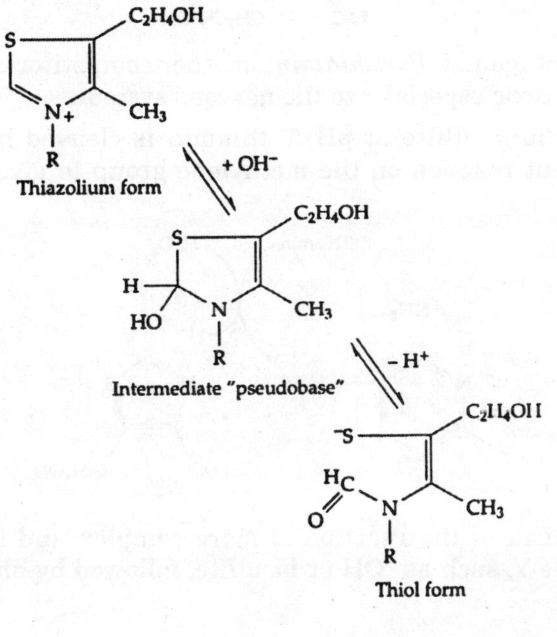

Thiazolium form

$+ OH^-$

Intermediate "pseudobase"

$- H^+$

Thiol form

Thiamin is unstable at high pH and is destroyed by the cooking of foods under mildly basic conditions. The thiol form undergoes hydrolysis and oxidation by air to a disulfide. The tricyclic form is oxidized to *thiochrome*, a fluorescent compound whose formation from thiamin by treatment with alka-line hexacyanoferrate (III) is the basis of a much used fluorimetric assay.

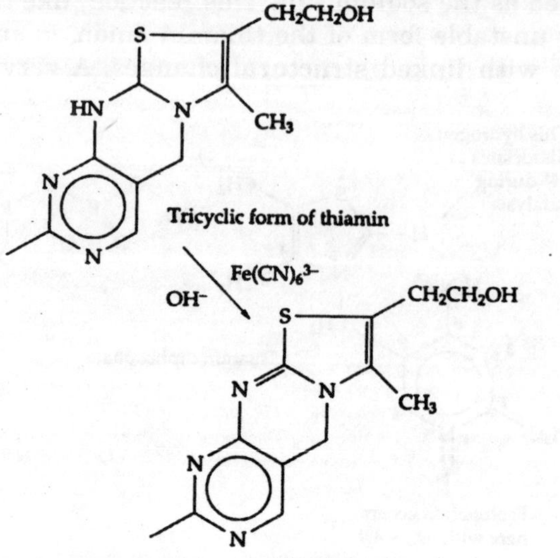

Tricyclic form of thiamin

$Fe(CN)_6^{3-}$, OH^-

Thiochrome

Treatment of thiamin with boiling 5N HC1 deaminates it to the hydroxy analogue

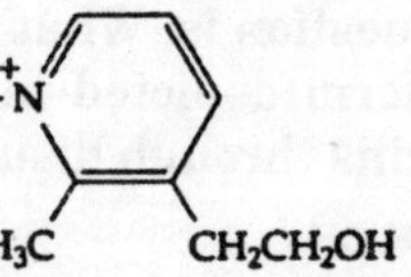

oxythiamin, a potent antagonist. *Pyrithiamin*, another competitor containing in place of the thiazolium ring, is very toxic especially to the nervous system.

In a solution of sodium sulfite at pH 5, thiamin is cleaved by what appears to be a nucleophilic displacement reaction on the methylene group to give the free thiazole and a sulfonic acid.

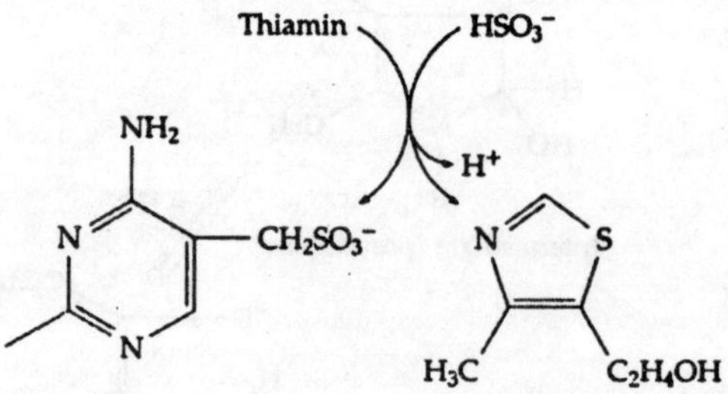

Thiamin + HSO_3^- → H^+ + (pyrimidine-$CH_2SO_3^-$) + (thiazole, H_3C, C_2H_4OH)

In fact the mechanism of the reaction is more complex and is evidently initiated by addition of *a* nucleophile Y, such as ^-OH or bisulfite, followed by elimination of the thiazole (eq. 8.18).

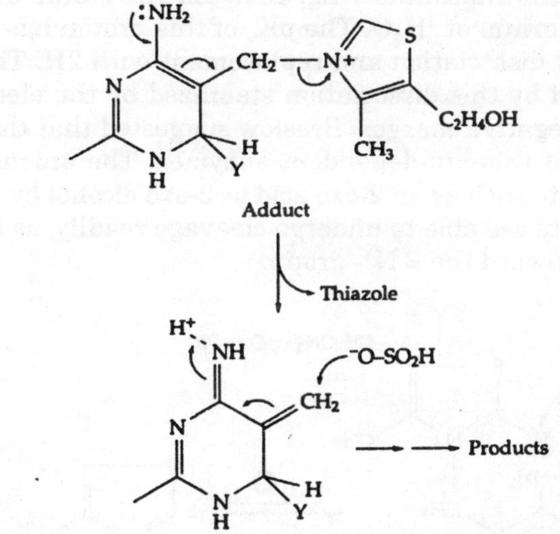

Adduct

Thiazole

Products

A similar cleavage is catalyzed by thiamin-degrading enzymes known as thiaminases which are found in a number of bacteria, marine organisms, and plants. In a bacterial thiaminase, group Y, of Eq. 8.18 is a cysteine -SH.

Thiamin is synthesized in bacteria, fungi, and plants from 1-deoxyxylulose 5-phosphate, which is also an intermediate in the nonmevalonate pathway of polyprenyl synthesis. However, thiamin diphosphate is a coenzyme for synthesis of this intermediate, suggesting that an alternative pathway must also exist. Each of the two rings of thiamin is formed separately as the esters 4-amino-5-hydroxy-methylpyrimidine diphosphate and 4-methyl-5-(β-hydroxyethyl) thiazole monophosphate. These precursors are joined with displacement of pyrophosphate to form thiamin monophosphate. In eukaryotes this is hydrolyzed to thiamin, then converted to thiamin diphosphate by transfer of a diphospho group from ATP. In bacteria thiamin monophosphate is converted to the diphosphate by ATP and thiamin monophosphate kinase.

Catalytic Mechanisms

The first real clue to the mechanism of thiamin-dependent cleavage came in about 1950 when Mizuhara showed that at pH 8.4 thiamin catalyzes the nonenzymatic conversion of

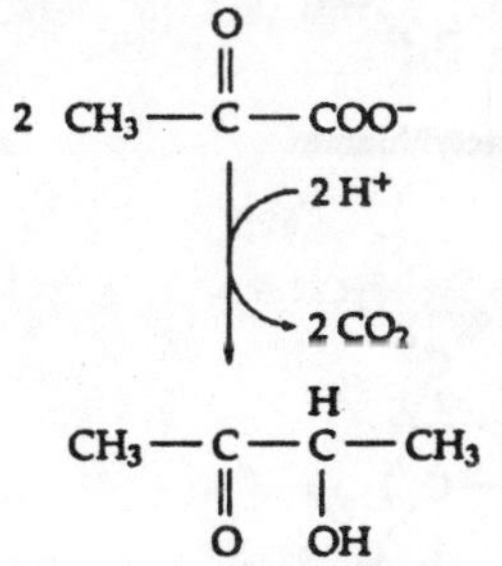

$$2\ CH_3-\overset{\overset{O}{\|}}{C}-COO^- \xrightarrow[\ 2\,CO_2\ ]{2\,H^+} CH_3-\underset{\underset{O}{\|}}{C}-\overset{H}{\underset{OH}{\underset{|}{C}}}-CH_3$$

pyruvate into acetoin Eq. 8.19. Following Mizuhara's lead, Breslow investigated the same reaction using the then new NMR method. He made the surprising discovery that the hydrogen atom in the 2 position of the thiazolium ring, between the sulfur and the nitrogen atoms, exchanged easily with deuterium of ^2H_2O. The pK_a of this proton has been estimated as ~18, low enough to permit rapid dissociation and replacement with 2H. The resulting *thiazolium dipolar ion* (or *ylid*) formed by this dissociation stabilized by the electrostatic interaction of the adjacent positive and negative charges. Breslow suggested that this dipolar ion is the key intermediate in reactions of thiamin-dependent enzymes. The anionic center of the dipolar ion can react with a substrate such as an 2-oxo acid or 2-oxo alcohol by addition to the carbonyl group. The resulting adducts are able to undergo cleavage readily, as indicated by the arrows showing the electron flow toward the = N^+- group.

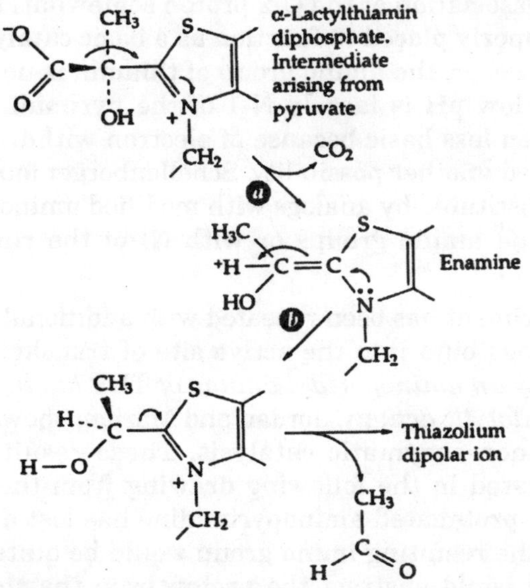

α-Lactylthiamin diphosphate. Intermediate arising from pyruvate

CO_2

Enamine

Thiazolium dipolar ion

Below the structures of the adducts in Eq. 8.20 are those of a 2-oxo acid and a β-ketol with arrows indicating the electron flow in decarboxylation and in the aldol cleavage. The similarities to the thiamin-dependent cleavage reaction are especially striking if one remembers that in some aldolases and decarboxylases the substrate carbonyl group is first converted to an N-protonated Schiff base before the bond cleavage.

We see that *the essence of the action of thiamin diphosphate as a coenzyme is to convert the substrate into a form in which electron flow can occur from the bond to be broken into the structure of the coenzyme.* Because of this alteration in structure, a bond breaking reaction that would not otherwise have been possible occurs readily. To complete the catalytic cycle, the electron flow has to be reversed again. The thiamin-bound cleavage product (an enamine) from either of the adducts can be reconverted to the thiazolium dipolar ion and an aldehyde as shown in step *b* of Eq. 8.21 for decarboxylation of pyruvate to acetaldehyde.

The adducts oc-lactylthiamin and α-lactylthiamin diphosphate have both been synthesized. As long as α-lactylthiamin is kept as a dry solid or at low pH, it is stable. However, it decarboxylates readily in neutral solution (Eq. 8.21). Decarboxylation is much more rapid in methanol, a fact that was predicted by Lienhard and associates. They suggested that decarboxylation is easier in a solvent of low polarity because the transition state has a lower polarity than does lactylthiamin. An enzyme could assist the reac-tion by providing a relatively nonpolar environment.

The crystal structures of thiamin-dependent α-zymes (see next section) as well as modeling suggest that lactylthiamin pyrophosphate has the conformation. If so, it would be formed by the addition of the ylid to the carbonyl of pyruvate in accord with stereoelectronic principles, and the carboxylate group would also be in the correct orientation for elimination to form the enamine in Eq. 8.21 step *b*. A transient 380- to 440-nm absorption band arising during the action of pyruvate decarboxylase has been attributed to the enamine.

What is the role of the pyrimidine portion of the coenzyme in these reactions? The pyrimidine ring has a large inductive effect on the basicity of the thiazolium nitrogen and may increase the rate of dissociation of the C-2 proton somewhat. More significant is the fact that the - NH_2 group is properly placed to function as a basic catalyst in the generation of the thiazolium dipolar ion. However, the amino group of thiamin is not very basic (pK_a ~4.9) and the site of protonation at low pH is largely N-1 of the pyrimidine ring. In the protonated form the -NH_2 group is even less basic because of electron withdrawal into the ring. Studies of thiamin analogs suggested another possibility. Schellenberger found pyruvate decarboxylase inactive when TDP was substituted by analogs with modified aminopyrimidine rings, *e.g.*, with methylated or dimethylated amino groups or with Nl of the ring replaced by carbon (an aminopyridyl analog).

More recently the experiment has been repeated with additional enzymes and X-ray studies have shown that the analogs bind into the active site of transketolase in a normal way. Of the compounds studied *only an aminopyridyl analog of* TDP *having a nitrogen atom at* 1′ (but CH at 3′) *had substantial catalytic activity.* Jordan and Mariam showed that N-1′-methylthiamin is a superior catalyst in non-enzymatic catalysis. These results are consistent with the speculative scheme illustrated in the following drawing from the first edition of this book. The - NH_2 group of the N1-protonated aminopyrimidine has lost a proton to form a normally *minor tautomer* in which the resulting imino group would be quite basic. Assisted by a basic group from the protein, it could abstract the proton from the thiazolium ring to form the ylid.

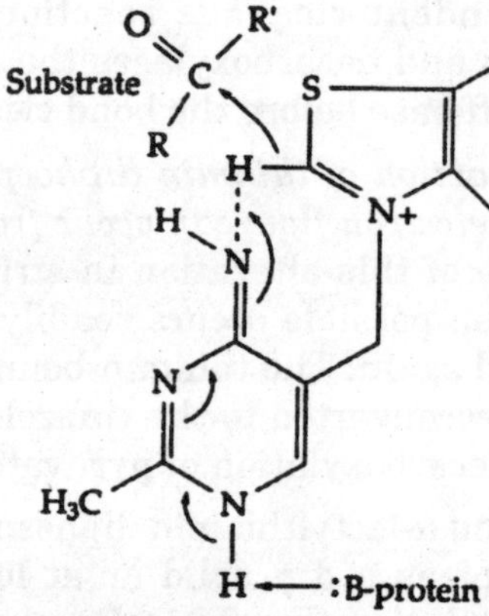

Crystallographic studies show that the catalytic base (B-protein) is the carboxylate group of a conserved glutamate side chain. Kern *et al.* used NMR spectroscopy of thiamin diphosphate present in native and mutant pyruvate decarboxylase and transketolase to monitor the exchange rates of the C2-H proton of the thiazolium ring. The results confirmed the importance of the conserved glutamate side chain for dissociation of the C2-H proton. Participation of other catalytic groups from the enzyme may also be important. However, these groups are not conserved in the whole family of enzymes.

For example, glutamine 122, which is within hydrogen-bonding distance of both the substrate and thiamin amino group, is replaced by histidine in transketolase. A variety of kinetic studies involving mutants, alternative substrates, and isotope effects in substrates and solvent have not yet resolved the details of the proton transfers that occur within the active site.

Structures of Thiamin-Dependent Enzymes

By 1998, X-ray structures had been determined for four thiamin diphosphate-dependent enzymes: a bacterial pyruvate oxidase, yeast and bacterial pyruvate decarboxylases, transketolase, and benzoylformate decarboxylase. The reactions catalyzed by these enzymes are all quite different, as are the sequences of the proteins. However, the thiamin diphosphate

A

B

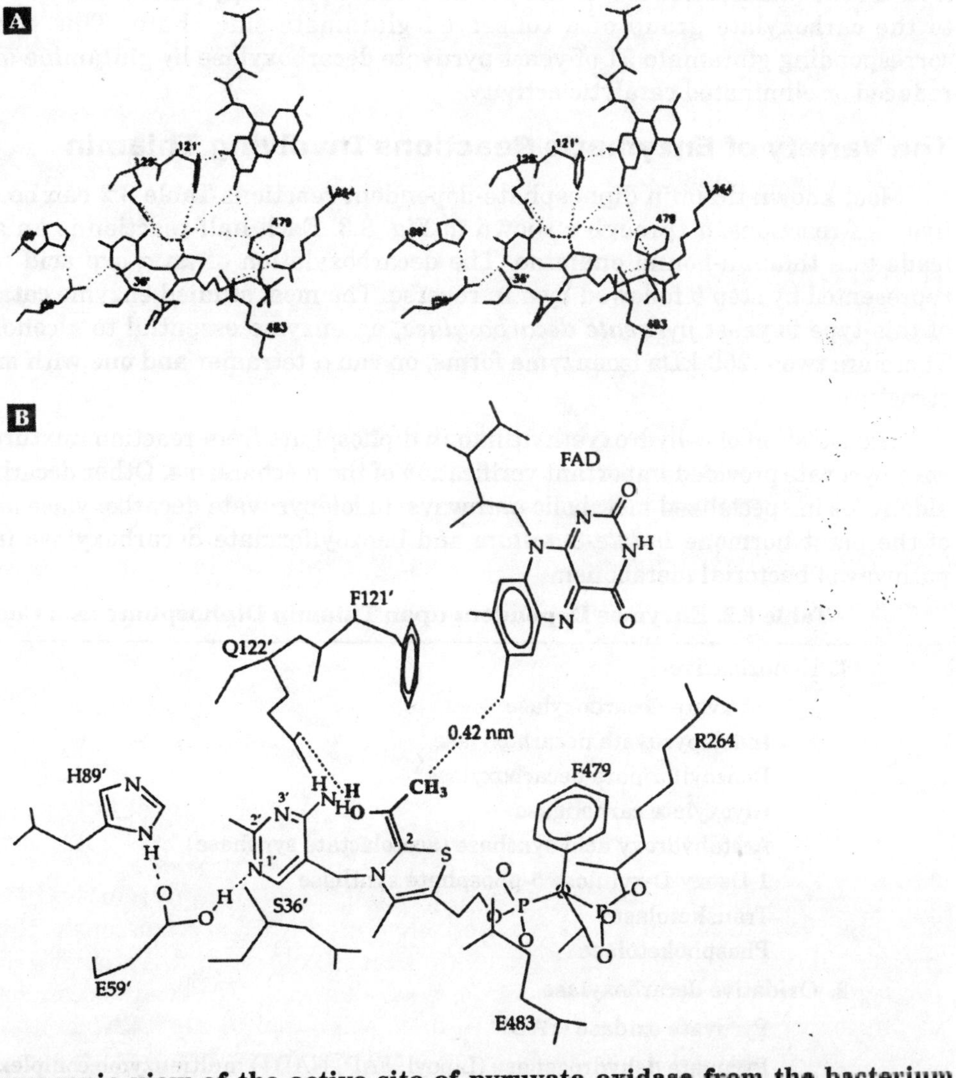

Fig. 8.2. (A) Stereoscopic view of the active site of pyruvate oxidase from the bacterium *Lactobacillus plantarium* showing the thiamin diphosphate as well as the flavin part of the bound FAD. The planar structure of the part of the intermediate enamine that arises from pyruvate is shown by dotted lines. Only some residues that may be important for catalysis are displayed: G35′, S36′, E59′, F121′, Q122′, R264, F479, and E483. Courtesy of Georg E. Schulz. (B) Simplified view with some atoms labeled and some side chains omitted. The atoms of the hypothetical enamine that are formed from pyruvate, by decraboxylation, are shown in green.

is bound in a similar way in all of them. A conserved pattern of hydrogen bonds holds the diphosphate group to the protein and also provides ligands to a metal ion. This is normally Mg^{2+}, which is held in nearly perfect octahedral coordination by two phosphate oxygen atoms, a conserved aspartate carboxylate, a conserved asparagine amide, and a water molecule. The thiamin rings are in a less polar region. The amino group of the pyrimidine is adjacent to the 2-CH of the thiazole and N1′ of the pyrimidine is apparently protonated and hydrogen bonded to the carboxylate group of a conserved glutamate side chain. This substitution of the corresponding glutamate 51 of yeast pyruvate decarboxylase by glutamine or alanine greatly reduced or eliminated catalytic activity.

The Variety of Enzymatic Reactions Involving Thiamin

Most known thiamin diphosphate-dependent reactions Table 8.2 can be derived from the five half-reactions, *a* through *e* shown in Fig. 8.3. Each half-reaction is an a cleavage which leads to a thiamin-bound enamine. The decarboxylation of an α-oxo acid to an aldehyde is represented by step *b* followed by *a* in reverse. The most studied enzyme catalyzing a reaction of this type is yeast *pyruvate decarboxylase*, an enzyme essential to alcoholic fermentation. There are two ~ 250-kDa isoenzyme forms, one an α tetramer and one with an $\alpha\beta_2$ quaternary structure.

The isolation of α-hydroxyethylthiamin diphosphate from reaction mixtures of this enzyme with pyruvate provided important verification of the mechanisms. Other decarboxylases produce aldehydes in specialized metabolic pathways: indolepyruvate decarboxylase in the biosynthesis of the plant hormone *indole*-3-*acetate* and benzoylformate decarboxylase in the maridelate pathway of bacterial metabolism.

Table 8.2. Enzymes Dependent upon Thiamin Diphosphate as a Coenzyme

1. Nonozidative
    - pyruvate decarboxylase*
    - Indolepyruvate decarboxylase
    - Benzoylformate decarboxylase*
    - Glyoxylate carboligase
    - Acetohydroxy acid synthase (acetolactate synthase)
    - 1-Deoxy-D-xylulose 5-phosphate synthase
    - Transketolase*
    - Phosphoketolase
2. Oxidative decarboxylase
    - Pyruvate oxidase (FAD)*
    - Pyruvate dehydrogenase (Lipoyl, FAD, NAD+) multienzyme complex
    - Pyruvate:ferredoxin oxidoreductase
    - Indolepyruvate:ferredoxin oxidoreductase

* Three-dimensional structures for these enzymes had been determined by 1998.

Formation of α-ketols from α-oxo acids also starts with but is followed by condensation with another carbonyl compound, in reverse. An example is decarboxylation of pyruvate and condensation of the resulting active acetaldehyde with a second pyruvate molecule to give *R*-α-acetolactate, a reaction catalyzed by *acetohydroxy acid synthase* (acetolactate synthase). Acetolactate is the precursor to valine and leucine. A similar ketol condensation, which is

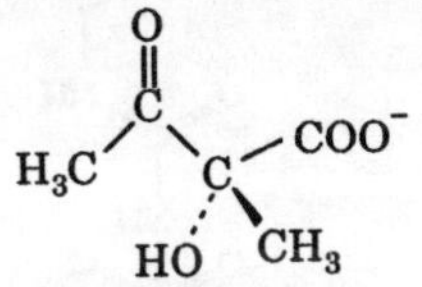

R-α-Acetolactate

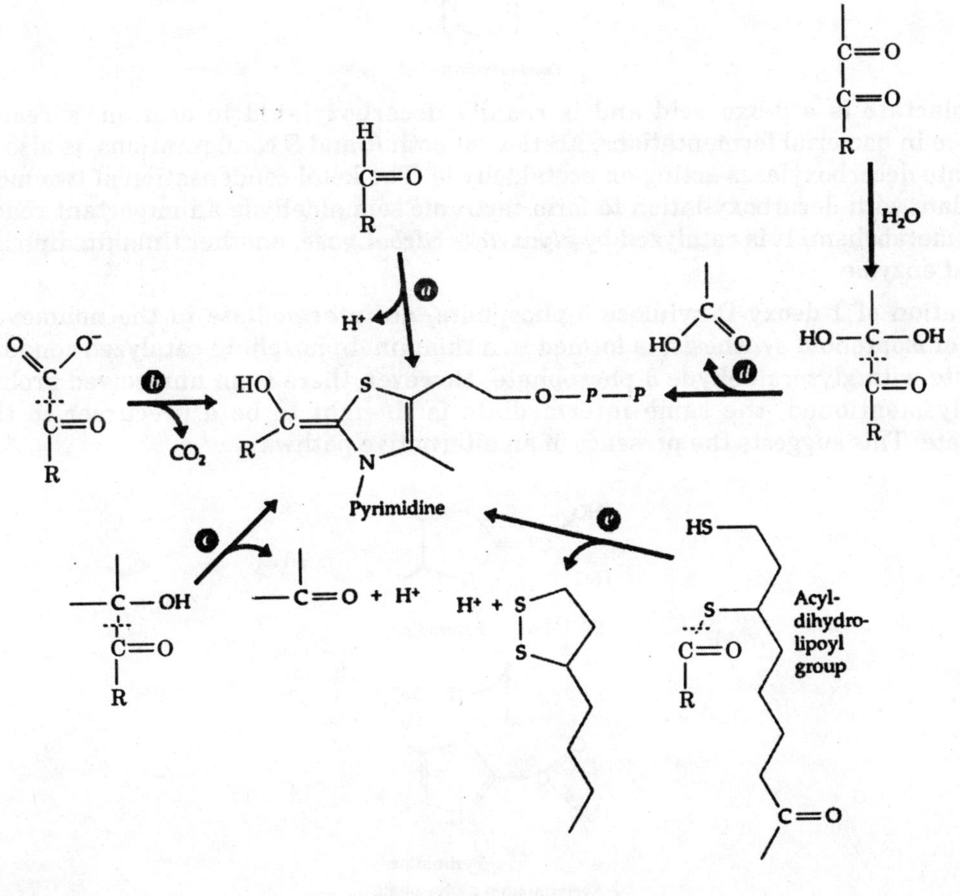

Fig. 8.3. Half-reactions making up the thiamin-dependent a cleavage and a condensation reactions.

catalyzed by the same synthase, is required in the biosynthesis of isolecucine. Since this synthase is not present in mammals it is a popular target for herbicides. It is inhibitied by many of the most widely used herbicides including sulfometuron methyl, whose structure is shown here.

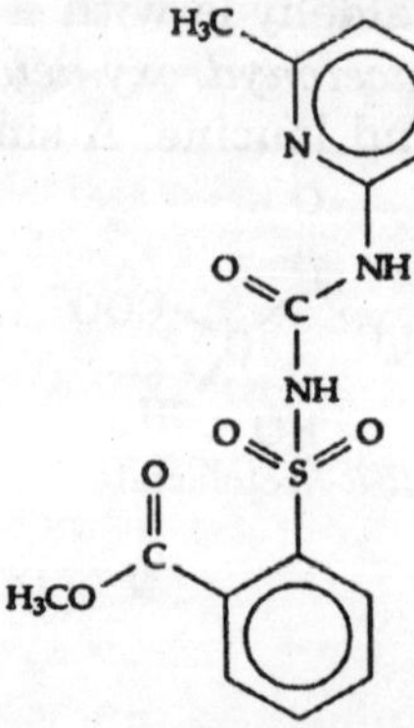

Sulfometuron methyl

Acetolactate is a β-oxo acid and is readily decarboxylated to acetion, a reaction of importance in bacterial fermentations. Acetion, of both *R* and *S* configurations, is also formed by pyruvate decarboxylases acting on acetaldehyde. The ketol condensation of two molecules of glyoxylate with decarboxy-lation to form tartronic semialdehyde an important reaction in bacterial metabolism. It is catalyzed by *glyoxylate carboligase*, another thiamin diphosphate-dependent enzyme.

Formation of 1-deoxy-D-xylulose 5-phosphate, an intermediate in the nonmevalonate pathway of isoprenoid synthesis, is formed in a thiamin diphosphate-catalyzed condensation of pyruvate with glyceraldehyde 3-phosophate. However, there is an unresolved problem. As previously mentioned, the same intermediate is thought to be a precursor to thiamin diphosphate. This suggests the presence of an alternative pathway.

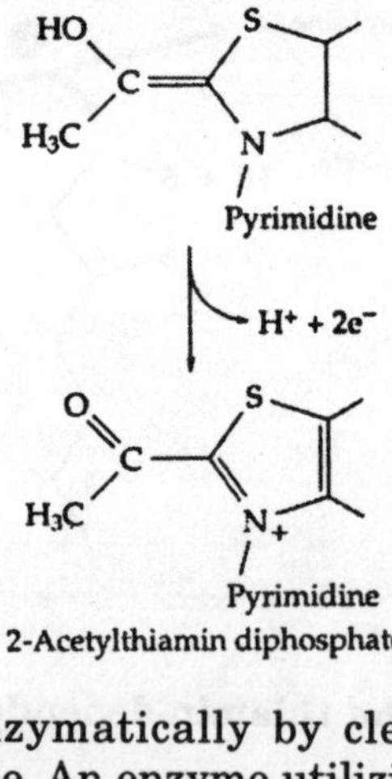

$H^+ + 2e^-$

2-Acetylthiamin diphosphate

Ketols can also be formed enzymatically by cleavage of an aldehyde followed by condensation with a second aldehyde. An enzyme utilizing these steps is *transketolase* which is essential in the pentose phosphate pathways of metabolism and in photosynthesis. α-Diketones can be cleaved to a carboxylic acid plus active aldehyde, which can react either via *a or c in* reverse. These and other combinations of steps are often observed as side reactions of such enzymes as pyruvate decarboxylase. A related thiamin-dependent reaction is that of pyruvate and acetyl-CoA to give the α-diketone, *diacetyl*, $CH_3COCOCH_3$. The reaction can be viewed as a displacement of the CoA anion from acetyl-CoA by attack of thiamin-bound active acetaldehyde derived from pyruvate.

Oxidative Decarboxylation and 2-Acetylthiamin Diphosphate

The oxidative decarboxylation of pyruvate to form acetyl-CoA or acetyl phosphate plays a central role in the metabolism of our bodies and of most other organisms. This reaction is usually formulated as the reverse of step *e*, which shows the cleavage of an *acyl-dihydrolipoyl* derivative. However, there is a possibility that the lipoyl group functions not but as an oxidant that converts the TDP enamine to 2-acetylthiamin diphosphate and only after that as an acyl group carrier. A related reaction that is known to proceed through acetyl-TDP is the previously mentioned bacterial pyruvate oxidase. As seen, this enzyme has its own oxidant, FAD, which is ready to accept the two electrons to produce bound acetyl-TDP. The electrons may be able to jump directly to the FAD, with thiamin and flavin radicals being formed at an intermediate stage.

The electron transfers as well as other aspects of oxidative decarboxylation are discussed. A reaction that is related to that of transketolase but is likely to function via acetyl-TDP is *phosphoketolase*, whose action is required in the energy metabolism of some bacteria. A product of phosphoketolase is acetyl phosphate, whose cleavage can be coupled to synthesis of ATP Phosphoketolase presumably catalyzes an a cleavage to the thiamin-containing enamine. A possible mechanism of formation of acetyl phosphate is elimination of H_2O from this enamine, tautomerization to 2-acetylthiamin, and reaction of the latter with inorganic phosphate.

CH_2OH / C=O / HO—C—H / R + P_i → CH_3 / C=O / OPO_3^{2-} (Acetyl phosphate) + O=C(H)—R

Acetyl phosphate + ADP → ATP + CH_3—COO^-

Thiamin Coenzymes in Nerve Action

The striking paralysis caused by thiamin deficiency together with studies of thiamin analogs as metabolites suggested a special action for this vitamin in nerves. The thiamin analog *pyrithiamin* both induces paralytic symptoms and displaces thiamin from nerve preparations. The nerve poison *tetrodotoxin* blocks nerve conduction by inhibiting inward diffusion of sodium, but it also promotes release of thiamin from nerve membranes. Evidence for a metabolic significance of thiamin triphosphate comes from identification of soluble and membrane-associated thiamin triphosphatases as well as a kinase that forms protein-bound thiamin triphosphate in the brain. Mono-, tri-, and tetraphosphates also occur naturally in smaller amounts. One might speculate about a possible role for the rapid interconversion of cationic and yellow anionic forms of thiamin via the tricyclic form in some aspect of nerve conduction.

E. PYRIDOXAL PHOSPHATE

The phosphate ester of the aldehyde form of vitamin Bg, *pyridoxal phosphate* (pyridoxal-P or PLP), is required by many enzymes catalyzing reactions of amino acids and amines. The reactions are numerous, and pyridoxal phosphate is surely one of nature's most versatile catalysts. The story begins with biochemical *transamination*, a process of central importance in nitrogen metabolism. In 1937, Alexander Braunstein and Maria Kritzmann, in Moscow, described the trans-amination reaction by which amino groups can be transferred from one carbon skeleton to another. For example, the amino group of glutamate can be transferred to the carbon skeleton of oxaloacetate to form aspartate and 2-oxoglutarate.

L-Glutamate + Oxaloacetate —Transamination→ 2-Oxoglutarate + L-Aspartate

This transamination reaction is a widespread process of importance in many aspects of the nitrogen metabolism of organisms. A large series of *transaminases (aminotransferases)*, for which glutamate is most often one of the reactants, have been shown to catalyze the reactions of other oxoacids and amino acids. In 1944, Esmond Snell reported the nonenzymatic conversion of pyridoxal into pyridoxamine by heating with glutamate. He recognized that this was also transamination and proposed that pyridoxal might be a part of a coenzyme needed for aminotrans-ferases and that these enzymes might act via two half-reactions that interconverted pyridoxal and pyridox-amine.

The hypothesis was soon verified and the coenzyme was identified as pyridoxal 5′-phosphate or pyridoxamine 5′-phosphate. At about the same time, Gunsalus and coworkers

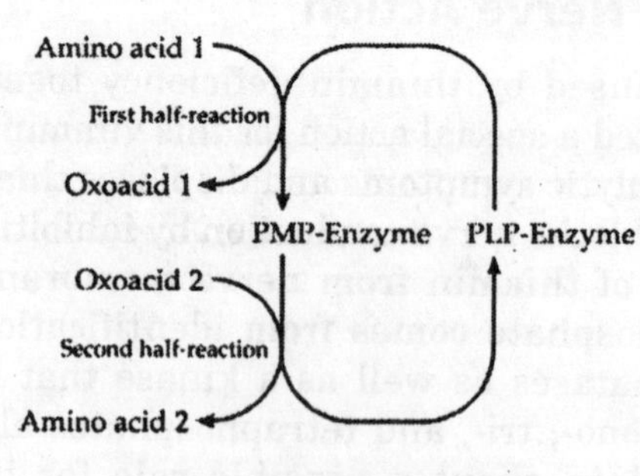

noticed that the activity of *tyrosine decarboxylase* produced by lactic acid bacteria was unusually low when the medium was deficient in pyridoxine. Addition of pyridoxal plus ATP increased the decarboxylase activity of cell extracts. PLP was synthesized and was found to be the essential coenzyme for this and a variety of other enzymes.

Nonenzymatic Models

Pyridoxal or PLP, in the complete absence of enzymes, not only undergoes slow transamination with amino acids but also catalyzes many other reactions of amino acids that are identical to those catalyzed by PLP-dependent enzymes. Thus, *the coenzyme itself can be regarded as the active site of the enzymes* and can be studied in nonenzymatic reactions. The latter can be thought of as *models* for corresponding enzymatic reactions. From such studies Snell and associates drew the following conclusions.

(*a*) The aldehyde group of PLP reacts readily and reversibly with amino acids to form Schiff bases which react further to give products.

(*b*) For an aldehyde to be a catalyst, a strong electron-attracting group, *e.g.*, the ring nitrogen of pyridine (as in PLP), must be *ortho* or *para* to the –CHO group. A nitro group, also strongly electron attracting, can replace the pyridine nitrogen in model reactions.

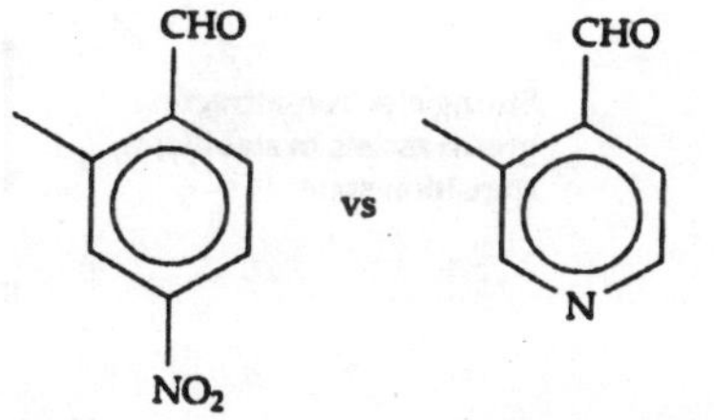

CHO CHO

vs

N

NO_2

(*c*) The presence of an – OH group adjacent to the – CHO group greatly enhances the catalytic activity. Since certain metal ions, such as Cu^{2+} and Al^{3+}, increase the rates in model systems and are known to chelate with Schiff bases of the type formed with PLP, it was concluded that either a metal ion or a proton formed a chelate ring and helped to hold the Schiff base in a planar conformation. *However, such a function for metal ions has not been found in PLP-dependent enzymes.*

(*d*) In model systems the 5-hydroxymethyl and 2-methyl groups are not needed for catalysis. However, in enzymes the 5-CH_2OH group is essential for attachment of the phosphate handle. The 2-CH_3 group is usually not necessary for coenzymatic activity.

Many investigations of nonenzymatic reactions of PLP and related compounds have been and are still being conducted.

A General Mechanism of Action of PLP

Based upon consideration of the various known PLP-dependent enzymes of amino acid metabolism, Braunstein and Shemyakin in 1952 proposed a general mechanism of PLP action which, in most details, was the same as the one proposed independently by Snell and associates on the basis of the nonenzymatic reactions. The general mechanism, which has been verified by studies of many enzymes, can be stated as follows: *Pyridoxal phosphate reacts to convert the amino group of a substrate into a Schiff base that is electronically the equivalent of an adjacent carbonyl.* However, a Schiff base of an amino acid with a simple aldehyde (for example, acetaldehyde) has the opposite polarity from that of C = O (see the following structures). Such an imine could not substitute for a carbonyl group in activating an α-hydrogen nor in facilitating C – C bond cleavage in the amino acid. It is necessary to have the strongly electron-attracting pyridine group conjugated with the C = N group in such a way that electrons can flow from the substrate into the coenzyme.

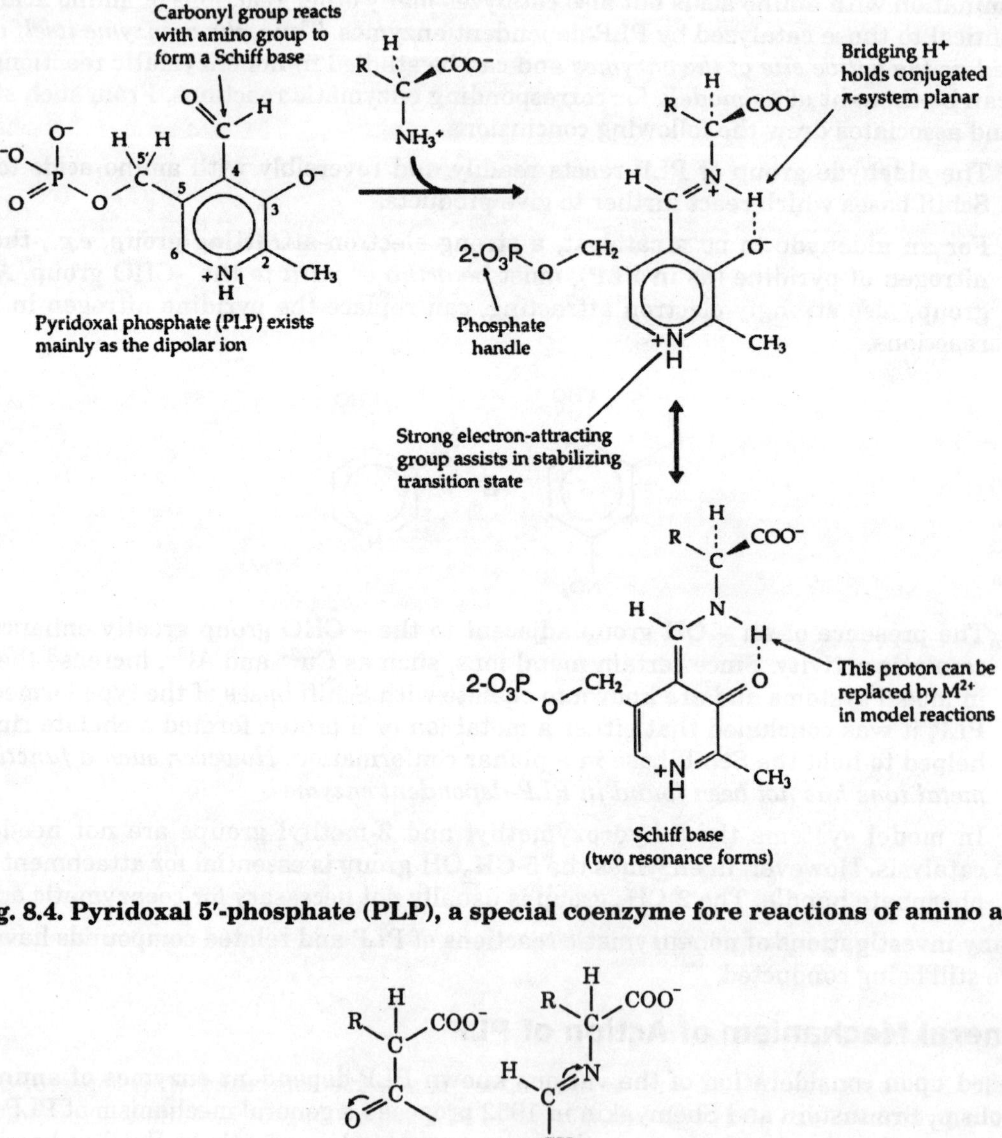

Carbonyl group reacts with amino group to form a Schiff base

Pyridoxal phosphate (PLP) exists mainly as the dipolar ion

Phosphate handle

Bridging H^+ holds conjugated π-system planar

Strong electron-attracting group assists in stabilizing transition state

This proton can be replaced by M^{2+} in model reactions

Schiff base (two resonance forms)

Fig. 8.4. Pyridoxal 5′-phosphate (PLP), a special coenzyme fore reactions of amino acids.

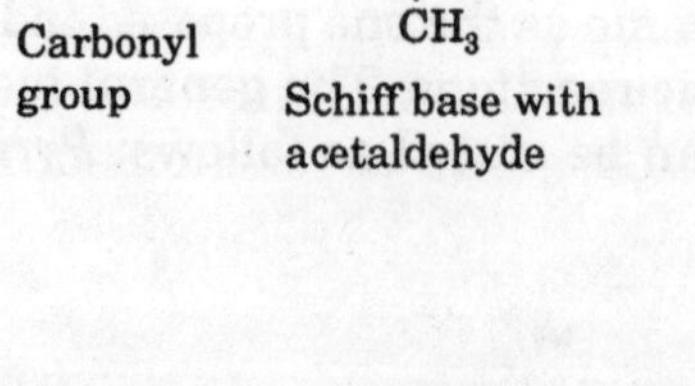

Carbonyl group

Schiff base with acetaldehyde

Before discussing the reactions of Schiff bases of PLP we should consider one fact that was not known in 1952. PLP is bound into an enzyme's active site as a Schiff base with a specific lysine side chain before a substate binds. This is often called the *internal aldimine*. When the substrate binds it reacts with the internal Schiff base by a two-step process called *transimination* to form the substrate Schiff base, which is also called the *external aldimine*.

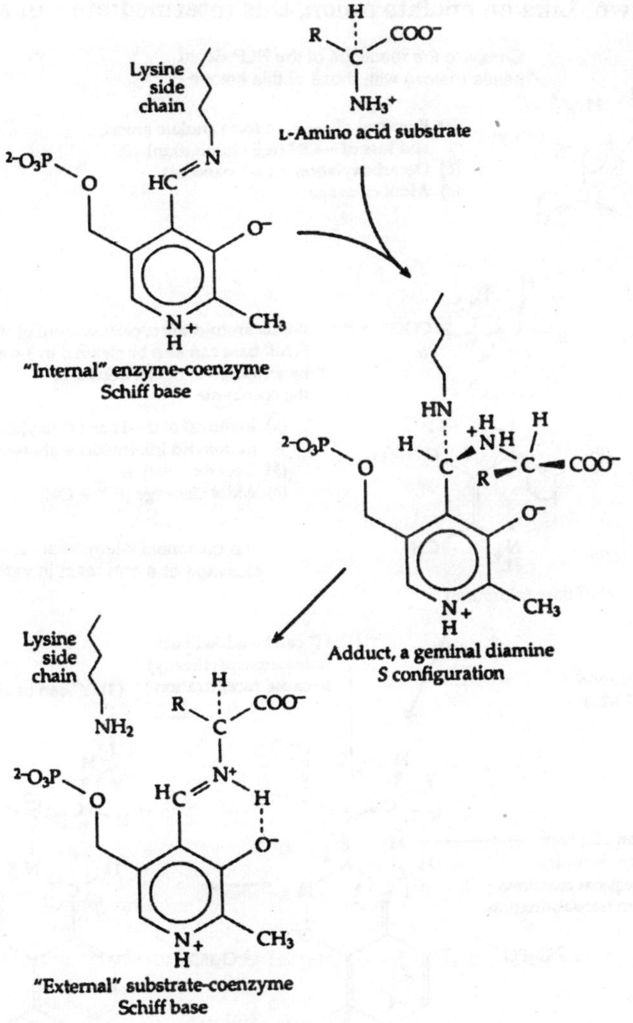

Lysine side chain

L-Amino acid substrate

"Internal" enzyme-coenzyme Schiff base

Adduct, a geminal diamine *S* configuration

Lysine side chain

"External" substrate-coenzyme Schiff base

The Variety of PLP-Dependent Reactions

In Fig. 8.5 reactions of PLP-amino acid Schiff bases are compared with those of β-oxo-acids. Beta-hydroxy-α-oxo acids and Schiff bases of PLP with β-hydroxy-α-amino acids can react in similar ways. The reactions fall naturally into three groups (*a, b, c*) depending upon whether the bond cleaved is from the α-carbon of the substrate to the hydrogen atom, to the carboxyl group, or to the side chain. A fourth group of reactions of PLP-dependent enzymes (*d*) also involve removal of the α-hydrogen but are mechanistically more complex. Some of the many reactions catalyzed by these enzymes are listed in Table 8.3.

Loss of the α-hydrogen (Group *a*)

Dissociation of the α-hydrogen from the Schiff base leads to a *quinonoid-carbanionic intermediate* whose structure in depicted. The name reflects the characteristics of the two resonance forms drawn. Like an enolate anion, this intermediate can react in several ways.

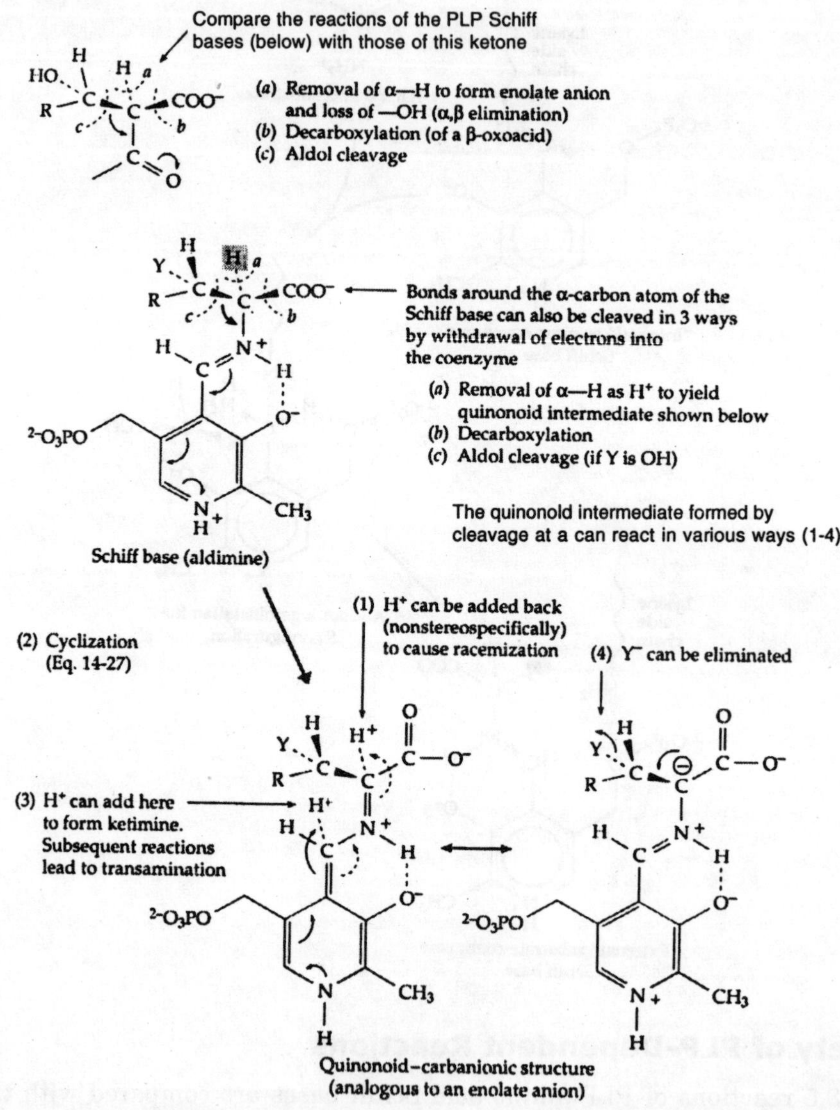

Fig. 8.5. Some reactions of Schiff bases of pyridoxal phosphate, (*a*) Formation of the quinonoid intermediate, (*b*) elimination of a β substituent, and (*c*) transamination. The quinonoid-carbanionic intermediate can react in four ways (1 – 4) if enzyme specificity and substrate structure allow.

(1) Racemization : A proton can be added back to the original alpha position but without stereospecificity. A racemase which does this is important to bacteria. They must synthesize D-alanine and D-glutamic acid from the corresponding L-isomers for use in formation of their peptidoglycan envelopes. The combined actions of alanine racemase plus D-alanine aminotransferase, which produces D-glutamate as a product, provide bacteria with both D amino acids. A fungal alanine racemase is necessary for synthesis of the immunosuppresant cyclosporin. High concentrations of free D-alanine are found in certain regions of the brain and also in various glands.

The carboxyl group of an amino acid can also activate the α-hydrogen. This may be the basis for an aspartate racemase and other racemases that are *not* dependent upon PLP.

(2) Cyclization : A second kind of reaction is represented by the conversion of S-adenosylmethionine to *aminocyclopropanecarboxylic acid*, a precursor to the plant hormone *ethylene.* The quinonoid intermediate cyclizes with elimination of methylthioadenosine to give a Schiff base of the product. The cyclization step appears to be a simple S_N 2-like reaction.

(3) Transamination : A proton can add to the carbon attached to the 4 position of the PLP ring (Fig. 8.5) to form a second Schiff base, often referred to as a *ketimine* (E. 8.28). The latter can readily undergo hydrolysis to *pyridoxamine phosphate* (PMP) and an α-oxo acid. This sequence represents one of thetwo half-reactions (Eq. 8.24 and 8.25) required for enzymatic transamination.

Transaminases participate in metabolism of most of the amino acids, over 60 different enzymes have been identified. Best studied are the *aspartate aminotransferases*, a pair of cytosolic and mitochondrial isoenzymes which can be isolated readily from animal hearts. Their presence in heart muscle and brain in high concentration is thought to be a result of their functioning in the malateaspartate shuttle. The sequences of the two proteins differ greatly, with only 50% of the residues being the same in both isoenzymes. However, these differences are largely on the outside surface, the folding pattern and internal structure are almost identical.

Three-dimensional structures of aspartate aminotransferases of *E. coli*, yeast, chickens, and mammals are extremely similar, even though sequence identity may be as low as 20%. Most other transaminases also use the L-gutamate-oxoglutarate pair as one of the product-reactant pairs but a few prefer smaller substrates with uncharged side chains. An example is serine: pyruvate (or alanine:glyoxylate) aminotransferase, an important mitochondrial and peroxisomal enzyme in both animals and plants. Other specialized aminotransferases act on aromatic amino acids, the branched chain amino acids valine, leucine, and isoleucine, and D-amino acids.

Many of them are highly specific for individual amino acids such as phosphoserine, ornithine *N*-acetylornithine, and 8-amino-7-oxononanoate An apparently internal transamination, which requires PMP and PLP, converts glutamate-1-semialdehyde into 8-aminolevulinate in the pathway of porphyrin bio-synthesis used by bacteria and plants.

(4) Elimination and β replacement : When a good leaving group is present in the β position of the amino acid it can be eliminated (Fig. 8.5 Eq. 8.29). A large number of enzymes catalyze such reactions. Among them are *serine and threonine dehydratases*, which eliminate OH^- as H_2O; *tryptophan indolelyase* (tryptophanase) of bacteria, which eliminates indole; *tyrosine phenollyase* (elimination of phenol); and *alliinase* of garlic (elimination of 1-propenylsulfenic acid). Cysta-thionine, a precursor to methionine, eliminates L-homocysteine through the action of *cystathionine* β *lyase* (cystathionase). Ammonia is eliminated from the β position of 2, 3-diaminopropionate by a bacterial lyase.

Table 8.3. Some Enzymes that Require Pyridoxal Phosphate as a Coenzyme

(*a*) Removing alpha hydrogen as H^+
- (1) Racemization
  - Alanine racemase*
- (2) Cyclization
  - Arninocyclopropane carboxylate synthase
- (3) Amino group transfer
  - Aspartate aminotransferase*
  - Alanine aminotransferase
  - D-Amino acid aminotransferase*
  - Branched chain aminotransferase
  - Gamma-aminobutyrate aminotransferase
  - ω-Amino acid:pyruvate aminotransferase*
  - Tyrosine aminotransferase Serineipyruvate aminotransferase
- (4) Beta elimination or replacement
  - D-and L-Serine dehydratases (deaminases)
  - Tryptophan indole-lyase (tryptophanase)*
  - Tyrosine phenol-lyase*
  - Alliinase
  - Cystathionine β-lyase (cystathionase)*
  - O-Acetylserine sulfhydrylase (cysteine synthase)
  - Cystathionine β-synthase Tryptophan synthase*

(*b*) Removal of alpha carboxylate as CO_2
- Diaminopimelate decarboxylase
- Glycine decarboxylase (requires lipoyl group)
- Glutamate decarboxylase
- Histidine decarboxylase
- Dopa decarboxylase Ornithine decarboxylase*
- Tyrosine decarboxylase
- Dialkylglycine decarboxylase (a decarboxylating transaminase)*

(*c*) Removal or replacement of side chain (or -H) by aldol cleavage
- Serine hydroxymethyltransferase
- Threonine aldolase
- δ-Aminolevulinate synthase
- Serine palmitoyltransferase
- 2-Amino-3-oxobutyrate-CoA ligase

(*d*) Reactions of ketimine intermediates
- Aspartate γ-decarboxylase
- Selenocysteine lyase
- *Nif S* protein of nitrogenase
- Gamma elimination and replacement
  - Cystathionine γ-synthase
  - Cystathionine γ-lyase
  - Threonine synthase

(*e*) Other enzymes
- Lysine 2,3-aminomutase
- Glycogen phosphorylase*
- Pyridoxamine phosphate (PMP) in synthesis of 3, 6-dideoxy hexoses

* The three-dimensional structures of these and other PLP-dependent enzymes were determined by 2000.

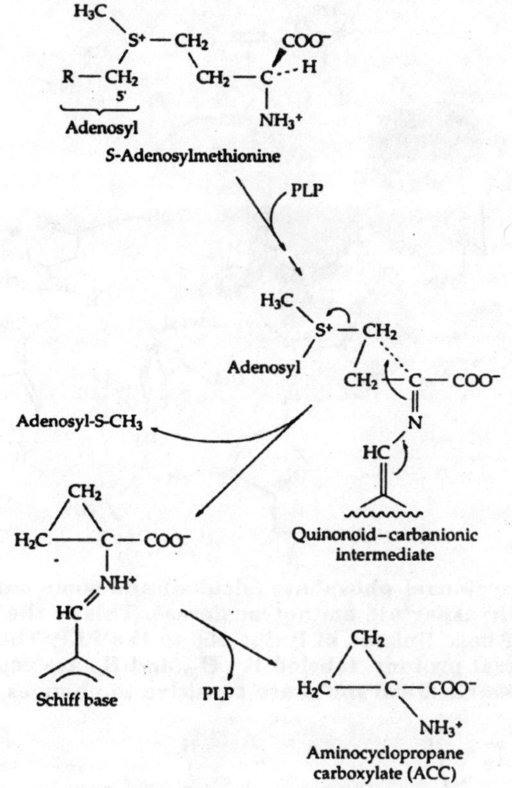

H_3C S^+ — CH_2 COO^- R — CH_2 5' CH_2 — C ··· H NH_3^+

Adenosyl

S-Adenosylmethionine

PLP

H_3C S^+ — CH_2 Adenosyl CH_2 — C — COO^- N HC

Adenosyl-S-CH_3

Quinonoid–carbanionic intermediate

CH_2 H_2C — C — COO^- NH^+ HC

Schiff base

PLP

CH_2 H_2C — C — COO^- NH_3^+

Aminocyclopropane carboxylate (ACC)

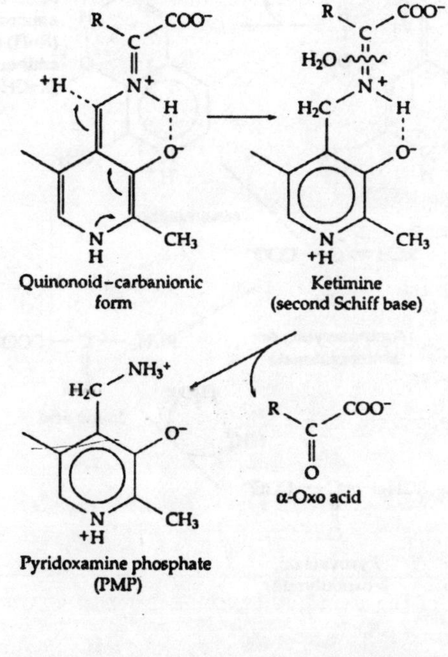

R COO^- C N^+ +H H O^- N H CH_3

Quinonoid–carbanionic form

R COO^- C H_2O N^+ H_2C H O^- N +H CH_3

Ketimine (second Schiff base)

NH_3^+ H_2C O^- N +H CH_3

Pyridoxamine phosphate (PMP)

R COO^- C O

α-Oxo acid

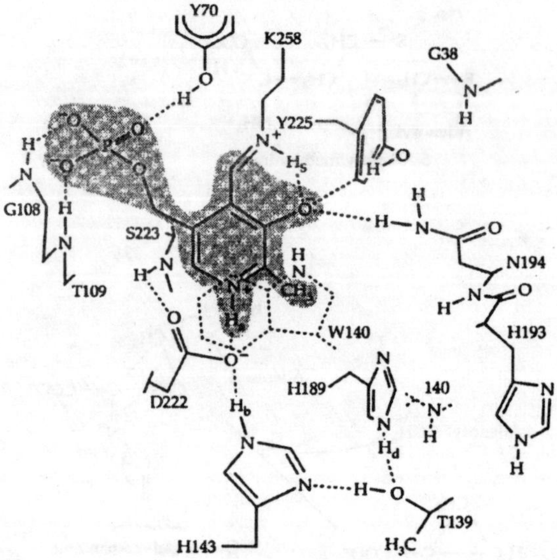

Fig. 8.6. Drawing showing pyridoxal phosphate (shaded) and some surrounding protein structure in the active site of cytosolic aspartate aminotransferase. This is the low pH form of the enzyme with an N-protonated Schiff base linkage of lysine 258 to the PLP. The tryptophan 140 ring lies in front of the coenzyme. Several protons, labeled H_a, H_b, and H_d are represented in H NMR spectra by distinct resonances whose chemical'shifts are sensitive to changes in the active site.

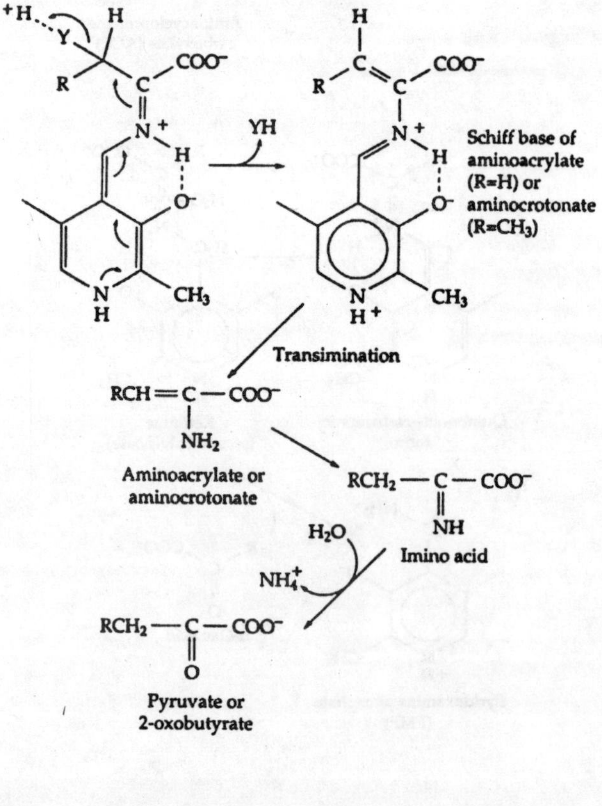

Schiff base of aminoacrylate (R=H) or aminocrotonate ($R=CH_3$)

Transimination

$RCH{=}C(NH_2){-}COO^-$

Aminoacrylate or aminocrotonate

$RCH_2{-}C({=}NH){-}COO^-$

Imino acid

H_2O, NH_4^+

$RCH_2{-}C({=}O){-}COO^-$

Pyruvate or 2-oxobutyrate

Beta replacement is catalyzed by such enzymes of amino acid biosynthesis as *tryptophan synthase O-acetylserine sulfhydrylase* (cysteine synthase), and *cystathionine β-synthase*. In both elimination and p replacement an unsaturated Schiff base, usually of aminoacrylate or aminocrotonate, is a probable intermediate. Conversion to the final products is usually assumed to be via hydrolysis to free aminoacrylate, tautomerization to an imino acid, and hydrolysis of the latter, *e.g.*, to pyruvate and ammonium ion. However, the observed stereospecific addition of a proton at the β-C atom of 2-oxobutyrate suggests that these steps may

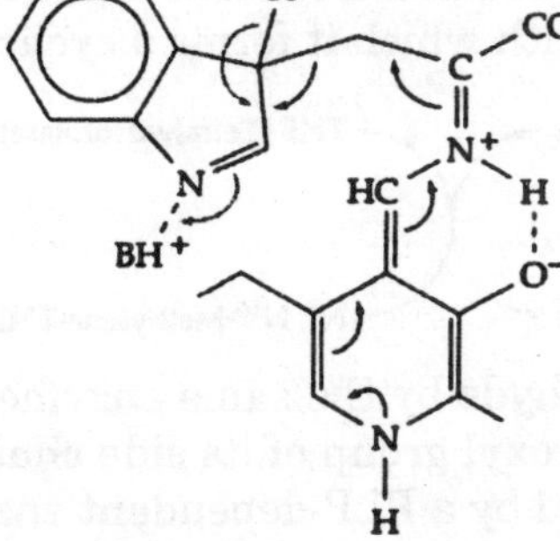

occur with the participation of groups from the enzyme. Before indole can be eliminated by tryptophan indolelyase the indole ring must presumably be tautomerized to the following form of the quinonoid intermediate. The same species may be created by tryptophan synthase upon addition of indole to the enzyme-bound aminoacrylate. The green arrows on the structure indicate the tautomerization that would occur to convert the indole ring to the structure found in tryptophan. The three-dimensional structure of tryptophan synthase. It is a complex of two enzymes with a remarkable tunnel through which the intermediate indole can pass. An unusual PLP-dependent β replacement is used to synthesize a transfer RNA ester of *selenocysteine* prior to its insertion into special locations in a few proteins.

Decarboxylation (*Group b*)

The bond to the carboxyl group of an amino acid substrate is broken in reactions catalyzed by *amino acid decarboxylases*. These also presumably lead to a transient quinonoid-carbanionic intermediate. Addition of a proton at the original site of decarboxylation followed by breakup of the Schiff base completes the sequence. Decarboxy-lation of amino acids is nearly irreversible and frequently appears as a final step in synthesis of amino compounds. For example, in the brain glutamic acid is decarboxylated to γ-*aminobutyric acid* (Gaba), while 3, 4-dihydroxyphenylalanine (dopa) and 5-hydroxytryptophan are acted upon by an *aromatic amino acid decarboxylase* to form, respectively, the neurotransmitters *dopamine* and serotonin. Histidine is decarboxylated to histamine.

Howcvcr, not all histidine decarboxylases use PLP as a coenzyme. Arginine is converted by a PLP-dependent decarboxylase to agmatine which is hydrolyzed to 1, 4-*diaminopropane*. This important cell constituent is also formed by hydrolysis of arginine to *ornithine* and decarboxylation of the latter. *Lysine* is formed in bacteria by decarboxylation of mesodiamino-pimelic acid. *Glycine* is decarboxylated oxidatively in mitochondria in a sequence requiring lipoic acid and tetrahydrofolate as well as PLP. A *methionine* decarboxylase has been isolated in pure form from a fern. The bacterial *dialkylglycine* decarboxylase is both a decarboxylase and an aminotransferase which uses pyruvate as its second substrate forming a ketone and L-alanine as products.

Side chain cleavage (Group c)

In a third type of reaction the side chain of the Schiff base of Fig. 8.5 undergoes aldol cleavage. Conversely, a side chain can be added by β condensation. The best known enzyme of this group is *serine hydroxymethyltransferase*, which converts serine to glycine and formaldehyde. The latter is not released in a free form but is transferred by the same enzyme specifically *to tetrahydrofolic acid*, with which it forms a cyclic adduct.

Threonine is cleaved to acetaldehyde by the same enzyme. In a more important pathway of degradation of threonine the hydroxyl group of its side chain is dehydrogenated to form 2-amino-3-oxobutyrate which is cleaved by a PLP-dependent enzyme to glycine and acetyl-CoA.

Conversely, ester condensation reactions join acyl groups from CoA derivatives to Schiff bases derived from glycine or serine. Succinyl-CoA is the acyl donor in second known pathway for biosynthesis of 5-*aminolevulinic acid*, an intermediate in porphyrin synthesis. The enzyme does not catalyze decarboxylation of glycine in the absence of succinyl-CoA, and the decarboxylation probably follows the condensation as indicated. In a similar reaction in the biosynthesis of *sphingosine* serine is condensed with palmitoyl-CoA and decarboxylated to form an aminoketone intermediate. 8-Amino-7-oxonanonoate synthase forms a precursor of biotin.

Ketimine Intermediate as Electron Acceptor (Group d)

The fourth group of PLP-dependent reactions are thought to depend upon formation of the ketimine intermediate of Eq. 8.28. In this form the original α-hydrogen of the amino acid has been removed and the C = NH$^+$ bond of the ketimine is polarized in a direction that favors electron withdrawal from the amino acid into the imine group. This permits another series of enzymatic reactions analogous to those of the β-oxo acid shown at the top of Fig. 8.5. Both elimination and C – C bond cleavage α, β to the C = N group of the ketimine can occur.

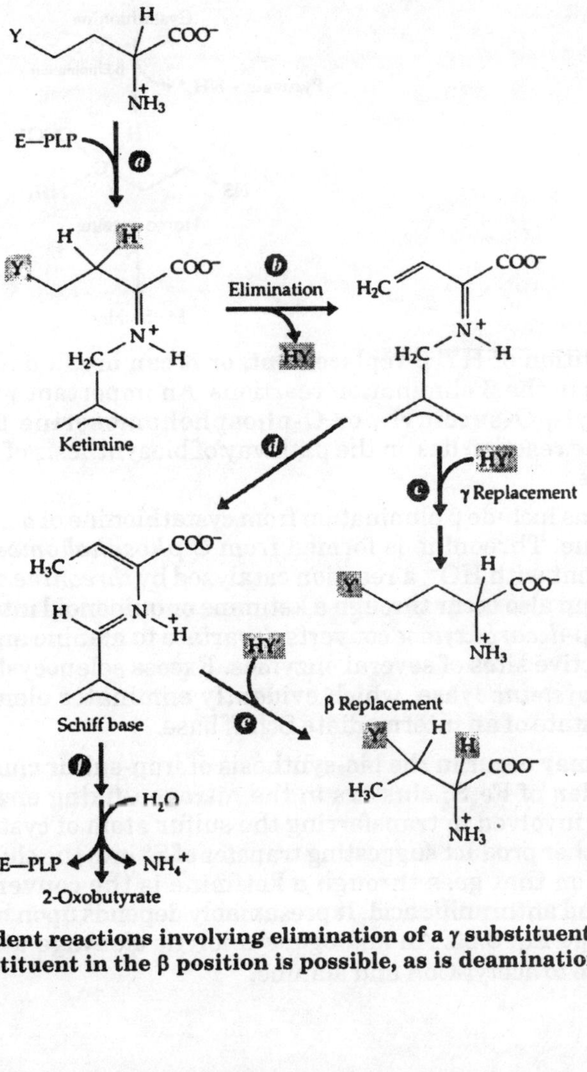

Fig. 8.7. Some PLP-dependent reactions involving elimination of a γ substituent. Replacement by another γ substiruent or by a substituent in the β position is possible, as is deamination to an α-oxo acid.

Enzymes of this group catalyze elimination of γ substituents from amino acids as illustrated in Fig. 8.7. Eliminated groups may be replaced by other substituents, either inthe a or the β positions. The ketimine formed initially by such an enzyme undergoes elimination of the γ substituent (β with respect to the C = N group) along with a proton from the β position of the original amino acid to form an unsaturated intermediate which can react in one of three ways, depending upon the enzyme. Addition of HY' leads to γ replacement, while addition of a proton at the a position leads, via reaction, to an α,β-unsaturated Schiff base. The

Cysteine + O-Succinylhomoserine —γ Replacement→ Succinate + Cystathionine —β Elimination→ Pyruvate + NH_4^+ + Homocysteine → Methionine

latter can react by addition of HY' β replacement, or it can break down to an α-oxo acid and ammonium ion, just as in the β elimination reactions. An important γ replacement reaction is conversion of *O*-acetyl-, *O*-succinyl-, or O-phosphohomoserine to cystathionine. This *cystathionine γ-synthase* reaction lies on the pathway of biosynthesis of methionine by bacteria, fungi, and higher plants.

Subsequent reactions include β elimination from cystathionine of *homocysteine* which is then converted to methionine. Threonine is formed from *O phosphohomoserine* via γ elimination followed by β replacement with HO^-, a reaction catalyzed by *threonine synthase*. The loss of a β-carboxyl group as CO_2 can also occur through a ketimine or quinonoid intermediate. For example, the bacterial *aspartate* β-*decarboxylase* converts aspartate to alanine and CO_2. *Selenocysteine* is utilized to create the active sites of several enzymes. Excess selenocysteine is degraded by the PLP-dependent *selenocysteine* lyase, which evidently eliminates elemental selenium from a ketimine or quinonoid state of an intermediate Schiff base.

A similar reaction may occur in the bio-synthesis of iron-sulfur clusters. The *Nif S* protein is essential for formation of Fe_4S_4 clusters in the nitrogen-fixing enzyme nitrogenase. This enzyme is in some way involved in transferring the sulfur atom of cysteine into an iron-sulfur cluster. Alanine is the other product suggesting transfer of S° into the cluster using the sequence. Another related reaction that goes through a ketimine is the conversion of the amino acid *kynurenine* to alanine and anthranilic acid. It presumably depends upon hydration of the carbonyl group prior to β cleavage Eq. 8.35. An analogous thiolytic cleavage utilizes CoA to convert 2-amino-4-ketopentanoate to acetyl-CoA and alanine.

Glycogen Phosphorylase

While PLP is ideally designed to catalyze reactions of amino compounds it was surprising to find it as an essential cofactor for glycogen phosphorylase. The PLP is linked as a Schiff base in the same way as in other PLP-dependent enzymes, but there is no obvious function for the coenzyme ring. The phosphate group probably acts as an acid-base catalyst. It has been estimated that 50% of the vitamin B_6 in our body is present as PLP in muscle phosphorylase. Studies of vitamin B_6-deficient rats suggest that PLP in phosphorylase serves as a reserve supply, much of which can be taken for other purposes during times of deficiency.

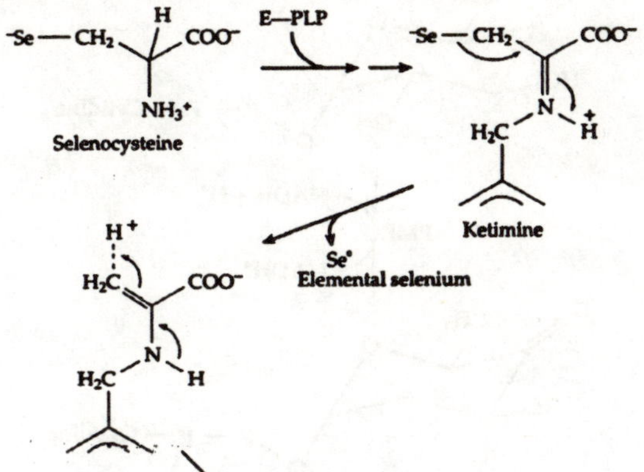

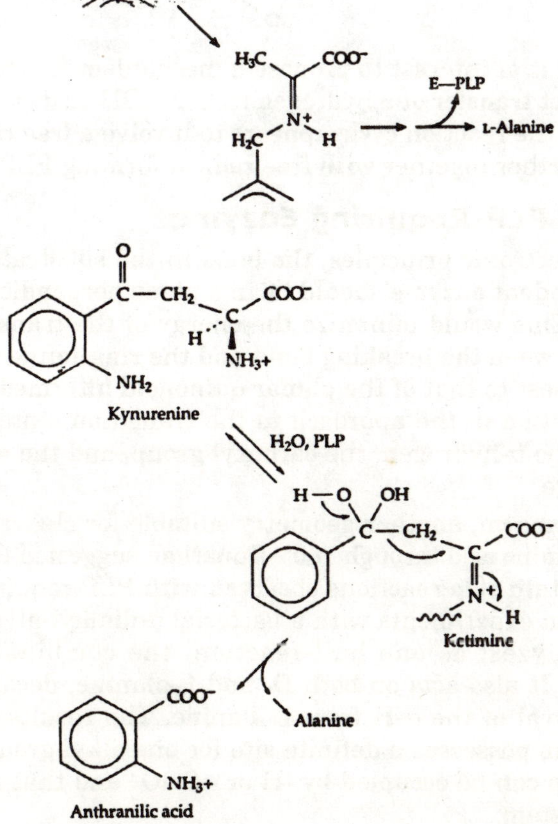

Pyridoxamine Phosphate as a Coenzyme

If PLP is a cofactor designed to react with amino groups of substrates, might not pyridoxamine phosphate (PMP) act as a coenzyme for reactions of carbonyl compounds? An example of this kind of function has been found in the formation of 3, 6-dideoxyhexoses needed for bacterial cell surface antigens. Glucose (as cytidine diphosphate glucose; CDP-glucose) is first converted to 4-oxo-6-deoxy-CDP-glucose. The conversion of the latter to 3, 6-dideoxy-CDP-glucose requires PMP as well as NADH or NADPH.

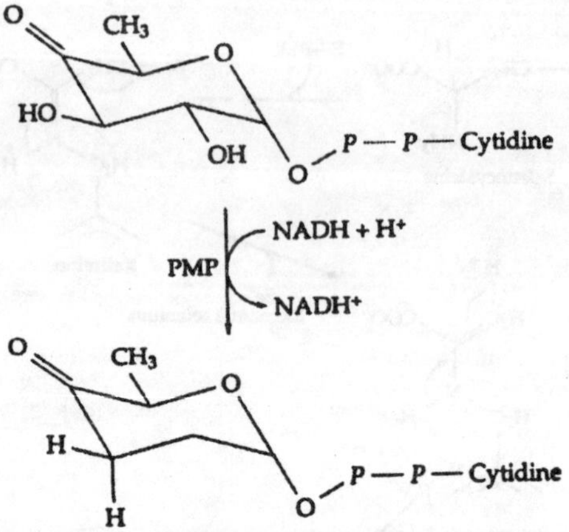

The student may find it of interest to propose a mechanism for this reaction, taking into account the expected direct transfer of a hydrogen from NADH as described, before consulting published papers. Part of the reaction cycle appears to involve a free radical derived from the PMP. This is discussed further together with free radical-forming PLP enzymes.

Stereochemistry of PLP-Requiring Enzymes

According to stereoelectronic principles, the bond in the substrate amino acid that is to be broken by a PLP-dependent enzyme should lie in a plane perpendicular to the plane of the cofactorimine π system. This would minimize the energy of the transition state by allowing maximum σ-π overlap between the breaking bond and the ring-imine π system. It also would provide the geometry closest to that of the planar quinonoid intermediate to be formed, thus minimizing molecular motion in the approach to the transition state. Three orientations of an amino acid in which the α-hydrogen, the carboxyl group, and the side chain, respectively, are positioned for cleavage.

For each orientation shown, another geometry suitable for cleavage of the same bond is obtained by rotating the amino acid through 180°. Dunathan suggested that this stereoelectronic requirement explains certain side reactions observed with PLP-requiring enzymes. The idea also received support from experiments with a bacterial α-dialkyl-glycinedecarboxylase. The enzyme ordinarily catalyzes, as one half-reaction, the combination decarboxylation-transamination reaction. It also acts on both D- and L-alanine, decarboxylating the former but catalyzing only removal of the α-H from L-alanine. The results can be rationalized by assuming that the enzyme possesses a definite site for one alkyl group but that the position of the second alkyl groxip can be occupied by -H or $-COO^-$ and that the group labilized lies perpendicular to the π system:

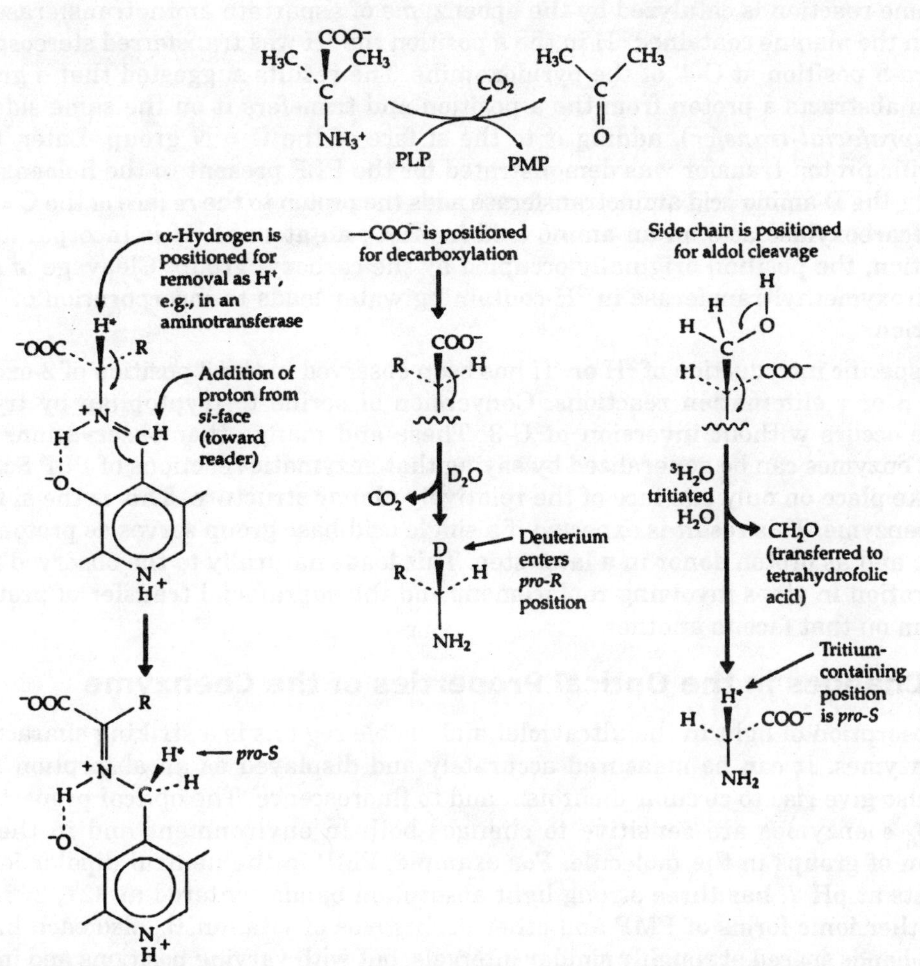

Fig. 8.8. Some stereochemical aspects of catalysis of PLP-requiring enzymes.

Glycine is unreactive, suggesting that occupation of the alkyl binding site is required for catalysis.

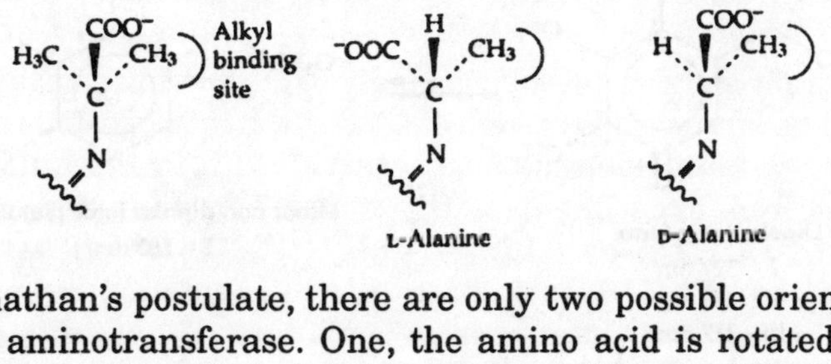

L-Alanine D-Alanine

According to Dunathan's postulate, there are only two possible orientations of the amino acid substrate in an aminotransferase. One, the amino acid is rotated 180° so that the a-hydrogen protrudes *behind* the plane of the paper. Dunathan studied *pyridoxamine:pyruvate amino-transferase*, an enzyme closely related to PLP-requiring aminotransferases and which catalyzes the trans-amination of pyridoxal with L-alanine to form pyridox-amine and pyruvate.

The same reaction is catalyzed by the apoenzyme of aspartate aminotransferase. In both cases, when the alanine contained ^2H in the a position the ^2H was transferred stereospecifically into the *pro-S* position at C-4′ of the pyridoxamine. The results suggested that a group from the protein abstracts a proton from the a position and transfers it on the same side of the π system (*suprafacial transfer*), adding it to the si face of the C = N group. Later, the same stereospecific proton transfer was demonstrated for the PLP present in the holoenzyme. Not surprisingly, the D-amino acid aminotransferase adds the proton to the *re* face of the C = N group. When a decarboxylase acts on an amino acid in ^2H_2O, an atom of ^2H is incorporated in the *pro-S* position, the position originally occupied by the carboxyl group. Cleavage of serine by serine hydroxymethyltransferase in ^3H-containing water leads to incorporation of ^3H in the *pro-S* position.

Stereospecific introduction of ^2H or ^3H has been observed in the β position of 2-oxobutyrate formed in β or γ elimination reactions. Conversion of serine to tryptophan by tryptophan synthetase occurs without inversion at C-3. These and many other observations on PLP-dependent enzymes can be generalized by saying that enzymatic reactions of PLP Schiff bases usually take place on only one face of the relatively planar structure. This is the si face at C-4′ of the coenzyme. This result is expected if a single acid-base group serves as proton acceptor in one step and as proton donor in a later step. This leads naturally to the observed retention of configuration in steps involving replacement and the suprafacial transfer of protons from one position on that face to another.

Seeing Changes in the Optical Properties of the Coenzyme

The absorption of light in the ultraviolet and visible regions is a striking characteristic of many coenzymes. It can be measured accurately and displayed as an absorption spectrum and may also give rise to circular dichroism and to fluorescence. The optical properties of the vitamin B_6 coenzymes are sensitive to changes both in environment and in the state of protonation of groups in the molecule. For example, PMP in the neutral dipolar ionic form, which exists at pH 7, has three strong light absorption bands centered at 327, 253, and 217 nm. The other ionic forms of PMP and other derivatives of vitamin B_6 also each have three absorption bands spaced at roughly similar intervals, but with varying positions and intensities.

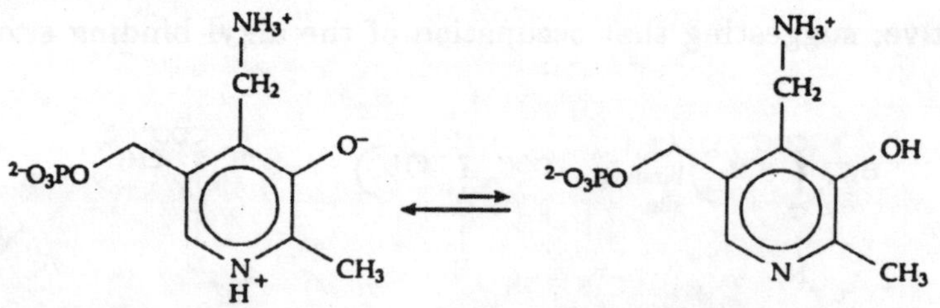

Dipolar ionic form
I 327 nm
II 253 nm
III 217 nm

Minor non-dipolar ionic tautomer
I 283 nm

The minor tautomer of PMP containing an uncharged ring has its low-energy (long-wavelength) band at 283 nm. When both the ring nitrogen and phenolic oxygen are protonated, the band shifts again to 294 nm and if both groups are deprotonated the resulting anion absorbs at 312 nm. Thus, observation of the absorption spectrum of the coenzyme bound to an enzyme surface can tell us whether particular groups are protonated or unprotonated. The peak of bound PMP at 330 nm in aspartate aminotransferase is indicative of the dipolar ionic form.

However, the 5-nm shift from the position of free PMP suggests a distinct change in environment. Pyridoxal phosphate exists in an equilibrium between the aldehyde and its covalent hydrate. The aldehyde has a yellow color and absorbs at 390, while the hydrate absorbs at nearly the same position as does PMP. The absorption bands of Schiff bases of PLP are shifted even further to longer wavelengths, with N-protonated forms absorbing at 415-430 nm. Forms with an unprotonated C = N group absorb at shorter wavelengths.

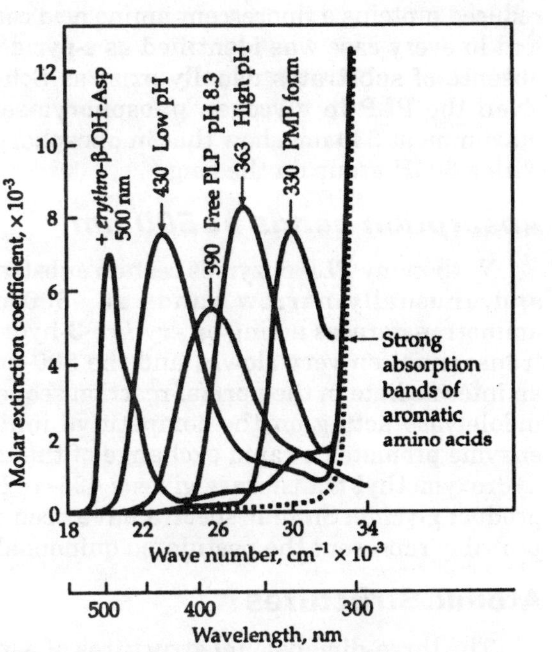

Fig. 8.9. Absorption spectra of various forms of aspartate aminotransferase compared with that of free pyridoxal phosphate. The low pH form of the enzyme observed at pH <5 is converted to the high pH form with pK_a ~6.3. Addition of *erythro*-3-hydroxyaspartate produces a quinonoid form whose spectrum here is shown only 1/3 its true height. The spectrum of free PLP at pH 8.3 is also shown. The spectrum of the apoenzyme (—) contains a small amount of residual absorption of uncertain origin in the 300- to 400-nm region.

Imine Groups in Free Enzymes

When, in 1957, W. T. Jenkins examined one of the first highly purified aspartate aminotransferase preparations he noted a surprising fact: The bound coenzyme, at pH 5, absorbed not at 390 nm, as does PLP, but at 430 nm, like a Schiff base. When the pH was raised, the absorption band shifted to 363 nm. The result suggests dissociation of a proton (with a pK_a of ~6.3) from the hydrogen-bonded position in a Schiff base of the type. It was quickly demonstrated for this enzyme and for many other PLP-dependent enzymes that reduction with sodium borohydride caused the spectrum to revert to one similar to that of PMP and fixed the coenzyme to the protein. After complete HC1 digestion of such borohydride-

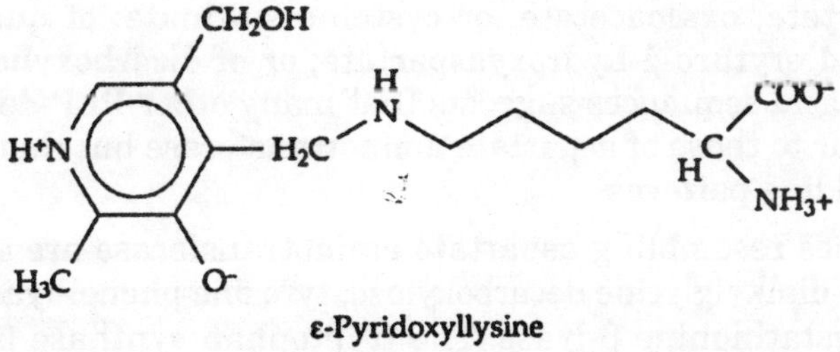

ε-Pyridoxyllysine

reduced proteins a fluorescent amino acid containing the reduced pyridoxyl group was obtained and in every case was identified as ε-pyridoxyl-lysine. Thus, PLP-containing enzymes in the absence of substrates usually exist as Schiff bases with lysine side chains of the proteins. Even the PLP in glycogen phosphorylase is joined in this way. However, its absorption maximum at 330 nm show that in phosphorylase it is present as the nondipolar ionic tautomer with a 3-OH group on the ring.

Absorption bands at 500 nm

With many PLP enzymes certain substrates and inhibitors cause the appearance of intense and unusually narrow bands at ~500 nm. Such a band is observed with aspartate aminotransferases acting on *erythro*-3-hydroxyaspartate (Fig. 8.9). This substrate undergoes transamination very slowly, and the 500-nm absorbing form which accumulates is probably an intermediate in the normal reaction sequence. A similar spectrum is produced by tryptophan indolelyase acting on the competitive inhibitor L-alanine. Under the same conditions the enzyme promotes a rapid exchange of the α-hydrogen of the alanine with ^2H of ^2H_2O. Serine hydroxymethyl-transferase gives a 495- to 500-nm band with both D-alanine and the normal product glycine. Similar spectra have been produced in nonenzymatic model reactions[242] and probably represent the postulated quinonoid-carbanionic intermediates.

Atomic Structures

The three-dimensional structures of aspartate aminotransferases from *E. coli* to humans are very similar. The folding pattern and active site structure are completely conserved. The major domain of the protein contains a central β sheet surrounded by helices with coenzyme attached to a lysine at the C terminus of one of the p strands. The protein is a dimer with the two major domains held together by both polar and nonpolar interactions. The two active sites are located at the interface between the subunits and residues from both subunits participate in forming the active site.

The internal Schiff base is formed with Lys 258. The protonated ring nitrogen of the dipolar ionic PLP forms an ion pair with the carboxylate of Asp 222 which protrudes from a central seven-stranded P sheet. The phenolic $-O^-$ forms a hydrogen bond with the –OH of Tyr 225. The interactions of Asp 222 and Tyr 225 fix the ring as the dipolar ionic tautomer. The phosphate group of the coenzyme, which ^{31}P NMR shows to be predominantly dianionic, forms an ion pair with the side chain of Arg 266 and hydrogen bonds to a backbone N- H at the N terminus of a long helix where it can interact with the positive end of the helix dipole.

In addition, the phosphate forms hydrogen bonds to four OH groups of Ser, Thr, and Tyr side chains. In front of the coenzyme ring are two guani-dinium groups from the side chains of Arg 386 and of Arg 292* (from the second subunit). These have been shown by X-ray crystallography to bind the two carboxylate groups of a substrate such as glutamate, 2-oxoglutarate, aspartate, oxaloacetate, or cysteine sulfinate; of quasi-substrates such as 2-methylaspartate and erythro-β-hydroxyaspartate; or of dicarboxylic inhibitors (Fig. 8.10). Comparison of amino acid sequences suggests that many other PLP-dependent enzymes have folding patterns similar to those of aspartate aminotransferase but that there are four or more additional different folding patterns.

Among the enzymes resembling aspartate aminotransferase are ω-amino acid: pyruvate aminotransferase, 2, 3-dialkylglycine decarboxylase, tyrosine phenol-lyase, a bacterial ornithine decarboxylase, and cystathionine β-lyase. The tryptophan synthase β subunit has a second folding pattern, while alanine racemase and eukaryotic ornithine decarboxylases have $(\alpha\beta)_8$-barrel structures resembling. A fourth structural pattern is that of D-amino acid aminotransferase. It is anticipated that the branched-chain aminotransferase will have a similar stucture. Glycogen phosphorylase has a fifth folding pattern.

Constructing a Detailed Picture of the Action of a PLP Enzyme

Consider the number of different steps that must occur in about one-thousanths of a second during the action of an aminotransferase. First, the substrate binds to form the "Michaelis complex." Then the transimination takes place in two steps and is followed by the removal of the α-hydrogen to form the quinonoid intermediate. An additional four steps are needed to form the ketimine, to hydrolyze it, and to release the oxoacid product to give the PMP form of the enzyme. The reaction sequences in some of the other enzymes are even more complex. How can one enzyme do all this? The first step in the sequence is the binding of the substrate to form the "Michaelis complex."

The positive charges on Arg 386 and Arg 292 doubtless attract the carboxylate groups of the substrate and aid in guiding it toward a correct fit. In a similar manner the $-O^-$ of the coenzyme, which is distributed by resonance into the $-C = N$ of the Schiff base linkage, attracts the $-NH_3^+$ of the substrate. When a substrate or inhibitor binds to the two guanidinium groups a small structural domain of the enzyme moves and closes around the substrate which now has very little contact with the external solvent. In the initial "Michaelis complex" the $-NH_3^+$ group of the substrate lies directly in front of the C-4′ carbon of the coenzyme, where it can initiate the transimination reaction. However, before this can happen a proton must be removed to convert the $-NH_3^+$ to $-NH_2$. Long before the three-dimensional structures were known, Ivanov and Karpeisky suggested that in the free enzyme the positively charged group (Arg 386) that binds the α-carboxylate of the substrate interacts electrostatically with the $-O^-$ of the coenzyme. This is one of the factors that keeps the pK_a of the $-CH=N^+H$-that is conjugated with this $-O^-$ at a low value of ~6.3 (at 0.1 M anion concentration).

However, in the Michaelis complex this interaction of the + charge of Arg 386 with the imine group must be weakened because of the pairing of the α-COO^- of the substrate with the + charge. This will increase the basicity of the imine nitrogen and will also cause a decrease in the pK_a of the substrate $-NH_3^+$, making is easier for a proton to jump from the $-NH_3^+$ to the imine group. This proton transfer, Thus, the nucleophilic $-NH_2$ group is generated by a process that at the same time increases the electrophilic properties of the carbon atom of the imine group. This favors the immediate addition of $-NH_2$ to $-C = N^+H-$ to give the adduct a *geminal diamine,* which is shown in three dimensions in Fig. 8.10 B. Notice that each step in the overall sequence changes the electronic or steric characteristics of the complex in a way that facilitates the next step. This is an important principle that is applicable throughout enzymology: *For an enzyme to be an efficient catalyst each step must lead to a change that sets the stage for the next.* These consecutive steps often require proton transfers, and each such transfer will influence the subsequent step in the sequence.

Some steps also require alterations in the conformation of substrate, coenzyme, and enzyme. One of these is the transimination sequence. On the basis of the observed loss of circular dichroism in the external aldimine, Ivanov and Karpeisky suggested that a rotation of the coenzyme occurs as the – NH_2 of the substrate adds to the C = N bond during transimination. This accomplishes the essential shortening by – 0.15 nm of the distance between C-4′ and the N atom from a van der Waals contact distance to a covalent bond distance while the carboxylate groups of the substrate remain bound in their initial positions. That the coenzyme really does change its orientation was suggested by a dramatic change in the absorption spectrum of a crystal recorded with plane polarized light (linear dichroism) when 2-methylaspartate was soaked into a crystal. X-ray crystallography confirmed rotation of the ring.

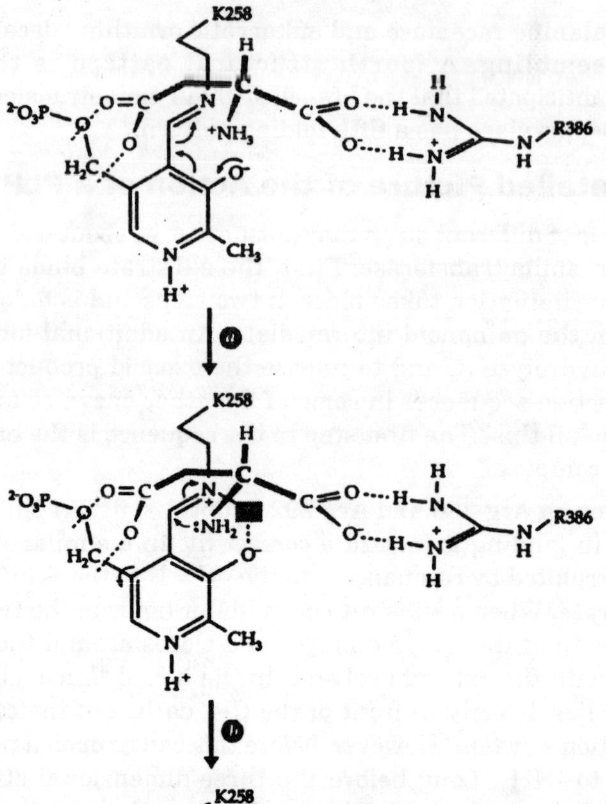

To complete the transimination sequence, which is shown only partially, another proton transfer is needed to move the positive charge on the substrate $-N^+H_2-$ to that of lysine 258, whose amino group is then eliminated. This requires additional tilting of the ring. The crystal structure of the external aldimine with α-methylaspartate has been determined as have those of ketimines with glutamate and aspartate, a carbinolamine, and quinonoid complexes of related enzymes.

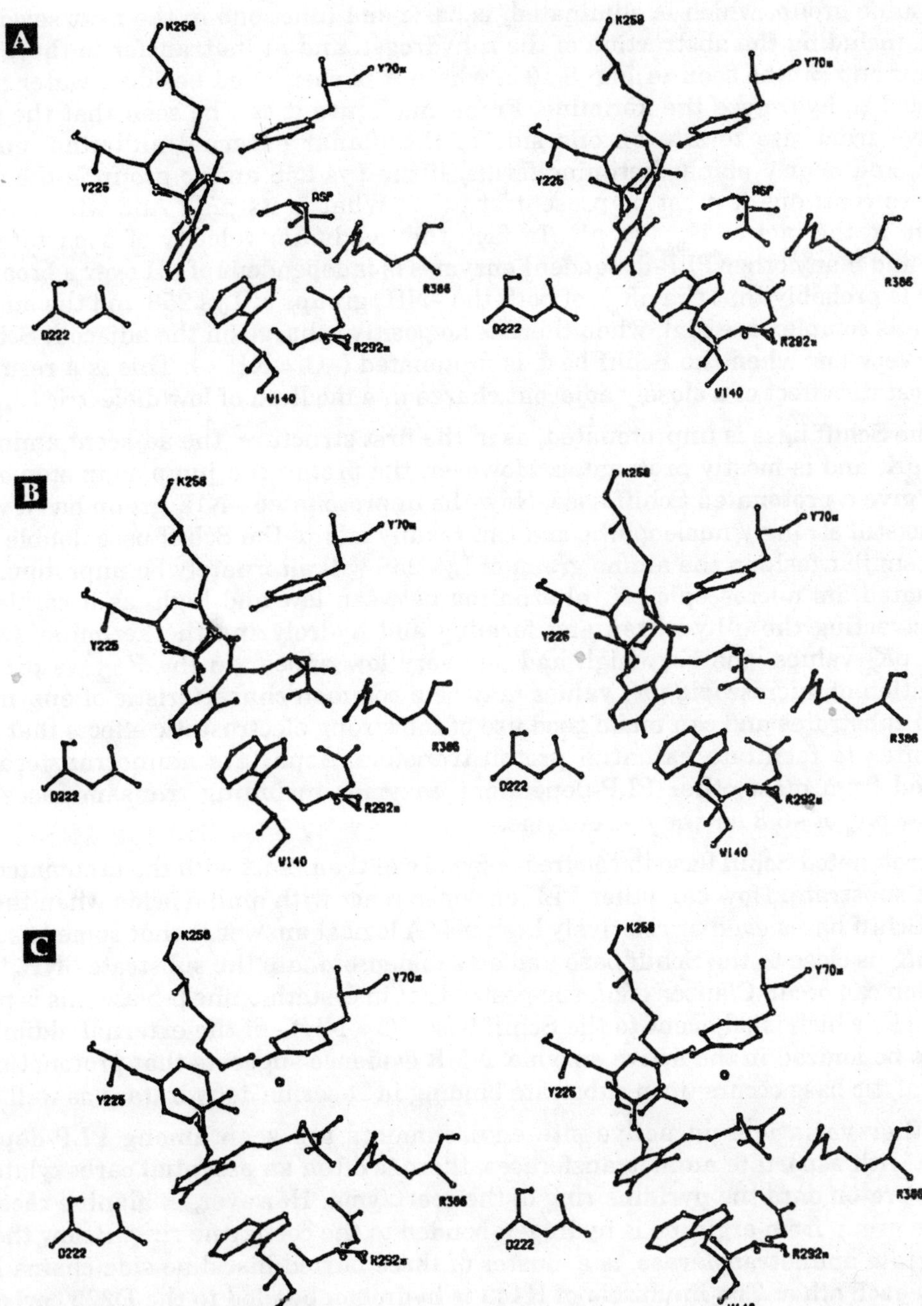

Fig. 8.10. Models of catalytic intermediates for aspartate aminotransferase in a half-transamination reaction from aspartate to oxalocetate. For clarity, only a selection of the active site groups are shown. (*A*) Michaelis complex of PLP enzyme with aspartate. (*B*) Geminal diamine. (*C*) Ketimine intermediate. The circle indicates a bound water molecule.

The ε-amino group, which is eliminated, is basic and functions in the next several steps of catalysis, including the abstraction of the α-hydrogen and in its transfer to the 4′-carbon. This amino group can be seen in Fig. 8.10 *o*, whoro it in poaitioncd bcside a water molecule that is needed to hydrolyze the ketimine. From this figure it can be seen that the group is able to move from site to site on one side of the planar external aldimine, quinonoid carbanionic, and nearly planar ketimine forms. If the Lys 258 amino group is the catalytic base for these reactions it must be present at pH 7. What is its pK_a? And why doesn't this pK_a show up in the plot of V_{max} vs pH. In fact, the maximum velocity of aspartate aminotransferase and many other PLP-dependent enzymes is independent of pH over a broad range. The answer is probably that the pK_a's of both the $-NH_2$ groups of Lys 258 and the amino acid in the Michalis complex are high when there is no positive charge on the adjacent Schiff base –C = N-but very low when the Schiff base is protonated ($-C = NH^+-$). This is a result of the large electrostatic effect of a closely adjacent charge in a medium of low dielectric constant.

When the Schiff base is unprotonated, as in the first structure, the adjacent amino group has a *high* pK, and is mostly protonated. However, the proton can jump as in step *a* of that equation to give a protonated Schiff base. Now the unprotonated $-NH_2$ group has a very *low* pK_a, but it is still strongly nucleophilic and can readily add to the Schiff base double bond in step *b*. In a similar fashion the amino group of Lys 258 will alternately be unprotonated and then protonated, its microscopic pK_a alternating between low and high, as it catalyzes the steps of abstracting the α-hydrogen and forming and hydrolyzing the ketimine. Only two microscopic pK_a values, one very high and one very low appear in the V_{max} vs pH profile. This alternation of microscopic pK_a values may be a common characteristic of enzymes that bind ionized substrates and can make good use of the strong electrostatic effects that arise in the active sites to facilitate essential proton transfers. Aspartate aminotransferases are distinguished from most other PLP-dependent enzymes including transaminases by the relatively low pK_a of ~6.3 for the *free* enzyme.

The unprotonated Schiff base in the free enzyme can then react with the protonated amino group of the substrate. How can other PLP enzymes react with amino acids when they have protonated Schiff bases even at relatively high pH? A logical answer is that some basic group with a low pK_a is close to the Schiff base and acts to deprotonate the substrate $-NH_3^+$ so that transimination can occur. Clausen *et al.* sug-gested that in cystathionine β-lyase this is probably tyrosine (Y111), which is adjacent to the Schiff base $-C = NH^+-$ of the external aldimine and is thought to be ionized in the active enzyme. NMR evidence suggests that protonation of an adjacent catalytic base occurs upon substrate binding in D-serine dehydratase as well.

Many other variations in active site environments are seen among PLP-dependent enzymes. As with aspartate aminotransferases, there is often an essential carboxylate group that holds a proton onto the pyridine ring of the coertzyme. However, in alanine racemase a guanidinium group from arginine is hydrogen bonded to the coenzyme ring. Below the active site of aspartate aminotransferase, is a cluster of three buried histidine side chains in close contact with each other. The imidazole of H143 is hydrogen bonded to the D222 carboxylate, the same carboxylate that forms an ion pair with the coenzyme. This system looks somewhat like the catalytic triad of the serine proteases in reverse.

As with the serine proteases, the proton-labeled H_b can be "seen" by NMR spectroscopy. So can the proton H_a on the PLP ring. These protons act as built-in sensors able to detect small changes in the electronic environment. For example, when the Schiff base proton dissociates around the pK_a of –6.2 the NMR resonance of H_a shifts upfield from 17.2 to 15.2 ppm as a result of donation of electons into the ring from the $-O^-$ of the coenzyme. This shift illustrates the reality of the strong electrostatic forces that operate across heterocyclic aromatic rings within active sites of proteins.

In alanine racemase a different histidine cluster is present beneath the active site and constitutes part of the "solvent" in which the catalyzed reaction takes place. At least in the case of aspartate aminotransferase, none of the histidines are absolutely essential for activity but the hydrogen-bonded network, which can be altered in mutant forms, may be important. The detailed description of a reaction sequence given here has to be altered for each specific enzyme. A vast amount of work, only a little of which is cited here, has been done on PLP enzymes. These studies involve calorimetry, kinetics, crystallography, optical spectroscopy, NMR, and genetic engineering and chemical modification.

F. PYRUVOYL GROUPS AND OTHER UNUSUAL ELECTROPHILIC CENTERS

A few enzymes that might be expected to have PLP at their active sites have instead a prosthetic group consisting of pyruvic acid bound by an amide linkage, a *pyruvoyl group*. These and several apparently related enzymes are the subject of this section.

Table 8.4. Some Pyruvoyl Enzymes

Decarboxylases	*Product*
Histidine (bacterial)	
S-Adenosylmethionine	
Aspartate α-decarboxylase	β-Alanine
Phosphatidylserine	Phosphatidylethanolamine
4′-Phosphopanthothenylcysteine	4′-Phosphopantetheine
Reductases (clostridial)	
Proline	
Glycine	

Adapted from van Peolje and Snell.

Decarboxylases

Mammalian *histidine decarboxylase* contains PLP but the enzyme from many bacteria contains a pyruvoyl group, as do a few other decarboxylases both of bacterial and eukaryotic origin. These enzymes are inhibited by carbonyl reagents and by borohydride. When ^{3}H-containing borohydride was used to reduce the histidine decarboxylase of *Lactobacillus*, ^{3}H was incorporated and was recovered in lactic acid following hydrolysis. This suggested the presence of a pyruvoyl group attached by an amide linkage and undergoing the chemical reactions. Reduction in the presence of the substrate histidine resulted in covalent binding of the histidine to the bound pyruvate. Thus, as with the PLP-containing decarboxylases, a Schiff base is formed with the substrate. Decarboxylation is presumably accomplished by using the electron-attracting properties of the carbonyl group of the amide:

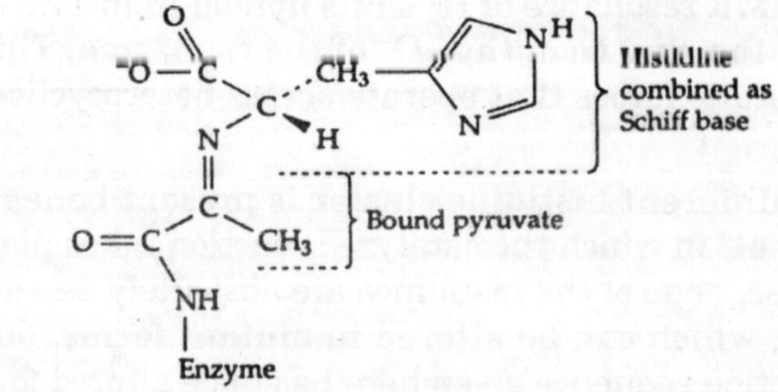

Histidine combined as Schiff base

Bound pyruvate

Enzyme

When ^{14}C-labeled serine was fed to organisms producing histidine decarboxylase, ^{14}C was incorporated into the bound pyruvoyl group. Thus, serine is a precursor of the bound pyruvate. The enzyme is manufactured in the cell as a longer 307-residue proenzyme which associates as hexamers (designated π_6). The active enzyme was found to be formed by cleavage of the π chains between Ser 81 and Ser 82 to form 226-residue oc chains and 81-residue p chains which associate as $(\alpha\beta)_6$. The a chains carry the N-terminal pyruvoyl group, which was formed from Ser 82. The activation occurs spontaneously by incubations of the proenzyme for 24-48 h at pH ~7 in the presence of divalent metal ions.

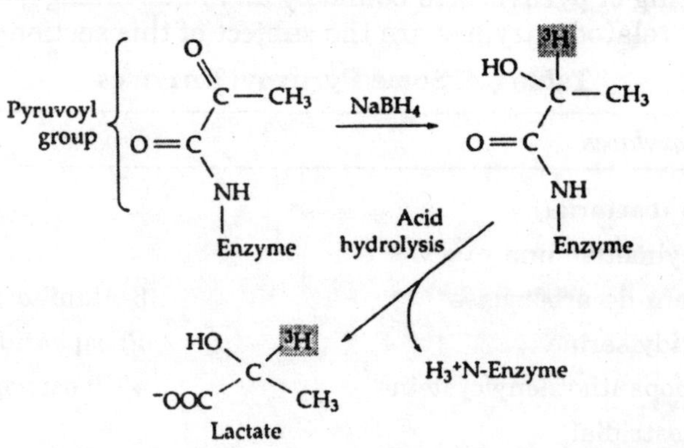

Pyruvoyl group — $NaBH_4$ — Enzyme — Acid hydrolysis — Enzyme — H_3^+N-Enzyme — Lactate

Substituted serine residues under mildly alkaline conditions readily undergo α,β elimination to form *dehydroalanine* residues. When prohistidine decarboxylase containing ^{18}O in its serine side chains was activated ^{18}O was found in the carboxylate group of Ser 81 of the β chains. It was shown that it had been transferred from the side chain of Ser 82. This suggested the formation of an intermediate oxygen ester of Ser 81 during formation of the pyruvoyl group. *S-Adenosylmethionine decarboxylase* is the first enzyme in the biosynthetic pathway to spermidine.

Whether isolated from bacteria, yeast, animals, or other eukaryotes, this enzyme always contains a bound pyruvoyl group. Both the mammalian enzyme and that from *E. coli* are $(\alpha\beta)_4$ tetramers formed in a manner similar to that of bacterial histidine decarboxylase. Other pyruvoyl decarboxylases are *phosphatidylserine decarboxylase*, an intrinsic membrane protein used to form phospha-tidylethanolamine, *aspartate α-decarboxylase*, which forms β alanine needed for biosynthesis of coenzyme A, and *4′-phosphopantethenoylcysteine decarboxylase*, the second of two decarboxylases required in the synthesis of coenzyme A. Because of the lack of a primary amino group, its mechanism must be somewhat different from that of other enzymes in this group.

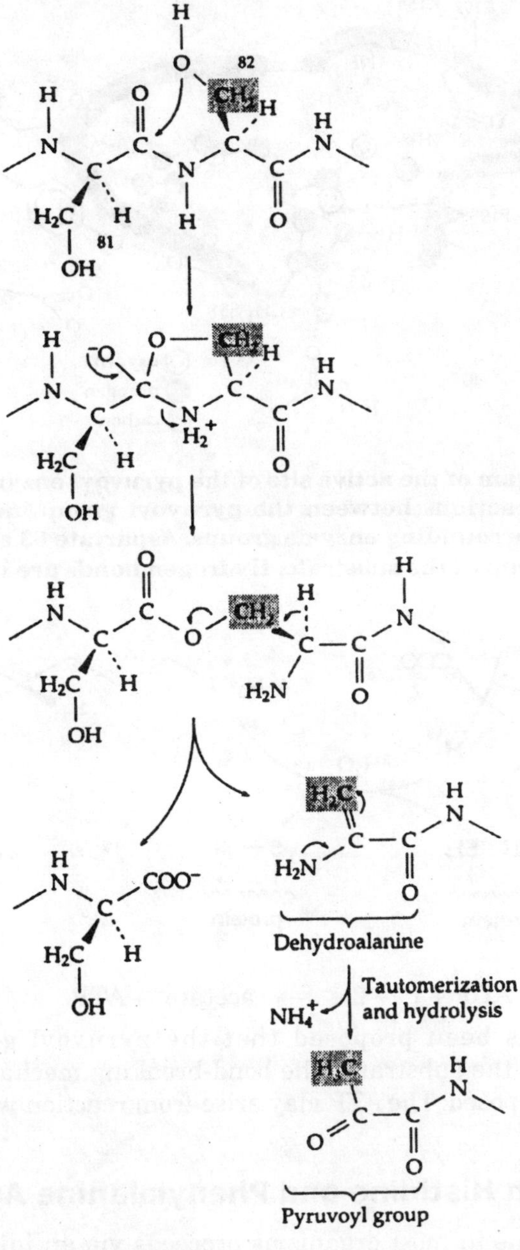

Proline and Glycine Reductases

An enzyme required in the anaerobic breakdown of proline by clostridia utilizes a dithiol-containing protein to reductively open the ring. This enzyme also contains an N-terminal pyruvoyl residue as does one subunit of a selenium-containing glycine reductase which utilizes a dithiol to convert glycine into acetate with coupled formation of ATP:

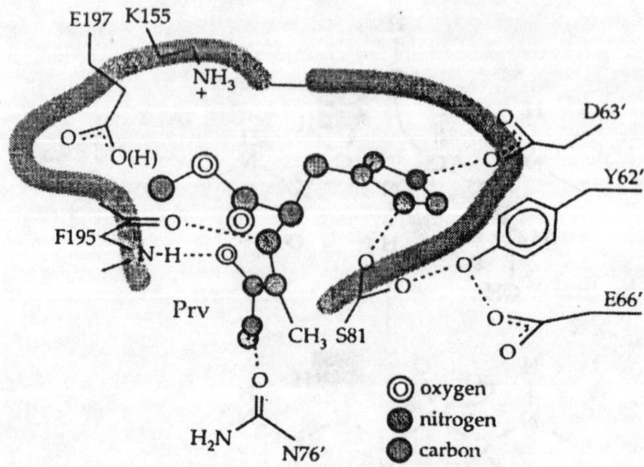

Fig. 8.11. Schematic diagram of the active site of the pyruvoyl enzyme histidine decarboxylase showing key polar interactions between the pyruvoyl group and groups of the inhibitor *O*-methylhistidine and surrounding enzyme groups. Aspartate 63 appears to form an ion pair with the imidazolium group of the substrate. Hydrogen bonds are indicated by dotted lines.

COO^- $N H_2^+$ H H_3^+N COO^- ^2H H

^2H_2O

SH SH — protein   →   S—S — protein

$$\text{Glycine} + \text{ADP} + P_i + 2\,e^- \rightarrow \text{acetate}^- + \text{ATP} \qquad ...(8.43)$$

In both cases it has been proposed that the pyruvoyl group forms a positively charged Scruff base with the substrate. The bond-breaking mechanisms are not obvious but some ideas have been proposed. The ATP may arise from reaction with an intermediate acetyl phosphate.

Dehydroalanine and Histidine and Phenylalanine Ammonia-Lyases

Catabolism of histidine in most organisms proceeds via an initial elimination of NH_3 to form *urocanic acid*. The absence of the enzyme *L-histidine ammonialyase* (histidase) causes the genetic disease histidinemia. A similar reaction is catalyzed by the important plant enzyme *L-phenylalanine ammonialyase*. It eliminates $-NH_3^+$ along with the *pro-S* hydrogen in the β position of phenylalanine to form *trans*-cinnamate. Tyrosine is converted to β-coumarate by the same enzyme. Cinnamate and coumarate are formed in higher plants and are converted into a vast array of derivatives.

The reactions unusual because the nucleophilic substituent eliminated is on the α-carbon atom rather than the β. There is nothing in the structures of the substrates that would permit an easy elimination of the α-amino group. Thus, it should not be surprising that both enzymes contain a special active center. When ^{2}H is introduced into the p position of phenylalanine, no isotope effect on the rate is observed. Rather, the rate-limiting step appears to be the release of ammonia from the coenzyme group. It appears that the enzyme must in some way make the amino group a much better leaving group than it would be otherwise.

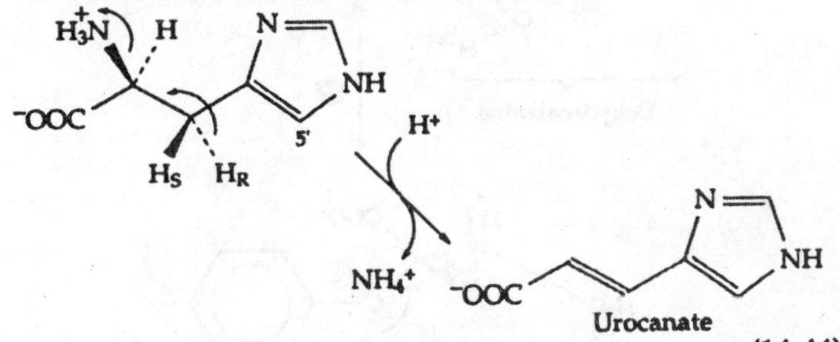

H_3N^+ H N NH ^-OOC 5' H_S H_R H^+ NH_4^+ ^-OOC N NH

Urocanate

(14-44)

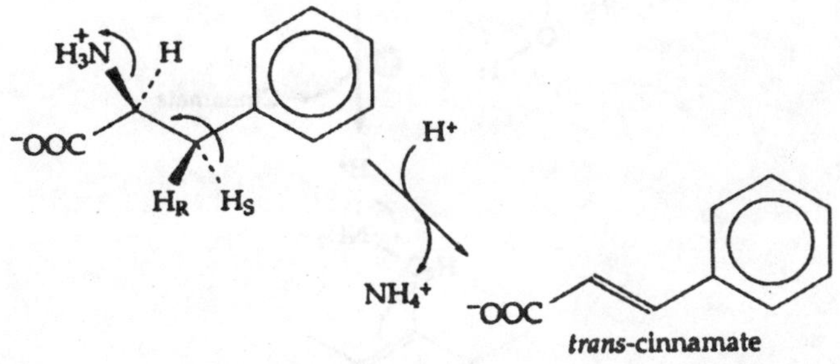

H_3N^+ H ^-OOC H_R H_S H^+ NH_4^+ ^-OOC

trans-cinnamate

Both enzymes are inhibited by sodium borohydride and also by nitromethane. After reduction with $NaB^{3}H_4$ and hydrolysis, ^{3}H-containing alanine was isolated. This suggested that they contain *dehydroalanine*, which could arise by dehydration of a specific serine residue. For phenylalanine ammonialyase from *Pseudomonas putida* this active site residue has been identified as S143. Replacement by cysteine in the S143C mutant also gave active enzyme while S143A was inactive. These results support the formation of dehydroalanine. One proposed mechanism of action of the enzymes involved addition of the substrate amino group to the C = C bond of the dehydroalanine followed by elimination.

Dissociation of the C - H bond in this step would be assisted by the electron-accepting properties of the adjacent aromatic ring. However, the proposed chemistry was not convincing. Noting that a 5-nitro substituent on the histidine greatly enhances the rate of reaction, Langer *et al.* proposed that the histidine reacts with the electrophilic carbon in the pyruvoyl center. However, this chemistry, too, was unprecedented in enzymology. Determination of the three-dimensional structure of histidine ammonialyase led to the discovery of a new prosthetic group and a solution to the problem. Two dehydration steps, convert an Ala-Gly-Ser sequence within the protein into 4-*methylidine-imidazole*-5-*one* (MIO), a modified dehydroalanine with enhanced electron-accepting properties. The proposed mechanism of action is portrayed.

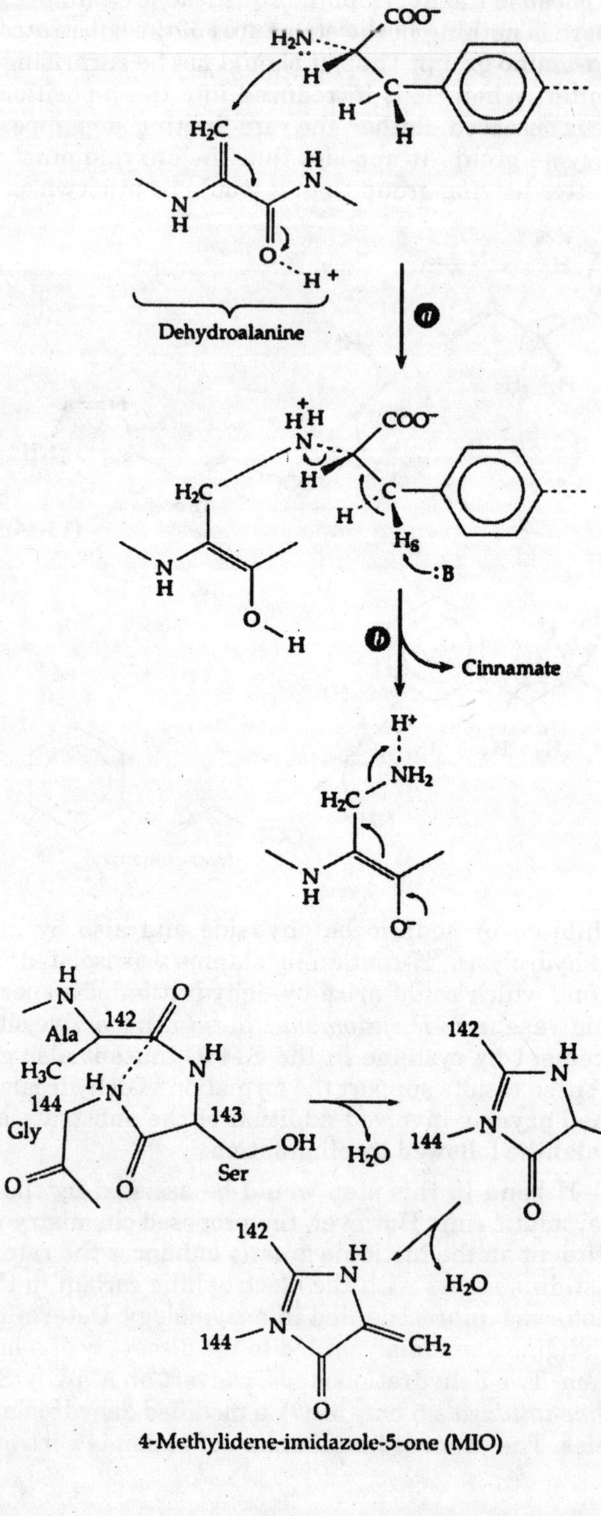

Dehydroalanine

a

b

Cinnamate

4-Methylidene-imidazole-5-one (MIO)

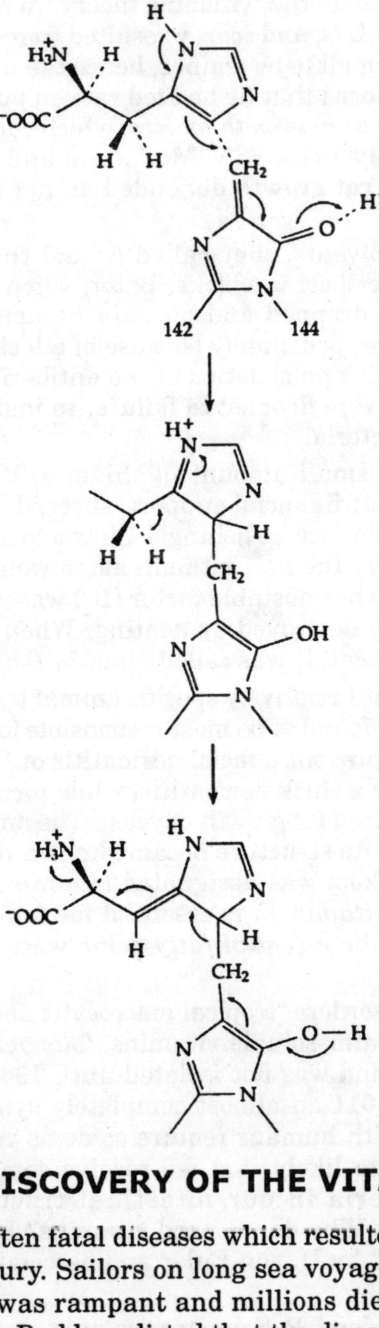

A. DISCOVERY OF THE VITAMINS

Several mysterious and often fatal diseases which resulted from vitamin deficiencies were prevalent until the past century. Sailors on long sea voyages were often the victims. In the Orient, the disease *beriberi* was rampant and millions died of its strange paralysis called "polyneuritis." In 1840, George Budd predicted that the disease was caused by the lack of some organic compound that would be discovered in "a not too distant future." In 1893, C. Eijkman, a Dutch physician working in Indonesia, observed paralysis in chicks fed white rice consumed by the local populace. He found that the paralysis could be relieved promptly by

feeding an extract of rice polishings. This was one of the pieces of evidence that led the Polish biochemist Casimir Funk to formulate the "vitamine theory" in about 1912. Funk suggested that the diseases *beriberi*, *pellagra*, *rickets*, and *scurvy* resulted from lack in the diet of four different vital nutrients. He imagined them all to be amines, hence the name *vitamine*. In the same year in England, F. G. Hopkins announced that he had fed rats on purified diets and discovered that amazingly small amounts of *accessory growth factors*, which could be obtained from milk, were necessary for normal growth. By 1915, E. V. McCollum and M. Davis at the University of Wisconsin had recognized that rat growth depended on not one but at least two accessory factors.

The first, soluble in fatty solvents, they called A, and the other, soluble in water, they designated B. Factor B cured beriberi in chicks. Later, when it was shown that vitamine A was not an amine, the *"e"* was dropped and *vitamin* became a general term. Progress in isolation of the vitamins was slow, principally because of a lack of interest. According to R. R. Williams, when he started his work on isolation of the antiberiberi factor in 1910 most people were convinced that his efforts were doomed to failure, so ingrained was Pasteur's idea that diseases were caused only by bacteria.

In 1926, Jansen isolated a small amount of thiamin, but it was not until 1933 that Williams, working almost without financial support, succeeded in preparing a large amount of the crystalline compound from rice polishings. Characterization and synthesis followed rapidly. It was soon apparent that the new vitamin alone would not satisfy the dietary need of rats for the B factor. A second thermostable factor (B_2) was required in addition to thiamin (B_1), which was labile and easily destroyed by heating. When it became clear that factor B_2 contained more than one component, it was called *vitamin B complex*.

There was some confusion until relatively specific animal tests for each one of the members had been devised. *Riboflavin* was found to be most responsible for the stimulation of rat growth, while *vitamin* B_6 was needed to prevent a facial dermatitis or "rat pellagra." *Pantothenic acid* was especially effective in curing a chick dermatitis, while *nicotinamide* was required to cure human pellagra. *Biotin* was required for growth of yeast. The antiscurvy (antiscorbutic) activity was called *vitamin C*, and when its structure became known it was called *ascorbic acid*. The fat-soluble factor preventing rickets was designated *vitamin D*. By 1922, it was recognized that another fat-soluble factor, *vitamin E*, is essential for full-term pregnancy in the rat. In the early 1930s *vitamin K and* the *essential fatty acids* were added to the list of fatsoluble vitamins.

Study of the human blood disorders "tropical macrocytic anemia" and "pernicious anemia" led to recognition of two more water-soluble vitamins, *folic acid and vitamin* B_{12}. The latter is required in minute amounts and was not isolated until 1948. Have all the vitamins been discovered? Rats can be reared 011 an almost completely synthetic diet. However, there is the possibility that for good health humans require some as yet undiscovered compounds in our diet. Furthermore, it is quite likely that we receive some essential nutrients that we cannot synthesize from bacteria in our intestinal tracts. An example may be the pyrroloquinoline quinone (PQQ). Why do we need vitamins? Early clues came in 1935 when nicotinamide was found in NAD^+ by H. von Euler and associates and in $NADP^+$ by Warburg and Christian.

Two years later, K. Lohman and P. Schuster isolated pure *cocarboxylase*, a dialyzable material required for decarboxylation of pyruvate by an enzyme from yeast. It was shown to be thiamin diphosphate. Most of the water-soluble vitamins are converted into coenzymes or are covalently bound into active sites of enzymes. Some lipid-soluble vitamins have similar functions but others, such as vitamin D and some metabolites of vitamin A, act more like hormones, binding to receptors that control gene expression or other aspects of metabolism.

B. THE BIOTIN-BINDING PROTEINS AVIDIN AND STREPTAVIDIN

A biochemical curiosity is the presence in egg white of the glycoprotein avidin. Each 68-kDa subunit of this tetrameric protein binds one molecule of biotin tenaciously with $K_f \sim 10\ M^{-1}$. Nature's purpose in placing this unusual protein in egg white is uncertain. Perhaps it is a storage form of biotin, but it is more likely an antibiotic that depletes the environment of biotin, A closely similar protein *streptavidin* is secreted into the culture medium by *Streptomyces avidinii.* Its sequence is homologous to that of avidin. It has a similar binding constant for biotin and the two proteins have similar three-dimensional structures. Biotin binds at one end of a p barrel formed from antiparallel strands and is held by multiple hydrogen bonds and a conformational alteration that allows a peptide loop to close over the bound vitamin. Historically, avidin was important to the discover of biotin.

The bonding between avidin and biotin is so tight that inclusion of raw egg white in the diet of animals is sufficient to cause a severe biotin deficiency. Avidin has also been an important tool to enzymologists interested in biotin-containing enzymes. Avidin invariably inhibits these enzymes and inhibition by avidin is diagnostic of a biotin-containing protein. Recently, avidin and streptavidin have found widespread application in affinity chromatography, in immunoassays, and in the staining of cells and tissues. These uses are all based on the ability to attach biotin covalently to side chain groups of proteins, polysaccharides, and other substances. The carboxyl group of the biotin "arm," which lies at the surface of the complex with avidin or streptovidin, can be converted to any of a series of reactive derivatives.

For example, β-nitrophenyl or N-hydroxysuccinimide esters of biotin can be used to attach biotin to amino groups of proteins to form biotinylated proteins. Other reactive derivatives can be used to attach biotin to phenolic, thiol, or carbonyl groups.

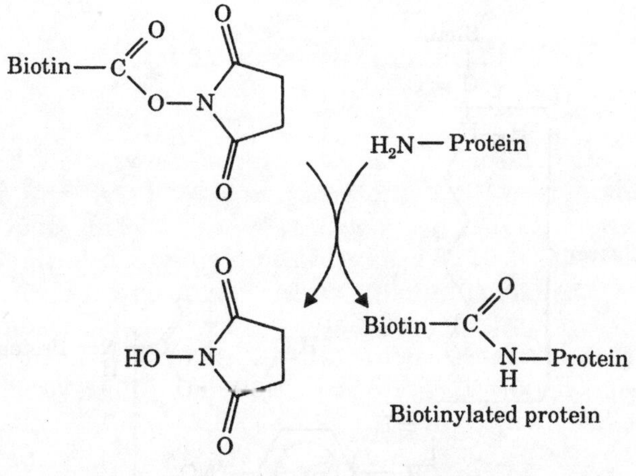

The affinity of avidin or streptavidin for the resulting biotinylated materials is still very high. This fact has permitted the application of this "biotinavidin technology" to numerous aspects of research and diagnostic medicine. For example, a specific antibody (IgG) can be utilized for immunoassay of a hormone or other ligand. A second antibody, specific for the

IgG-ligand complex, can be produced in a biotinylated form. A steptavidin complex of a biotinylated enzyme such as alkaline phosphatase, β-galactosidase, or horseradish peroxidase, for which a sensitive colorimetric assay is available, is allowed to react with the biotin•anti-IgG•IgG•ligand complex. The streptavidin now releases enzyme in proportion to the amount of ligand originally present. Since the biotinylated anti IgG and streptavidin•biotinylated enzyme can be stored as a stable mixture, the assay is simple and fast.

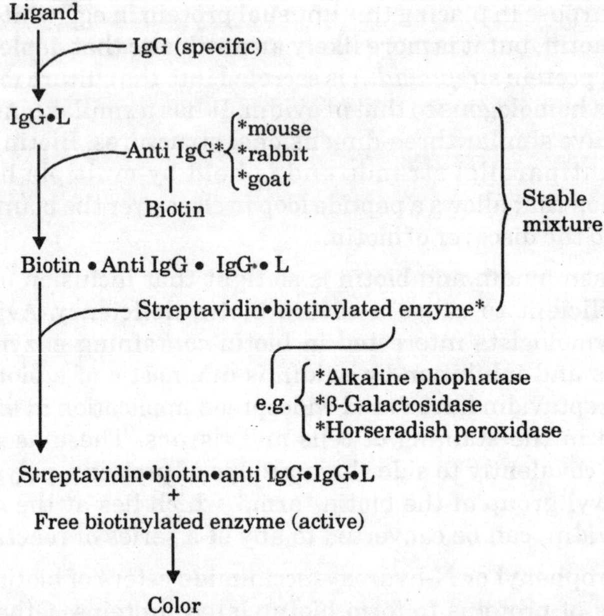

Avidin technology can also be applied to the isolation of proteins and other materials from cells. Because the irreversibility of the binding of biotin may be a problem, photocleavable

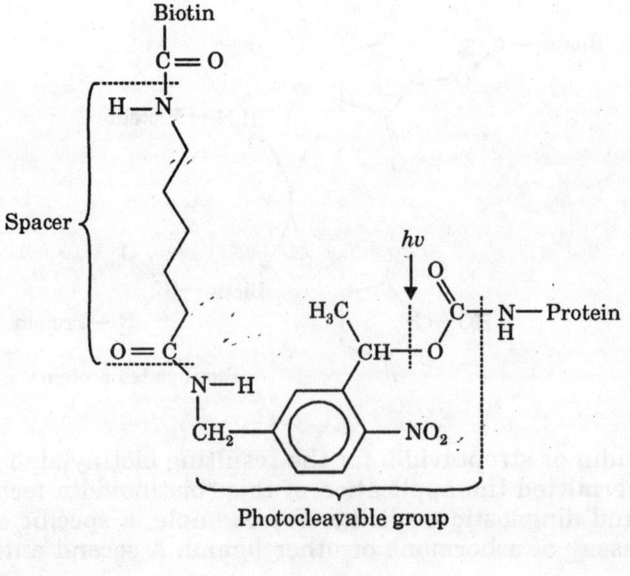

Biotin

C=O

H—N

Spacer

hν

O

H_3C

C—N—Protein

H

O=C

CH—O

N—H

CH_2

NO_2

Photocleavable group

biotin derivatives have been developed. In the following structure, the biotin derivative has been joined to a protein (as in the first equation in this box) and is ready for separation, perhaps on a column containing immobilized avidin or streptavidin. After separation the biotin together with the linker and photocleavable group are cut off by a short irradiation with ultra violet light, leaving the protein in a free form.

C. THE VITAMIN B_6 FAMILY: PYRIDOXINE, PYRIDOXAL AND PYRIDOXAMINE

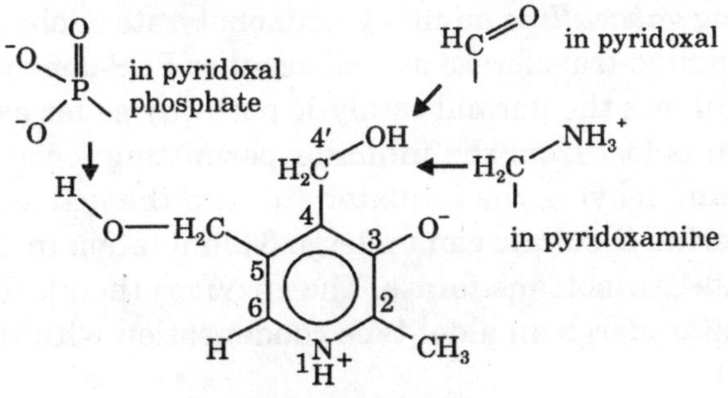

in pyridoxal phosphate

in pyridoxal

in pyridoxamine

Pyridoxine (pyridoxol)
Vitamin B_6 alcohol

Pyridoxine, the usual commercial form of vitamin B_6, was isolated and synthesized in 1938. However, studies of bacterial nutrition soon indicated the existence of other naturally occurring forms of the new vitamin which were more active than pyridoxine in promoting growth of certain lactic acid bacteria. The amine *pyridoxamine* and the aldehyde *pyridoxal* were identified by Snell, who found that pyridoxal could be formed from pyridoxine by mild oxidation and that pyridoxamine could be formed from pyridoxal by heating in a solution with glutamic add via a transamlnation reaction. These simple experiments also suggested the correct structures of the new forms of vitamin B_6. Animal tissues contain largely pyridoxal, pyridoxamine, and their phosphate esters.

The lability of the aldehyde explains the ease of destruction of the vitamin by excessive heat or by light. On the other hand, plant tissues contain mostly pyridoxine, which is more stable. Kinases use ATP to form the phosphate esters, which are interconvertible within cells. Pyridoxine 5'-phosphate can be oxidized to PLP and the latter may undergo transamination to PMP. The acid-base chemistry and tautomerism of pyridoxine were discussed. Many poisonous substances as well as useful drugs react with PLP-requiring enzymes. Thus, much of the toxic effect of the "carbonyl reagents" hydroxylamine, hydrazine, and semicarbazide stems from their formation of stable derivatives analogous to Schiff bases with PLP.

$$H-O-NH_2,\ R-O-NH_2$$

Hydroxylamines

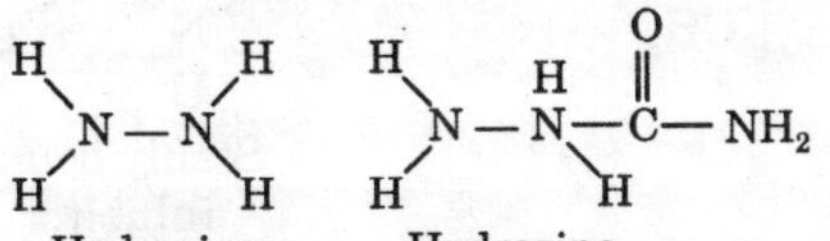

$$H_2N-NH_2 \qquad H_2N-NH-C(=O)-NH_2$$

Hydrazine Hydrazine

Isonicotinyl hydrazide (INH), one of the most effective drugs against tuberculosis, is inhibitory to pyridoxal kinase, the enzyme that converts pyridoxal to PLP. Apparently, the drug reacts with pyridoxal to form a hydrazone which blocks the enzyme. Pyridoxal kinase is not the primary target of INH in mycobacteria. However, patients on long-term isonicotinyl hydrazide therapy sometimes suffer symptoms of vitamin B_6 deficiency. PLP-dependent enzymes are inhibited by a great variety of enzyme-activated inhibitors that react by several distinctly different chemical mechanisms. Here are a few.

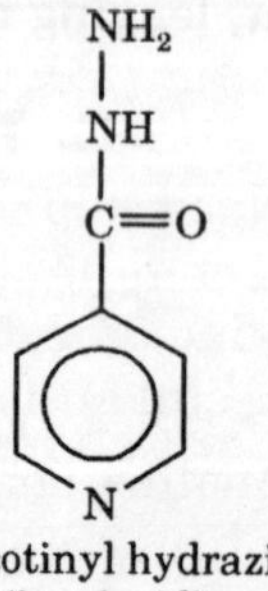

NH_2—NH—C=O (pyridine ring, N)

Isonicotinyl hydrazide (isoniazid)

The naturally occurring *gabaculline* mimics y-aminobutyrate (Gaba) and inhibits γ-aminobutyrate amino-transferase as well as other PLP-dependent enzymes. The inhi-bitor follows the normal catalytic pathway as far as the ketimine. There, a proton is lost from the inhibitor permitting formation of a stable benzene ring and leaving the inhibitor stuck in the active site: Beta-chloroalanine and serine O-sulfate can undergo β elimination in active sites of glutamate decarboxylase or aspartate arninotrans-ferase. The enzymes then form free aminoacrylate, a reactive molecule that can undergo an aldol-type condensation with the external aldimine to give the following product.

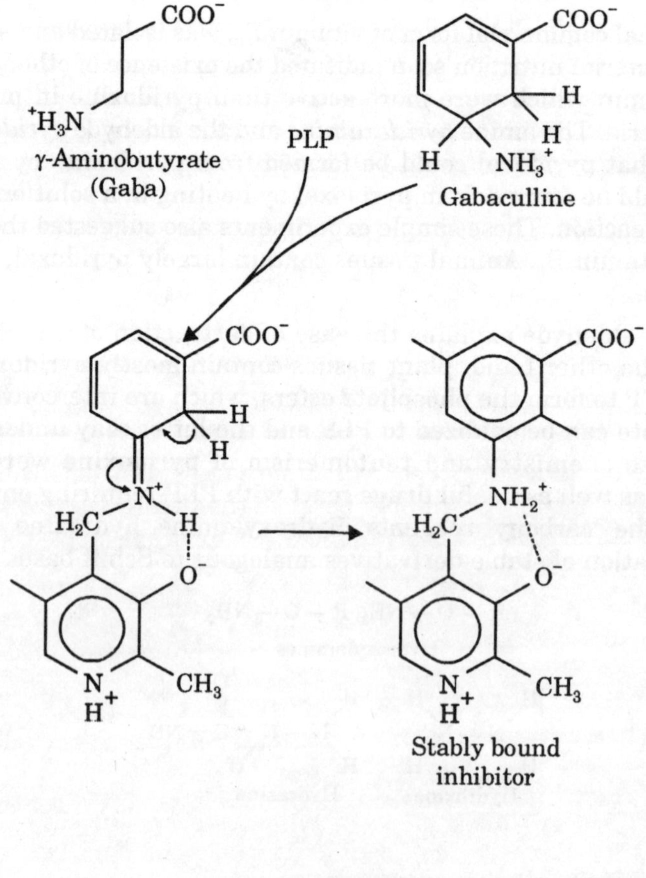

COO^- ^+H_3N

γ-Aminobutyrate (Gaba)

PLP

COO^- H H H NH_3^+

Gabaculline

COO^- H H N^+ H_2C H O^- N H CH_3 $\longrightarrow$ COO^- NH_2^+ H_2C O^- N H CH_3

Stably bound inhibitor

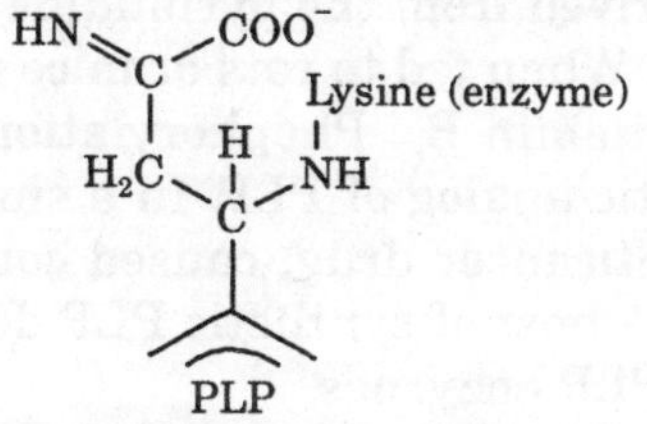

HN=C(COO⁻)–CH₂–C(H)(NH–Lysine (enzyme))–C=PLP

PLP

Nucleophilic groups from enzymes can add to double bonds, *e.g.*, in an aminoacrylate Schiff base, or to multiple bonds present in the inhibitor. An example is γ-vinyl γ-aminobutyrate (4-amino-5-hexenoic acid), another inhibitor of brain γ-amino-butyrate aminotransferase which is a useful anti-convulsant drug.

Another enzyme-activated inhibitor is the streptomyces antibiotic *D-cycloserine* (oxamycin), an antitubercular drug that resembles D-alanine in structure. A potent inhibitor of alanine racemase, it also inhibits the non-PLP, ATP-dependent, *D-alanyl-D-alanine synthetase* which is needed in the bio-synthesis of the peptidoglycan of bacterial cell walls.

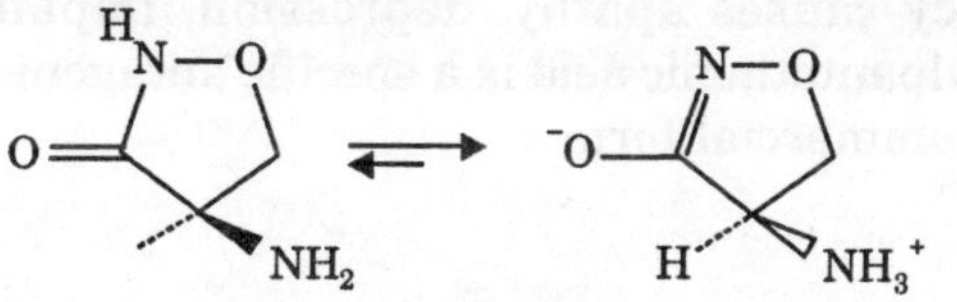

D-Cycloserine

L-Cycloserine inhibits many PLP enzymes and is toxic to humans. This observation led Khomutov *et al.* to synthesize the following more specific "cyclo-glutamates," structural analogs of glutamic acid with fixed conformations. Nature apparently anticipated the synthetic chemist, because it has been reported that the mushroom *Tricholoma muscariun* contains one of the same compounds. It is said to impart two interesting properties to the mushroom: a delicious flavor and a lethal action on flies that alight on the mushroom's surface!

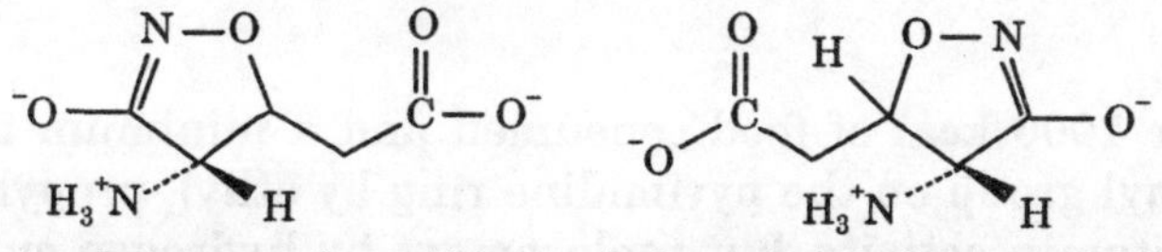

A cycloglutamate that inhibits aspartate aminotransferase

An isomeric cycloglutamate trichlomic acid found in certain mushrooms

The substituted cysteine derivative L-penicillamine causes convulsions and low glutamate decarboxylase levels in the brain, presumably because the Schiff base formed with PLP can then undergo cyclization, the SH group adding to the C = N to form a stable thiazolidine ring.

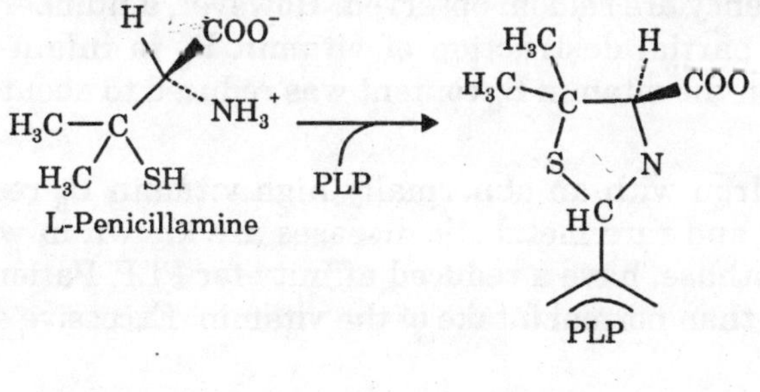

L-Penicillamine

PLP

Toxopyrimidine the alcohol derived from the pyrimidine portion of the thiamin molecule, is a structural analog of pyridoxal. When fed to rats or mice it induces running fits which can be stopped by administration of vitamin B_6. Phophorylation of toxopyrimidine by pyridoxal kinase may produce an antagonistic analog of PLP. In a similar fashion, 4-deoxypyridoxine, which was tested as a possible anticancer drug, caused convulsions and other symptons of vitamin B_6 deficiency in humans. A host of synthetic PLP derivatives have been made, some of which are effective in blocking PLP enzyrnes.

D. DIETARy REQUIREMENTS FOR B VITAMINS

It is difficult to determine the amounts or vitamins needed for good health and they may differ considerably from one individual to another. The quantities listed below are probably adequate for most young adults but must be increased during pregnancy and lactation and after very strenuous exercise.

Pantothenic Acid

10-15 mg/day. Deficiency causes apathy, depression, impaired adrenal function, and muscular weakness, ω-Methylpantothenic acid is a specific antagonist. The calcium salt, calcium pantothenate, is the usual commercial form.

Biotin

0.15 – 0.3 mg/day. The discovery that biotin deficiency in young chickens can lead to sudden death resulted in a recommendation to supplement infant formulations with biotin. *Desthiobiotin*, in which the sulfur has been removed and replaced by two hydrogen atoms, can replace biotin in some organisms and appears to lie on one pathway of biosynthesis. *Oxybiotin*, in which the sulfur has been replaced by oxygen, is active for many organisms and partially active for others. No evidence for conversion to biotin itself has been reported, and oxybiotin may function satisfactorily in at least some enzymes.

Thiamin

0.23 mg or more per 1000 kcal of food consumed and a minimum total of 0.8 mg/day. Replacement of the methyl group on the pyrimidine ring by ethyl, propyl, or isopropyl gives compounds with some vitamin activity, but replacement by hydrogen cuts activity to 5% of the original. The butyl analog is a competitive inhibitor.

Vitamin B_6

1.5 – 2 mg/day; 0.4 mg/day for infants. Vitamin B_6 is widely distributed in foods, and symptoms of severe deficiency are seldom observed. However, a number of cases of convulsions have been attributed to partial destruction of vitamin B_6 in infant liquid milk formulas. Conculsions occurred when the vitamin B_6 content was reduced to about one-half that normally present in human milk.

Several cases of children with an abnormally high vitamin B_6 requirement (2 – 10 mg/day) have been reported, and rare metabolic diseases are known in which specific enzymes, such as cystathionine synthase, have a reduced affinity for PLP. Patients with these diseases also benefit from a higher than normal intake of the vitamin. Excessive excretion of the vitamin may also occur, an example being provided by a strain of laboratory mice that require twice the normal amount of vitamin B_6 and which die in convulsions after a brief period of vitamin B_6 depletion. Dietary supplementation with large amounts of vitamin B_6 for treatment of medical conditions such as carpal tunnel syndrome has been controversial/Amounts of pyridoxine over 50 mg per day may damage peripheral nerves, probably because of the chemical reactivity of pyridoxal and PLP.

Nicotinamide : *About 7.5 mg/day. Tryptophan can substitute to some extent.*

Riboflavin : *About 1.5 mg/day.*

Folic acid : *About 0.2-0.4 mg/day.*

Vitamin B_{12} : *About 2 ng/day.*

Vitamin C (ascorbic acid) : 50-200 mg/day.

CHAPTER 9 Population Nutrition, Health Promotion and Government Policy

It is now generally accepted that lifestyle—diet, tobacco use, exercise—has a major impact on health, especially Western diseases. However, there is a world of difference between awareness of these facts and their translation into preventive action. This chapter not only focuses on nutrition in relation to health promotion, it also examines other areas, especially smoking and exercise. This is necessary because most health promotion campaigns take a broad lifestyle approach and simultaneously tackle nutrition, exercise, and smoking. Trends toward a healthier lifestyle over the last 20 yr have been inconsistent. In the United States, deaths from coronary heart disease (CHD) have fallen by half since their peak in the late 1960s. Contrariwise, from 1976-1980 and 1988-1994 about 7% of American adults became newly obese.

A rising prevalence of obesity has also been reported from other Western countries. In the United States, per capita intake of fruit and vegetables increased by about 20% from 1970 to 1994. Nevertheless, only about two Americans in five consume the recommended five or more servings per day of these vital foods. Moreover, half of Americans eat no fruit on any given day. This poor rate of progress in the area of diet should be seen as part of a more general problem that large sections of the population give a low priority to a healthy lifestyle. For instance, the proportion of middle-aged adults in England engaged in at least moderate exercise, such as a brisk walk for at least 30 min 5 or more days each week, is no more than one half of men and one quarter of women.

In the United States about one third of adults achieve this level of exercise, and another one third report no leisure time physical activity. There has been an impressive fall by about half in smoking rates in men in many Western countries over the last 30 yr. The percentage of American men who smoke dropped from 44 to 28% between 1970 and 1993. However, recent trends indicate that smoking rates have been creeping up in recent years, especially in young people. For instance, 37% of American students in grade 10 are smokers.

HEALTH PROMOTION CAMPAIGNS

During the 1970s, the intimate connection between lifestyle and health became increasingly apparent. As a result, many people assumed that the next step was to disseminate this information to the public and exhort lifestyle changes, action deemed sufficient to bring about the necessary changes. However, a review of 24 evaluations of the effectiveness of using the mass media across a range of health topics found little evidence of behavior change as a result of education alone. Here, we look at various types of health promotion campaigns, most of them focused on risk factors for cardiovascular disease.

Campaigns in Communities

A number of community interventions have used the mass media combined with various other methods to reach the target population. Three major projects were carried out in the United States during the 1980s. Their aims were to lower elevated levels of blood cholesterol, blood pressure, and weight, to cut smoking rates, and to persuade more people to exercise. Each program lasted 5 – 8 yr and succeeded in implementing its intervention on a broad scale, involving large numbers of programs and participants. In the Stanford Five-City Project, conducted by Farquhar and colleagues in California, two intervention cities received health education via TV, radio, newspapers, other mass-distributed print media, direct education, and schools. On average each adult was exposed to 26 h of education, achieved at the remarkably low per capita cost of $4 per year (*i.e.,* about 800 times less than total health-care costs).

A similar project was the Minnesota Heart Health Program, which included three intervention cities and three control cities in the upper Midwest. A third project was the Pawtucket Heart Health Program in which the population of Pawtucket, RI, received intensive education at the grass roots level: schools, local government, community organizations, supermarkets, and so forth, but without involving the media. A recent analysis combined the results of the three studies so as to increase the sample size to 12 cities.

Improvements in blood pressure, blood cholesterol, body-mass index (BMI), and smoking were of very low magnitude and were not statistically significant; the estimated risk of CHD mortality was unchanged. These results are mirrored by two additional community projects: the Heart To Heait Project in Florence, South Carolina and the Bootheel Heart Health Project in Missouri also showed little success. One factor contributing to the lack of effect may have been that the projects took place at a time when American lifestyles were becoming generally more healthy and CHD rates were falling. This suggests that when a population starts receiving health education, even if little more than reports in the mass media and government policy pronouncements, large numbers of people will be induced to make changes in lifestyle.

A health promotion campaign superimposed on such secular trends may have little *additional* benefit. However, we cannot discount the possibility that different types of intervention might be successful where those just described were not. Fortunately, we have some examples of reasonably successful community projects for heart disease prevention. One of the earliest and most informative such projects was conducted in North Karelia, a region of eastern Finland with an exceptionally high rate of the disease. The intervention began in 1972 before much health information had reached the population. Nutrition education was an important component of the intervention.

Over the next few years, CHD rates in North Karelia fell sharply. Later, an intensive educational campaign spread to the rest of the country leading to a national drop in CHD rates. Two other European studies also achieved some success. Positive results were seen in the German Cardiovascular Prevention Study, which took place from about 1985-1992, when there was no particular favourable trend in risk factors for the population as a whole. It was carried out in six regions of the former West Germany using a wide-ranging approach similar to that used in the American community studies.

The intervention caused a small decrease in blood pressure and serum cholesterol (about 2%) and a 7% fall in smoking, but had no effect on weight. Action Heart was a community-based health promotion campaign conducted in Rotherham, England. After 4 yr, 7% fewer people smoked and 9% more drank low-fat milk, but there was no change in exercise habits, obesity, or consumption of wholemeal bread. Recent American and Australian community campaigns are of particular interest because each was narrowly focused on changing only one aspect of lifestyle and used paid advertising as a major intervention strategy. The 1% Or Less campaign aimed to persuade the population of two cities in West Virginia to switch from whole milk to low-fat milk (1 % or less). Advertising in the media was a major component of the intervention (at a cost of slightly less than a dollar per person) together with super-market campaigns (taste tests and display signs), education in schools, as well as other community education activities. Low-fat milk sales, as a proportion of total milk sales, increased from 18 to 41% within just a few weeks.

The intervention campaign was repeated in another city in West Virginia; this time only paid adver-tising was used. Low-fat milk sales increased from 29 to 46% of total milk sales. An Australian intervention campaign also used paid advertising as a major component. The campaign ran in the State of Victoria from 1992-1995 and aimed to increase consumption of fruit and vegetables. Significant increases in consumption of these foods was reported (fruit by 11% and vegetables by 17%). Taken together, the community intervention studies indicate that small changes in cardiovascular risk factors can be made by the methods used to date. The evidence is suggestive that interventions focusing on a small number of changes and using paid advertising can be more successful.

Worksite Health Promotion

As an alternative to health promotion using a community intervention approach, other interventions have focused on the worksite. A pioneering project of this type, which started in 1976, was carried out in Europe by the World Health Organization. The project was conducted over 6 yr in 80 factories in Belgium, Italy, Poland, and the UK with the aim of preventing CHD. The trial achieved modest risk factor reductions (1.2% for plasma cholesterol, 9% for smoking, 2% for systolic blood pressure, and 0.4% for weight); these were associated with a 10%' reduction in CHD. At around the same time, "Live for Life" was carried out by the Johnson & Johnson company in the United States. This comprehensive intervention was started in 1979 and lasted 2 yr. Employees exposed to the program showed significant improvements in smoking behavior, weight, aerobic capacity, incidence of hypertension, days of sickness, and health-care expenses. Another worksite project took place in New England. Employees were encouraged to increase their intake of fiber and to reduce their fat intake.

Compared with the control sites, the program had no effect on fiber intake, but fat intake fell by about 3%. A few years later, the research team reported that they succeeded in increasing employees' intake of fruit and vegetables by 19% (half a serving per day) using an approach that targeted employees and their families. A similar project in Minnesota offered employees weight control and smoking cessation programs. The program had no effect on weight, but the prevalence of smoking was reduced by 2% more than occured in the control worksites.

Health Promotion in the Physician's Office

In 1994, two British studies reported the effects of intervention carried out by nurses in the offices of family physicians. The aim was to improve cardiovascular risk factors. Each study was a randomized trial aimed at cardiovascular screening and lifestyle intervention. Both studies achieved only modest changes despite intensive intervention. The OXCHECK study reported no significant effect on smoking or excessive alcohol intake but did observe small significant improvements in exercise participation, weight, dietary intake of saturated fat, and serum cholesterol.

The Family Heart Study achieved a 12% lowering of risk of CHD (based on a risk factor score). These results are consistent with the results of a meta-analysis of 17 randomized controlled intervention trials where dietary advice was given to healthy adults. Changes over 9 – 18 mo included small decreases in serum cholesterol and blood pressure, leading to a 14% fall in the estimated risk of CHD. A variation of the foregoing trials is the targeting of patients at high risk of CHD, probably the most cost-effective form of intervention. A study from Sweden exemplifies this approach. Subjects at relatively high risk of cardiovascular disease received either simple advice from their physician or intensive advice (five 90-min sessions plus an all-day session).

The intensive advice had a modest impact; it reduced the risk of CHD by approx 6%. The major deficiency of the high-risk approach, as Rose has pointed out, is that it only affects a minority of future cases: the 15% of men at "high risk" of CHD account for only 32% of future cases. Therefore, to achieve a major effect on CHD, it is necessary to target the entire population. This logic also applies to other diseases related to diet and lifestyle practices such as stroke and cancer.

Health Promotion and the Individual

What the foregoing projects teach us is that appealing to individuals to change their lifestyles will be effective in some instances but not in others and can therefore be frustratingly difficult. Although some projects have achieved a 'moderate degree of success, typically progress has amounted to no more than a few percentage points. This might be expected to reduce the risk of CHD by about 5 – 15%. Although this is certainly beneficial, it will not affect the majority of people at risk, however. Thus, exhortations to the individual, whether via the media, in the community, at the worksite, or in the physician's office, are most unlikely to turn the tide of Western diseases. Myriad factors influence people's lifestyle behavior besides concerns about how to protect health. Social factors, such as housing, employment, and income also shape people's attitudes, as does education.

Advertising directly affects what people want and prices determine whether they can afford it. We are also creatures of habit and custom; resistance may therefore be expected when lifestyle modification demands changes in longstanding behavior and goes against fashion or peer pressure. We must also bear in mind that individuals have little control over many aspects of their physical environment, such as pollution and food contamination. It is probably naive, therefore, to expect dramatic results from interventions that merely exhort the individual to lead a healthier lifestyle. Indeed, this has sometimes been characterized as "victim blaming." This in no way dismisses interventions aimed at encouraging people to improve their lifestyle. Quite the contrary.

Minor changes can make valuable contributions to public health that more than justify the expense and effort involved. For instance, Jeffery and associates concluded that a smoking cessation program at a worksite costs about $100 to $200 per smoker who quits, whereas the cost to the employer for each employee who smokes is far greater. Similarly, Action Heart estimated that the cost per year of life gained was a mere 31 (British) pounds.

Each reduction in blood cholesterol of 1% yields a 2% decrease in coronary risk, and reductions of risk in the range 10 – 15% would have a huge beneficial impact if achieved in entire populations. Health promotion, therefore, can be a cost-effective way to improve lifestyles and thereby improve the health of large numbers of people. This is emphasized by the fact that in the United States poor dietary practices cost an estimated $71 billion per year in lost productivity, premature deaths, and medical costs. More research is required to determine why different health promotion projects have achieved such varying levels of success. Would campaigns be more successful if the focus was on one lifestyle change rather than many? Is paid advertising the best means to utilize scarce resources?

GOVERNMENT POLICY

Effective interventions may need to tackle the factors that determine how people make food choices. Such interventions require the implementation of policies, especially by governments. In the words of Davey Smith and Ebrahim : "...even with the substantial resources given to changing people's diets the resulting reductions in cholesterol concentrations is disappointing. [Health promotion programs] are of limited effectiveness. Health protection—through leg-islative and fiscal means—is likely to be a better investment." Governments have a variety of powers at their disposal that can be put into service. One approach, which relies entirely on voluntary cooperation, is to issue statements of policy.

However, these can easily amount to no more than hollow declarations as is illustrated by government policies on tobacco in many countries. On the other hand, policy statements can serve as a clarion call to action. For instance, British and American government policy on diet and disease, in conjunction with the media and medical science, helped change the climate of opinion so that it is now widely accepted that diets should preferably be much lower in fat and richer in fiber.

The Effect of Price on Sales

Prominent among available government powers are legislation and the use of taxation and subsidies. Action on tobacco control most graphically illustrates the necessity for placing these powers as a tool of health promotion. Educational efforts over the last three decades have been enormously important in persuading millions of people to quit smoking. Nevertheless, smoking rates are still well over half of their level of 30 yr ago. There is convincing evidence that price hikes are an effective means to reduce smoking rates (*i.e.*, there is price elasticity). It has been estimated that a 10% increase in price reduces tobacco consumption by about 5%, especially among the lower socioeconomic groups.

The Canadian experience is particularly illuminating. The prevalence of smoking in young Canadians fell by half during the 1980s in tandem with a doubling of the price. This trend was reversed in the early 1990s when the price was slashed in an attempt to reduce smuggling from the United States. Price increases appear to be a far more effective means of tobacco control than education or media campaigns. Alcohol intake shows a similar price elasticity to tobacco intake: a price rise of 10% causes a decrease in consumption by 3 – 8%. Studies in Eastern Europe, especially Poland and the former USSR, have demonstrated that pricing, sometimes in combination with rationing, sharply reduces consumption and associated mortality. The lesson we learn from tobacco and alcohol is, first and foremost, that price increases are an effective vehicle to lower consumption. What applies to cigarettes and alcohol no doubt also applies to food. By means of taxes and subsidies fruit, vegetables, and whole grain cereals might become more attractively priced in comparison with less healthy choices. This would most likely induce many people to shift their diets in a healthier direction.

Recommendations along these lines in the area of food and nutrition policy were advocated by the World Health Organization at the Adelaide Conference in 1988. The policy recommendation given was: "Taxation and subsidies should discriminate in favor of easy access for all to healthy food and improved diet." Jeffery and colleagues in the United States carried out a series of studies that have demonstrated the potential of policy interventions, especially of low prices, to increase the consumption of healthy food choices. In one study, the price of low-fat snacks sold in worksite vending machines was halved. Purchases of these foods increased from 26% of total sales before intervention to 46% after.

In a worksite cafeteria, the range of fruit and salad ingredients was increased at the same time as the price was halved. As a result, purchases tripled. In a similar study conducted in a high school cafeteria, prices for fruit, carrots, and salads was halved. This led to a fourfold increase in sales of fruit, a twofold increase for carrots, and a slight increase for salads.

Advertising, Marketing, and Labeling of Food

Another area where policy interventions could positively affect food choices concerns food advertising. The annual advertising budget for Coca-Cola Classic alone is over $130 million a year, and the budget for McDonalds is over $600 million. In stark contrast, the education component of the National Cancer Institute-sponsored 5-A-Day campaign to promote fruit and vegetable consumption receives well under 1 million. The extent to which these huge imbalances in advertising budgets affect people's actual diets is not known, but is almost certainly significant. Common sense dictates that if advertising did not work, the advertisers would not be wasting their money. A particular issue is food advertising on children's TV.

A study of advertisements appearing on Saturday morning TV in the United States found that 44% were for fats, oils, and sugar, 23% were for highly sugared cereals, and 11 % were for fast food restaurants. None were for fruit and vegetables. The authors concluded that: "The diet that is presented on Saturday morning television is the antithesis of what is recommended for healthful eating for children." Similar findings were reported for Canadian TV. Advertising is but one part of the wider production and marketing strategy of the food industry.

James and Ralph pointed out that in response to demand, manufacturers sell foods with less fat, but the missing fat often reappears in "added value" foods, which may be little more than concoctions of fat, sugar, and salt. James made the compelling point that the food industry promotes high-fat food because it is so profitable, while at the same time, food labeling is "completely confusing" (with particular reference to Britain). The system is, in theory, based on "consumer choice" but, in reality, choices become largely uninformed decisions.

Government Policy and Food

The foregoing discussion suggests that government policies concerning food prices and, to a lesser extent, food advertising and labeling may be an effective means to induce desirable changes in eating patterns. Here, we offer some specific suggestions as to how existing government policies could be modified along the foregoing lines so as to encourage diets higher in fruit and vegetables and lower in fat.

1. Subsidies paid to milk producers could be changed to favor low-fat milk. Similarly, by the use of such means as subsidies, grading regulations, and labeling, and perhaps even taxation, the sale of low-fat meat could be encouraged over high-fat varieties.
2. There is always scope for improved food labels to facilitate purchase of foods with a low content of fat, especially saturated fat. In addition, labeling and nutrition information should be extended to areas presently outside the system, especially restaurant menus, soft drinks, snack foods, and fresh meat.
3. By means of regulations and rewards, schools could be encouraged to sell meals of superior health value while restricting the sale of junk food. Similar policies could be applied to other institutions under government control, such as the military, prisons, and cafeterias in government offices.
4. Television advertising could be regulated so as to control the content, duration, and frequency of commercials for unhealthy food products, especially when the target audience is children.

Jeffery argued that this approach be used as a strategy against obesity: "I believe that it is time to think about alternatives to the traditional individual-focused strategies for obesity control. I believe we should also view our food supply as a potential environmental hazard that promotes obesity and to consider public health policy strategies for improving it." He suggested that perhaps caloriedense products should be packaged in small serving sizes and that taxation should be based on the fat and sugar content of food. James also made proposals for how policies could be used to help combat obesity.

The approach discussed above was well put by *Blackburn* :

...even the newer community-based lifestyle strategies continue to assign much of the burden of change to the individual. A shift of focus to reducing, by policy change, many widespread practices that are life-threatening, while enhancing life-supportive practices, should redirect the currently misplaced emphasis on achieving "responsible" behavior and its purported difficulty. For example, local communities may more appropriately be considered to have a "youth tobacco access problem," approachable in part by regulation, than a "youth smoking problem," approachable mainly by education. Policy interventions may also be designed to .... make preventive practice more economical, as well as to encourage the development of more healthy products by industry. They may be a partial answer to another major paradox: while unhealthy personal behavior is medically discouraged for individuals, the whole of society legalizes, tolerates, and even encourages the same practices in the population.

Schmid *et al.*, summed up the approach discussed here :

Health departments that support disincentives for high-fat foods, tax breaks for cafeterias that offer healthy food choices, policies that require zoning ordinances to include sidewalks, or school facilities open to the public might be labeled radical or experimental today; tomorrow, however, they may be considered prudent stewards of the public health.

The problem of lead pollution is an excellent illustration of what can be achieved by governmental action. In the 1970s, regulations implemented by the American government forced major reductions or removal of lead from gasoline, paint, water, and consumer products. As a result, the blood lead level of the average American child is now less than one quarter of what it had been in the late 1970s.

Barriers Against Public Health Policies

Although many might consider the policies discussed here to be worthy of implementation, it must be appreciated that barriers exist. In particular, industry profits enormously from the sale of highly processed food and has often shown itself to be resistant to change. In this regard, industry often secures government support. The history of attempts to enact legislative control over tobacco illustrate how effective an industry can be when it utilizes a large budget in attempts to delay, dilute, or stop laws. There is clear evidence as to the likely reason why the US Congress has been so lethargic when it comes to antismoking legislation.

In 1991 and 1992, the average senator received $ 11,600 per year from the tobacco industry. In the opinion of the researchers who carried out this study: "The money that the tobacco industry donates to members of Congress ensures that the tobacco industry will retain its strong influence in the federal tobacco policy process." Similarly, researchers looked at the California legislature and concluded: "Legislative behavior is following tobacco money rather than reflecting constituents' prohealth attitudes on tobacco control". If the tobacco industry can achieve so many successes, then it will likely be much easier for the food industry to thwart interventions that threaten its profits. This is because the relationship between diet and disease is far less clear than is the case with tobacco. Indeed, there is ample evidence that governments are sympathetic to the wishes of the agricultural and food industries.

Typically, although the health arm of governments encourages people to eat less fat, the departments responsible for the agricultural and food industries are largely concerned with maintaining high sales. James and Ralph asserted that: "Analysis of different policies suggest that health issues are readily squeezed out of discussion by economic and vested interests." There is considerable evidence of how industry has successfully pressured governments to bow to their wishes on questions of nutrition policy.

As discussed by Nestle, the meat industry has been particularly effective in rewriting dietary guidelines. In the late 1970s, the goal was "eat less meat." This then became "choose lean meat." By 1992, people were encouraged to consume at least two or three servings daily. There is also evidence that the 1992 version of Canada's Food Guide was similarly modified under pressure from the food industry.

Discussing the question of salt, Goodlee, assistant editor of the British Medical Journal, put it as follows:

> *...some of the world's major food manufacturers have adopted desperate measures to try to stop governments from recommending salt reduction. Rather than reformulate their products, manufacturers have lobbied governments, refused to cooperate with expert working parties, encouraged misinformation campaigns, and tried to discredit the evidence... The tactics over salt are much the same as those used by other sectors of industry. The Sugar Association in the United States and the Sugar Bureau in Britain have waged fierce campaigns against links between sugar and obesity and dental caries.*

National Nutrition Policies: Examples

One pioneering project was the Norwegian Nutrition and Food Policy. Implemented in 1976, it recognized the need to integrate agricultural, economic, and health policy. The policy included consumer and price subsidies, marketing measures, consumer information, and nutrition education in schools. Unfortunately, the policy clashed with policies aiming to stimulate agriculture. As a result, subsidies went to pork, butter, and margarine rather than to potatoes, vegetables, and fruit.

Despite these setbacks, the policy has achieved some suc-cess in moving the national diet in the intended direction. Another noteworthy effort which implemented several of the policies discussed here was Heartbeat Wales. This project was carried out in Wales with the aim of preventing CHD. Specific measures included better food labeling, price incentives, and greater availability of healthier food. The active support was enlisted of catering departments and a food retailer. Unfortunately, little information appears to be available as to the degree of success of this policy.

Are Nutrition Policies Acceptable to the Public?

An important question concerns the extent to which the public would accept the suggested policies. The issues of seat belt use and drunk driving illustrate that when legislation is implemented and the public is educated as to their importance, there is a high degree of acceptance. A study by Jeffery and colleagues in the upper midwest of the United States indicated widespread support for regulatory controls in the areas of alcohol, tobacco and, to a lesser extent, high-fat foods, especially with respect to children and youths. If such policies are acceptable to Americans, then they are also likely to be acceptable in other countries.

SOCIOECONOMIC STATUS AND HEALTH

One area of importance is the relationship between socioeconomic status (SES) and health. Low SES is strongly and consistently associated with a raised mortality rate. This applies to total mortality, as well as to death from CHD and cancer. The risk ratios are in the range 1.5 – 4, clearly making SES a major determinant of health. Various measures of SES have been examined—income, social status of job, being unemployed, area of residence, and education—and each seems to manifest a similar relationship to mortality. Another important finding is that the degree of inequality within a country predicts morbidity and mortality.

In other words, the key factor predicting health and life expectancy of a country is not average income or average level of education, but the gap between the rich and poor. Various studies have investigated why SES is associated with increased mortality. In general, lower SES is associated with higher rates of smoking and a diet of lower nutritional quality. Is SES merely a proxy measure of lifestyle? Or does SES affect health by a more direct mechanism? This question is of much more than mere theoretical importance and has a bearing on health strategies. If people of low SES are unhealthy because they lead an unhealthy lifestyle, then the solution lies in encouraging changes in their lifestyles. But, if a low SES is intrinsically unhealthy, then the solution lies elsewhere. Our best evidence is that both possibilities are partially correct.

After correcting for confounding variables, especially smoking, exercise, blood cholesterol, blood pressure, and weight, most studies have found that the strength of the association between SES and mortality is reduced by about a quarter or a half. This *indicates that people with lower SES tend to lead a less healthy lifestyle and this partl*y explains their poorer health.

But this still leaves half to three-quarters of the association between SES and mortality unexplained. None of the studies we cited examined diet in relation to SES. If people of low SES tend to eat a Jess nutritious diet, this would help explain why such conditions as hypercholesterolemia, hypertension, and overweight are associated with low SES. Nevertheless, it appears that much of the association between SES and mortality cannot be explained by lifestyle and must therefore be a more direct consequence of low SES. Psychological factors appear to play an important role in explaining the association between SES and mortality.

The psychological factor most closely associated with risk of poor health is lack of control at work. We can speculate that other psychological factors, such as resentment, frustration, and a feeling of disempowerment, all contribute to poor health among low-income groups. Whatever the precise mechanisms, there is little doubt that structural elements of inequality within Western societies—economic, educational, social status—lead to reduced health. What should be done about this? An effective strategy to deal with the challenge of low SES may have to include efforts to reduce socioeconomic inequalities. If people of lower SES could be persuaded to adopt the same lifestyle, including diet, as those of higher SES, perhaps as much as half of the problem would likely disappear. Therefore, dietary advice is still worth the effort.

CONCLUSIONS

Based on the close association between income inequalities in a society and health, an essential component of enhancing a population's health must be government measures to make societies more egalitarian. Unfortunately, the legacy of Thatcherism and Reagonomics means that many governments are ideologically opposed to taking such action. At the same time, governments have been even more business friendly and non-interventionist than ever.

This was well put by James with specific regard to obesity:

The needed transformation in thinking on transport, environment, work facilities, education, health and food policies, and perhaps in social and economic policies is unlikely when governments are wedded to individualism, but without these changes to enhance physical activity and alter food quality, societies are doomed to escalating obesity rates.

This viewpoint applies to the relationship between nutrition and all diseases related to it.

The primary force driving government policy is economics rather than a desire to improve the national health. However, the weight of evidence strongly suggests that until governments reorientate their nutrition policies so that national health becomes the first priority, we will be losing great opportunities for the prevention of such diseases as cancer and CHD. The philosophy discussed here need not stop at nutrition: what applies to nutrition certainly applies to other areas of lifestyle, especially to smoking. Exercise also lends itself to policy initiatives. What is the point in telling people to exercise if there is a lack of appropriate facilities? What is the point in telling people to cycle if the roads are too dangerous for bikes? What is needed is a comprehensive view of human health that takes all such factors into consideration. In the 21 st century, no doubt, people will look back with incredulity on today's world where narrow commercial interests and government laissez-faire predominate while the national health founders.

CHAPTER 10 Coenzymes of Oxidation-Reduction Reactions

The *dehydrogenation* of an alcohol to a ketone or aldehyde (Eq. 10.1) one of the most frequent biological oxidation reactions. Although the hydrogen atoms removed from the substrate are often indicated simply as 2[H], it was recognized early in the twentieth century that they are actually transferred to hydrogen-carrying coenzymes such as NAD^+, $NADP^+$, FAD, and *riboflavin*

$$H-\overset{|}{\underset{|}{C}}\longrightarrow OH-\overset{|}{\underset{|}{C}}=O+2[H] \qquad \text{...(10.1)}$$

5′-phosphate (FMN). This chapter deals with these coenzymes and also with a number of other organic oxidation-reduction coenzymes. They may be considered either as carriers of hydrogen or as carriers of electrons ($H = H^+ + e^-$) in metabolic reactions.

When NAD^+ becomes reduced by dehydrogenation of an alcohol, one of the hydrogen atoms removed from the alcohol becomes firmly attached to the NAD^+, converting it to NADH. The other is released as a proton (Eq. 10.2). Study of ^2H-labeled alcohols and their oxidation by NAD^+ has shown that *dehydrogenases catalyze direct transfer to NAD^+ of the hydrogen that is attached to carbon in the alcohol.* There is never any exchange of this hydrogen atom with protons of the medium. At the same time, the hydrogen attached to the oxygen of the alcohol is released into the medium as H^+ :

$$NAD^+ + H-\overset{|}{\underset{|}{C}}-O-H\longrightarrow \overset{|}{C}=O+NADH+H^+ \qquad \text{...(10.2)}$$

This hydrogen in release as a proton, H^+

This H is transferred directly to NAD^+

The foregoing observations suggested that these biological dehydrogenations may be viewed as removal of a hydride ion (H^-) together with a proton (H^+) rather than as removal of two hydrogen atoms. NAD^+ and NADP+ are regarded as hydride ion-accepting coenzymes. However, it has been impossible to establish conclusively that the hydrogen atom and electron are transferred simultaneously as H^-. Transfer of the hydrogen atom to or from these coenzymes may conceivably be followed by or preceded by transfer of an electron. The situation is even less clear for the flavin coenzymes FAD and riboflavin phosphate for which intermediate *free radical* oxidation states are known to exist.

However, regardless of the actual mechanism, it is convenient to classify most hydrogen transfer reactions of metabolism as if they occurred by transfer of a hydride ion. *The hydride*

ion can be regarded as a nucleophile which can add to double bonds or can be eliminated from substrates in reactions of types that we have already considered. **Some of these reactions are listed in Table 10.1.**

Table 10.1. Biochemical Hydrogen Transfer Reactions[a]

Reaction	*Example (oxidant)*
A Dehydrogenation of an Alcohol $HC{-}OH \rightarrow H^- + H^+ + {>}C{=}O$	**Alcohol dehydrogenase (NAD^+)**
B Dehydrogenation of an amine $HC{-}NH_2 \rightarrow H^- + H^+ + {>}C{=}NH$; $+H_2O \rightarrow NH_4^+ + {>}C{=}O$	**Amino acid dehydrogenase, amine oxidase (NAD^+ or flavin)**
C Dehydrogenation of adduct of thiol and aldehyde $R{-}SH + {-}C(=O){-}H \rightleftharpoons {-}C(OH)(H){-}SR \rightarrow H^+ + H^- + {-}C(=O){-}SR$	**Gyceraldehyde 3-phosphate dehydrogenase (NAD^+ or $NADP^+$)**
D Dehydrogenation of acyl-CoA, acyl-ACP, or arboxylic acid $R{-}CH_2{-}CH_2{-}C(=O){-}Y \rightarrow H^- + H^+ + R{-}CH{=}CH{-}C(=O){-}Y$ (Y = —S—CoA or —OH)	**Acyl-CoA dehydrogenases (Flavin)** **Succinic dehydrogenase (Flavin)** **Opposite: enoyl reductase (NADPH)**
E. Reduction of desmosterol to cholesterol by NADPH $H^- + R{-}CH{=}C{<} + H^+ \rightarrow R{-}CH_2{-}CH{<}$	**Reduction of demonsterol to cholesterol by NADPH**

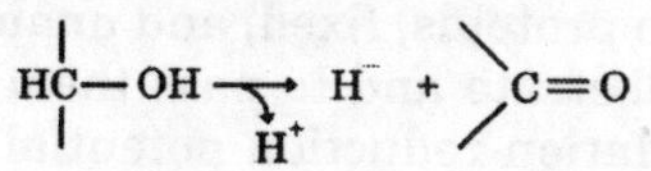

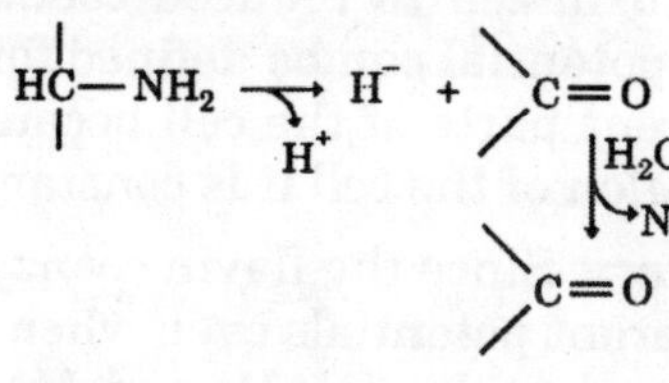

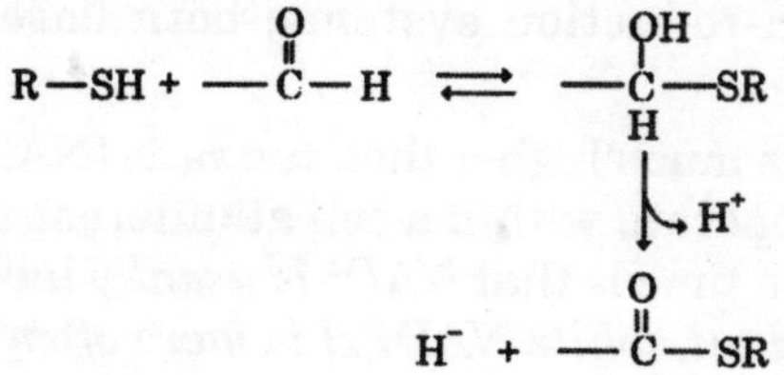

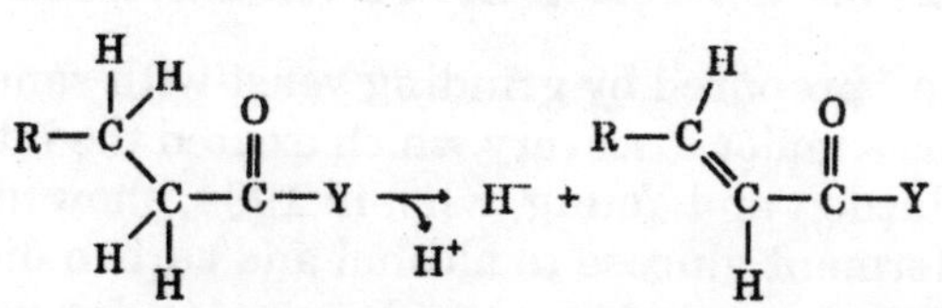

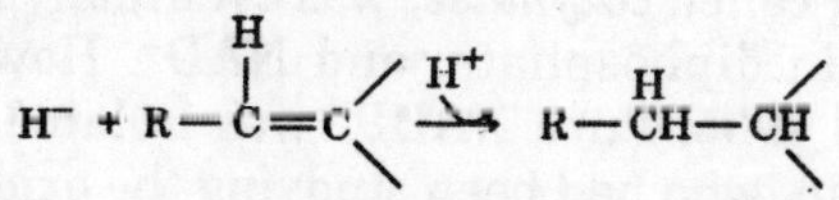

*** Reaction type 9 A,B of Table on the inside cover at the end of the book: A hypothetical hydride ion H^- is transferred from the substrate to a conenzyme of suitable reduction potential such as NAD^+, $NADP^+$, FAD, or riboflavin 5'-phosphate. The reverse of hydrogenation is shown for E. Many of the reactions are reversible and often go spontaneously in the reverse direction from that shown here.**

Why are there *four* major hydrogen transfer coenzymes, NAD^+, $NADP^+$, FAD, and riboflavin phosphate (FMN), instead of just one? Part of the answer is that the *reduced pyridine nucleotides* NADPH and NADH are more powerful reducing agents than are reduced *flavins*. Converscly, flavin coenzymes are more powerful oxidizing agents than are NAD+ and $NADP^+$. This difference reflects the chemical difference between the vitamins *riboflavin and nicotinamide* which form the oxidation-reduction centers of the coenzymes.

Another difference is that NAD^+ and $NADP^+$ tend to be present in free forms within cells, diffusing from a site on one enzyme to a site on another. These coenzymes are *sometimes* tightly bound but flavin coenzymes are *usually* firmly bound to proteins, fixed, and unable to move. Thus, they tend to accept hydrogen atoms from one substrate and to pass them to a second substrate while attached to a single enzyme. The oxidation-reduction potential of a pyridine nucleotide coenzyme sýstem is determined by the standard redox potential for the free coenzyme together with the ratio of concentrations of oxidized to reduced coenzyme ($[NAD^+]/[NADH]$. If these concentrations are known, a redox potential can be defined for the NAD^+ system within a cell. This potential may vary in different parts of the cell because of differences in the $[NAD^+]/[NADH]$ ratio, but within a given region of the cell it is constant.

On the other hand, the redox potentials of flavoproteins vary. Since the flavin coenzymes are not dissociable, two flavoproteins may operate at very different potentials even when they are physically close together. Why are there two pyridine nucleotides, NAD^+ and $NADP^+$, differing only in the presence or absence of an extra phosphate group? One important answer is that they are members of two different oxidation-reduction systems, both based on nicotinamide but functionally independent.

The experimentally measured ratio $[NAD^+]/[NADH]$ is much higher than the ratio [NADP+]/[NADPH]. Thus, these two coenzyme systems also can operate within a cell at different redox potentials. A related generalization that holds much of the time is that *NAD^+ is usually involved in pathways of catabolism, where it functions as an oxidant, while NADPH is more often used as a reducing agent in biosynthetic processes.*

A. PYRIDINE NUCLEOTIDE COENZYMES AND DEHYDROGENASES

In 1897, Buchner discovered that "yeast juice," prepared by grinding yeast with sand and filtering, catalyzed fermentation of sugar. This was a major discovery which excited the interest of many other biochemists. Among them were Harden and Young, who, in 1904, showed that Buchner's cell-free yeast juice lost its ability to ferment glucose to alcohol and carbon dioxide when it was dialyzed. Apparently, fermentation depended upon a low-molecular-weight substance that passed out through the pores of the dialysis membrane. Fermentation could be restored by adding back to the yeast juice either the concentrated dialysate or boiled yeast juice (in which the enzyme proteins had been destroyed).

The heat-stable material, which Harden and Young called *cozymase,* was eventually found to be a mixture of inorganic phosphate ions, thiamin diphosphate, and NAD^+. However, characterization of NAD^+ was not accomplished until 1935. Pure $NADP^+$ was isolated from red blood cells in 1934 by Otto Warburg and W. Christian, who had been studying the oxidation of glucose 6-phosphate by erythrocytes. They demonstrated a requirement for a dialyzable coenzyme which they characterized and named *triphosphopyridine nucleotide* (TPN^+, but now officially $NADP^+$; Fig. 10.1). Thus, even before its recognition as an important vitamin in human nutrition, nicotinamide was identified as a component of $NADP^+$.

Warburg and Christian recognized the relationship of NADP+ and NAD^+ (then called DPN^+) and proposed that both of these compounds act as hydrogen carriers through alternate reduction and oxidation of the pyridine ring. They showed that the coenzymes could be reduced either enzymatically or with sodium dithionite $Na_2S_2O_4$.

$$S_2O_4^{2-} + NAD^+ + H+ \rightarrow NADH + 2\ SO_2 \quad ...(10.3)$$

The reduced coenzymes NADH and NADPH were characterized by a new light-absorption band at 340 nm. This is not present in the oxidized forms, which absorb maximally at 260 nm (Fig. 10.2). The reduced forms are stable in air, but their reoxidation was found to be catalyzed by certain yellow enzymes which were later identified as flavoproteins.

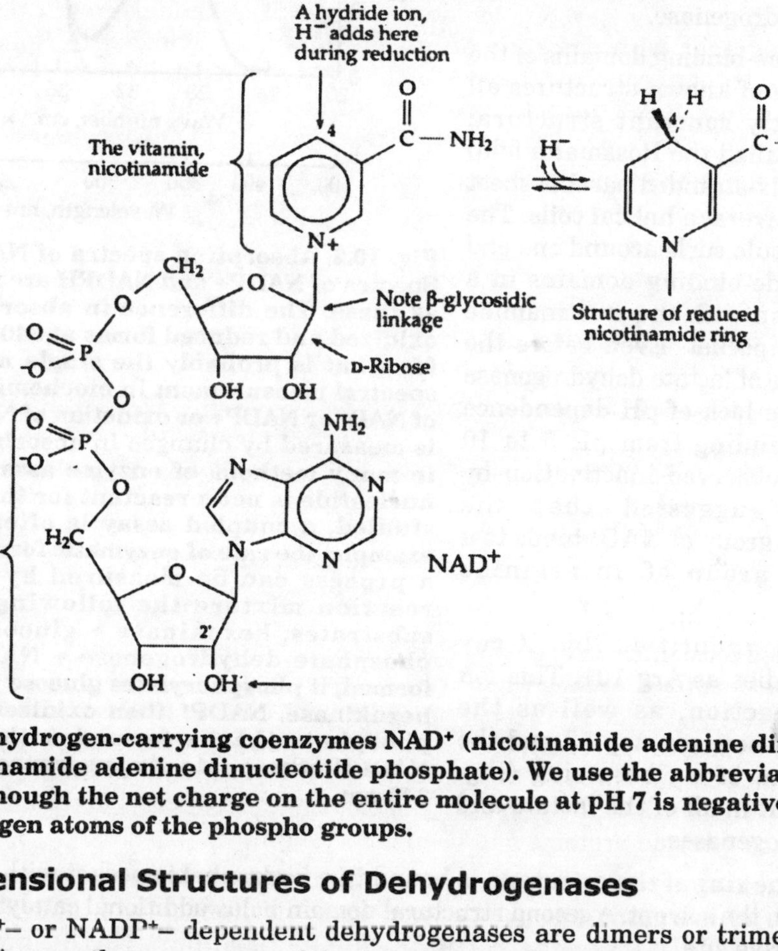

Fig. 10.1. The hydrogen-carrying coenzymes NAD^+ (nicotinanide adenine dinucleotide) and $NADP^+$ (nicotinamide adenine dinucleotide phosphate). We use the abbreviations NAD^+ and $NADP^+$, even though the net charge on the entire molecule at pH 7 is negative because of the charges on oxygen atoms of the phospho groups.

Three-Dimensional Structures of Dehydrogenases

Most NAD^+– or $NADP^+$– dependent dehydrogenases are dimers or trimers of 20- to 40-kDa subunits. Among them are some of the first enzymes for which complete structures were determined by X-ray diffraction methods. The structure of the 329-residue per subunit muscle (M_4) isoenzyme of *lactate dehydrogenases* from the dogfish was determined to 0.25 nm resolution by Rossmann and associates in 1971. More recently, structures have been determined for mammaliam muscle and heart type (H_4) isoenzymes, for the testicular (C_4)

isoenzyme from the mouse, and for bacterial lactate dehydrogenases. In all of these the polypeptide is folded nearly identically. The structures of the homologous cytosolic and mitochondrial isoenzymes of *malate dehydrogenase* are also similar, as are those of the bacterial enzyme. All of these proteins consist of two structural domains and the NAD^+ is bound to the nucleotide-binding domain in a similar manner for glyceraldehyde phosphate dehydrogenase.

The coenzyme-binding domains of the dehydrogenases of known structures all have this nearly constant structural feature (often called the Rossmann fold) consisting of a six-stranded parallel sheet together with several a helical coils. The coenzyme molecule curls around one end of the nucleotide-binding domains in a "C" conformation with the mcotinamide ring lying in a pocket. Even before the crystal structure of lactate dehydrogenase was known, the lack of pH dependence of coenzyme binding from pH 5 to 10 together with observed inactivation by butanedione suggested that the pyrophosphate group of NAD^+ binds to a guanidinium group of an arginine residue.

This was identified by X-ray diffraction studies as Arg 101. This ion pairing interaction, as well as the hydrogen bond between Asp 53 and the 2′ oxygen atom of a ribose ring (Fig. 10.3), is present in all of the lactate and malate dehydrogenases.

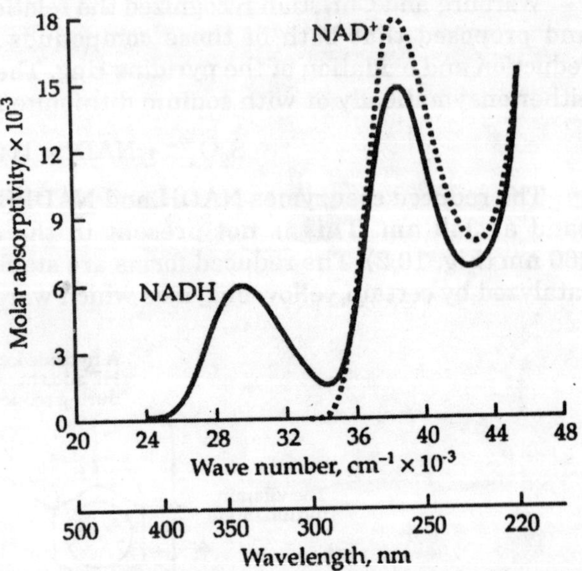

Fig. 10.2. Absorption spectra of NAD^+ and NADH. Spectra of NADP+ and NADPH are nearly the same as these. The difference in absorbance between oxidized and reduced forms at 340 nm is the basis for what is probably the single most often used spectral measurement in biochemistry. Reduction of NAD^+ or NADP+ or oxidation of NADH or NADPH is measured by changes in absorbance at 340 nm in many methods of enzyme assay. If a pyridine nucleotide is not a reactant for the enzyme being studied, a coupled assay is often possible. For example, the rate of enzymatic formation of ATP in a process can be measured by adding to the reaction mixture the following enzymes and substrates: hexokinase + glucose + glucose-6-phosphate dehydrogenase + $NADP^+$. As ATP is formed, it phosphorylates glucose via the action of hexokinase. NADP* then oxidizes the glucose 6-phosphate that is formed with production of NADPH, whose rate of appearance is monitored at 340 nm.

The adenine ring of the coenzyme is bound in a hydrophobic pocket with its amino group pointed out into the solvent. A second structural domain holds additional catalytic groups needed to form the active site.

Stereospecificity and Mechanism

When N AD^+ is reduced in ^2H_2O by dithionite an atom of ^2H is introduced into the reduced pyridine. Chemical degradation showed the ^2H to be present at the 4 position of the ring para to the nitrogen atom (see Fig. 10.1). As shown by Westheimer and associates, during enzymatic

reduction of NAD+ by deuterium-containing ethanol, $CH_3 - C^2H_2OH$, one of the ^2H atoms is transferred into the NADH formed, thus establishing the *direct transfer of a hydrogen atom.* When the NAD^2H formed in this way is reoxidized enzymatically with acetaldehyde, with regeneration of NAD+ and ethanol, the ^2H is completely removed.

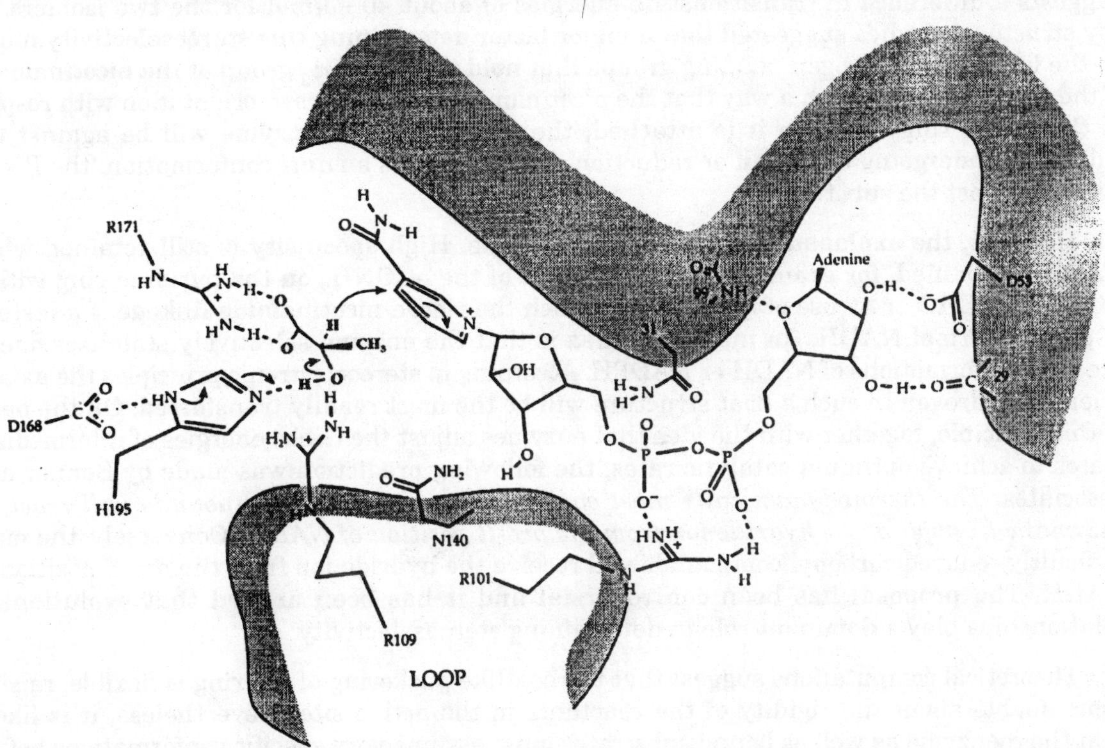

Fig. 10.3. Diagrammatic structure of NAD+ and L-lactate bound into the active site of lactate dehydrogenase.

This was one of the first recognized examples of the ability of an enzyme to choose between two identical atoms at a prochiral center. The two sides of the nicotinamide ring of NAD were

pro-R ⟶ H_A H_B ⟵ pro-S

CONH₂

designated A and B and the two hydrogen atoms at the 4 position of NADH as H_A (now known as pro-R) and H_B (pro-S). Alcohol dehydrogenase always removes the pro-R hydrogen. Malate, isocitrate, lactate, and D-glycerate dehydrogenases select the same hydrogen. However, dehydrogenases acting on glucose 6-phosphate, glutamate, 6-phosphogluconate, and 3-phosphoglyceraldehyde remove the *pro-S* hydrogen. By 1979 this stereo-specificity had been determined for 127 dehydrogenases, about half of which were found A-specific and half B-specific.

Isotopically labeled substrates have usually been used for this purpose but a simple NMR method has been devised. How complete is the stereospecificity? Does the enzyme sometimes make a mistake? *Lactate dehydrogenase displays nearly absolute specificity,* transferring a proton into the "incorrect" *pro-S* position no more than once in 5×10^7 catalytic cycles. This suggests a difference in transitionstate energies of about 40 kJ/mol for the two isomers. X-ray structural studies suggested that a major factor determining this stereoselectivity might be the location of hydrogen-bonding groups that hold the $-CONH_2$ group of the nicotinamide. If these are located in such a way that the nicotinamide ring has, a *syn* orientation with respect to the ribose ring to which it is attached, the A face of the coenzyme will be against the substrate undergoing oxidation or reduction. If the ring has an *anti* conformation, the B side will be against the substrate.

However, the explanation cannot be this simple. High specificity is still retained when NADH is modified, for example, by replacement of the $-CONH_2$ on the pyridine ring with $-COCH_3$ or $-CHO$ or by use of α-NADH, in which the ribose-nicotinamide linkage is a instead of β as in normal NADH. An intriguing idea is that the enzyme selectively stabilizes one of the boat conformations of NADH or NADPH. According to stereoelectronic principles the axially oriented hydrogen in such a boat structure will be the most readily transferred. On the basis of this principle, together with the idea that enzymes adjust the Gibbs energies of intermediate states to achieve optimum catalytic rates, the following prediction was made by Benner and associates: *The thermodynamically most easily reduced carbonyl compounds will react by enzymatic transfer of the hydride ion from the pro-R position of NADH.* Conversely, the most difficultly reduced carbonyl compounds will receive the hydride ion from the *pro-S* position of NADH. The proposal has been controversial and it has been argued that evolutionary relationships play a dominant role in determining stereoselectivity.

Theoretical computations suggest that the boatlike puckering of the ring is flexible, raising some doubts about the rigidity of the reactants in the active site. Nevertheless, it is likely that the coenzyme as well as bound substrates must assume very specific conformations before the enzyme is able to move to the transition state for the reaction. A great deal of effort has been expended in trying to understand other factors that may explain the high stereospecificity of dehydrogenases. When a hydrogen atom is transferred by an enzyme from the 4 position of NADH or NADPH to an aldehyde or ketone to form an alcohol, the placement of the hydrogen atom on the alcohol is also stereospecific. Thus, alcohol dehydrogenase acting on NAD^2H converts acetaldehyde to (R)-mono-[^2H] ethanol.

Pyruvate is reduced by lactate dehydrogenase to L-lactate, and so on. Even mutations that disrupt the binding of the carboxylate of lactate, *e.g.,* substitution of arginine 171 by tryptophan or phenylalanine and introduction of a new arginine on the other side of the active site, do not alter the L-stereospecificity. However, the specificity for lactate is lost.

One Step or Two-step Transfer?

Another major question about dehydrogenases is whether the hydrogen atom that is transferred moves as a hydride ion, as is generally accepted, or as a hydrogen atom with separate transfer of an electron and with an intermediate NAD or NADPH free radical. In one study parasubstituted benzaldehydes were reduced with NADH and NAD^2H using yeast alcohol dehydrogenase as a catalyst. This permitted the application of the Hammett equation to the rate data. For a series of benzaldehydes for which σ^+ varied widely, a value of $\rho = +2.2$ was observed for the rate constant with both NADH and NAD^2H. Thus, electron-accepting substituents in the para position hasten the reaction. While the significance of this observation is not immediately obvious, the relatively low value of ρ is probably incompatible with a mechanism that requires complete transfer of a single electron from NADH to acetaldehyde in the first step. A primary isotope effect on the rates was $k_H/k_D = 3.6$, indicating that the C – H bond in NADH is broken in the ratelimiting step.

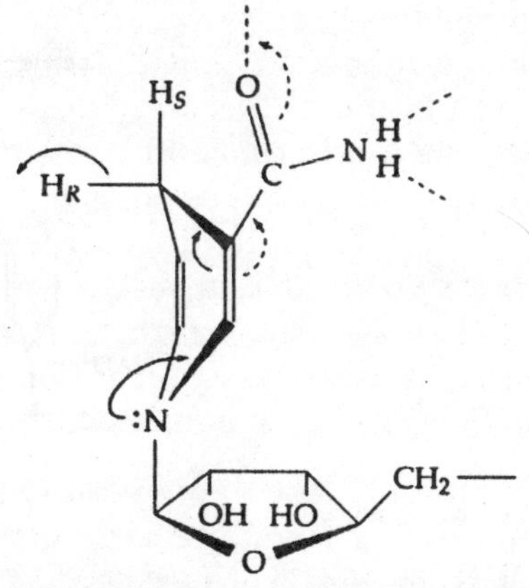

The fact that the isotope effect is the same for all of the substituted benzaldehydes also argued in favor of a hydride ion transfer. Studies of the kinetics of nonenzymatic model reactions of NADH with quinones have also been interpreted to favor a single-step hydride ion transfer. Application of Marcus theory to data from model systems also supports the hydride transfer mechanism. Quantum mechanical tunneling may be involved in enzymatic transfer of protons, hydride ions, and electrons. Tunneling is often recognized by unusually large primary or secondary kinetic isotope effects. According to semi-classical theory, the maximum effects for deuterium and tritium are given by the following ratios:

$$(k_{2_H}/k_{1_H}) = (k_{3_H}/k_{2_H})^{3.3} = 7$$

Higher ratios, which suggest tunneling, are frequently observed for dehydrogenases. Tunneling is apparently coupled to fluctuations in motion within the enzyme-substrate complex. Study of effects of pressure on reactions provides a new approach that can aid interpretation.

Fig. 10.4. The nicotinamide ring of NADH in a *syn* boat conformation suitable for transfer of an axially oriented pro-R hydrogen atom from its A face as H⁻. The flow of electrons is shown by the solid arrows. The dashed arrows indicate competing resonance which favors planariry of the ring and opposes the H⁻ transfer. Hydrogen bonds from the protein to the carboxamide group (dashed lines) affect both this tendency and the conformation of the nucleotide.

Coenzyme and Substrate Analogs

The structures of enzyme•NAD^+•substrate complexes (see Fig. 10.3) may be studied by X-ray crystallography under certain conditions or can be inferred from those of various stable enzyme-inhibitor complexes or from enzyme reconstituted with NAD^+ that has been covalently linked to the substrate. In the following diagram, NAD^+ (I) is shown with L-lactate lying next to its A face, ready to transfer a hydride ion to the *pro-R* position in NADPH. Also shown (II) is 5-(S)-lac-NAD^+, with the covalently linked lactate portion in nearly the same position as in diagram I. This NAD derivative was used to obtain the first 0.27-nm structure with a bound substrate-like molecule in lactate dehydrogenase.

Since then the structures of many complexes with a variety of dehydrogenases have been studied. In coenzyme analog III the ring is bound to ribose with a C – C bond and it lacks the positive charge on the nicotinamide ring in NAD^+. It does not react with substrate. However, it binds to the coenzyme site of alcohol dehydrogenase and forms with ethanol a ternary complex whose structure has been solved.

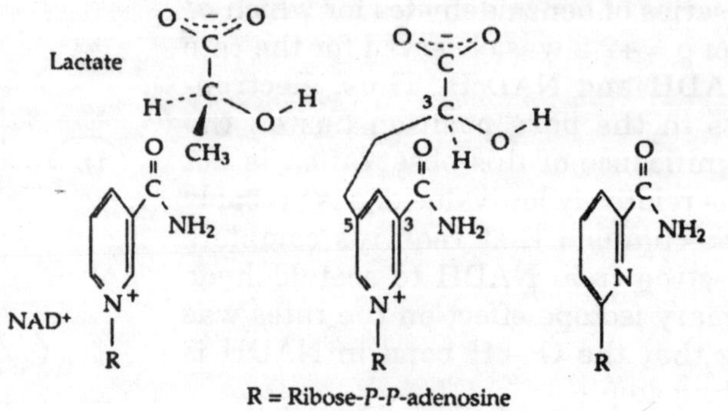

Lactate

NAD+

R = Ribose-*P*-*P*-adenosine

I II III

5-(*S*)-lac-NAD+,
(3*S*)-5-(3-carboxy-3
hydroxypropyl)-NAD+

A related approach is to study complexes formed with normal NAD^+ but with an unreactive second substrate. An example is oxamate, which binds well to lactate dehydrogenase to form stable ternary complexes for which equilibrium isotope effects have been studied.

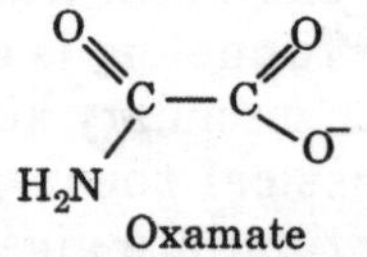

Oxamate

In the structure of the lactate dehydrogenase active site shown in Fig. 10.3, the lactate carboxylate ion is held and neutralized by the guanidinium group of Arg 171, and the imidazole group of His 195 is in position to serve as a general base catalyst to abstract a proton from the hydroxyl group of the substrate. This imidazole group is also hydrogen bonded to the carboxylate of Asp 168 as in the "charge-relay" system of the serine proteases. The same features are present in the active site of malate dehydrogenases and have been shown essential by study of various mutant forms. The His:Asp pair of the dehydrogenases is not part of the nucleotide-binding domain but is present in the second structural domain, the "catalytic domain." This is another feature reminiscent of the serine proteases.

A bacterial D-glycerate dehydrogenase also has a similar structure and the same catalytic groups. However, the placement of the catalytic histidine and the arginine that binds the α-carboxylate group of the substrate is reversed, allowing the enzyme to act on the D-isomer.

Conformational Changes During Dehydrogenase Action

Dehydrogenases bind coenzyme and substrate in an ordered sequence. The coenzyme binds first, then the oxidizable or reducible substrate. The binding of the coenzyme to lactate dehydrogenase is accompanied by a conformational change by which a loop, involving residues 98 – 120 and including one helix, folds over the coenzyme like a lid. This conformational change must occur during each catalytic cycle, just as in the previously discussed cases of citrate synthase and aspartate aminotransferase. One effect of folding of the loop is to bring the side chain of Arg 109 into close proximity to His 195 and to the OH group of the bound lactate. The closing of the loop also forces the positively charged nicotinamide ring more deeply into

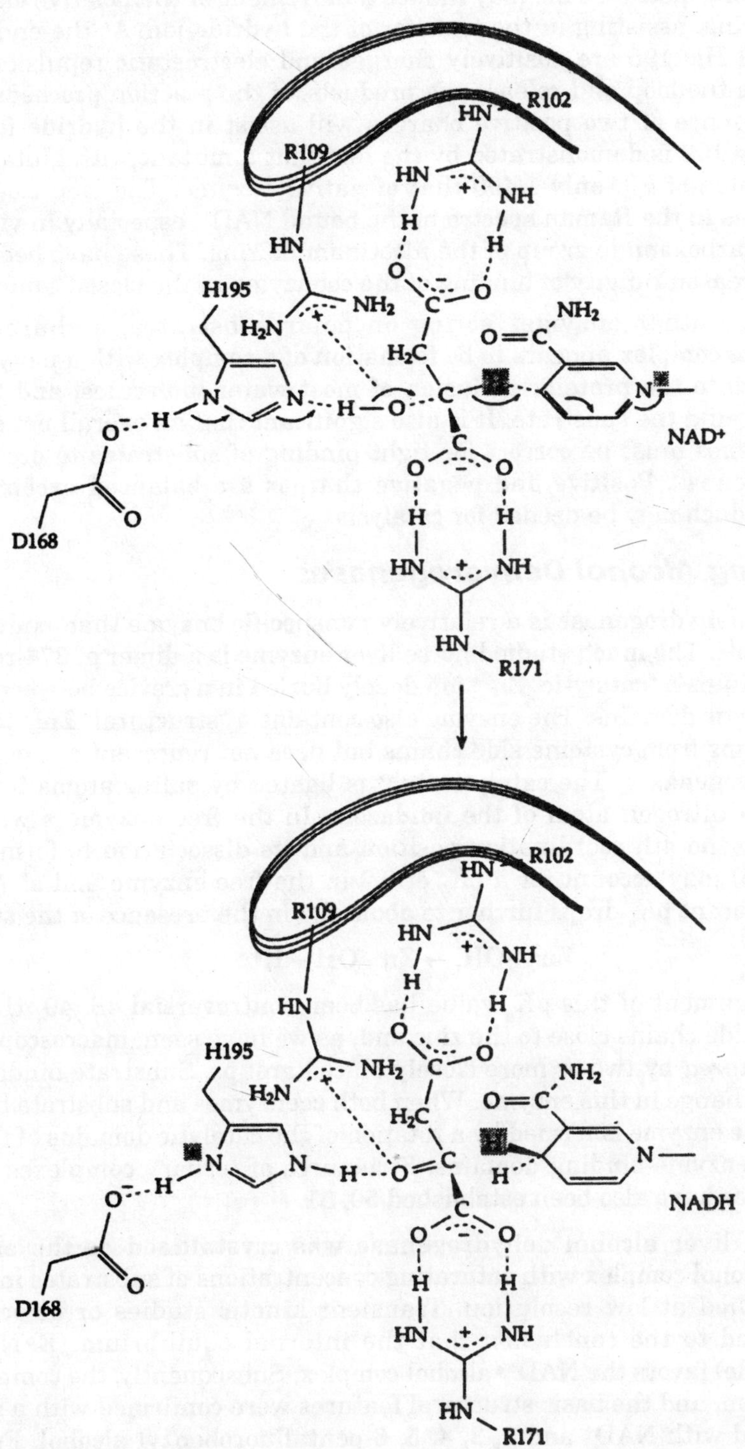
R102
R109
H195
NAD+
D168
R171
R102
R109
H195
NADH
D168
R171

a relatively nonpolar pocket. This may induce a movement of the positive charge toward the 4′ carbon of the ring, assisting in the transfer of the hydride ion. At the end of this transfer both Arg 109 and His 196 are positively charged and electrostatic repulsion between them may help to move the loop and release the products. If the reaction proceeds in the opposite direction the presence of two positive charges will assist in the hydride ion transfer. The importance of Arg 109 is demonstrated by the fact that a mutant with glutamine in place of Arg 109 has a value of k_{cat} only 1/400 that of native enzyme. The loop closure also causes significant changes in the Raman spectra of the bound NAD^+, especially in vibrational modes that involve the carboxamide group of the nicotinamide ring. These have been interpreted as indicating an increased ridigity of binding of the coenzyme in the closed conformation.

As with many other enzymes acting on polar substrates, a characteristic of the pretransition state complex appears to be formation of a complex with a network of hydrogen bonds extending into the protein, exclusion of most water molecules, and tight packing of protein groups around the substrate. It is also significant that an overall net electrical charge on the ES complexes must be correct for tight binding of substrates to occur for lactate or malate dehydrogenase. Positive and negative charges are balanced except for one excess posi-tive charge which may be needed for catalysis.

Zinc-containing Alcohol Dehydrogenases

Liver alcohol dehydrogenase is a relatively nonspecific enzyme that oxidizes ethanol and many other alcohols. The much studied horse liver enzyme is a dimer of 374-residue subunits, each of which contains a "catalytic" Zn^{2+} ion deeply buried in a crevice between the nucleotide-binding and catalytic domains. The enzyme also contains a "structural" Zn^{2+} ion that is bound by four sulfur atoms from cysteine side chains but does not represent a conserved feature of all alcohol dehydrogenases. The catalytic Zn^{2+} is ligated by sulfur atoms from cysteines 46 and 174 and by a nitrogen atom of the imidazole. In the free enzyme a water molecule is thought to occupy the 4th coordination position and its dissociation to form the $Zn^+ - HO^-$ complex (Eq. 10.5) may account for a pK_3 of 9.2 in the free enzyme and of 7.6 in the NAD^+ complex. The apparent pK_a drops further to about 6.4 in the presence of the substrate

$$Zn^{2+}\text{–}OH, \rightarrow Zn^{+}\text{–}OH + H^{+} \qquad ...(10.5)$$

ethanol. The assignment of this pK_a value has been controversial 48, 49. Histidine, Glu 68 and Asp 49 have side chains close to the zinc and, as we have seen, macroscopic pK_a values of proteins can be *shared* by two or more closely placed groups. Substrate binding also induces a conformational change in this enzyme. When both coenzymes and substrate bind the "closed" conformation of the enzyme is formed by a rotation of the catalytic domains of the two subunits relative to the coenzyme-binding domains. Structures of ternary complexes with inhibitors and with substrates have also been established 50, 51.

For example, liver alcohol dehydrogenase was crystallized as the enzyme•NAD^+• *p*-bromobenzyl alcohol complex with saturating concentrations of substrates in an equilibrium mixture and studied at low resolution. Transient kinetic studies or direct spectroscopic determinations led to the conclusion that the internal equilibrium (E•NAD^+•alcohol = E•NADH•aldehyde) favors the NAD^+•alcohol complex. Subsequently, the complex was studied at higher resolution, and the basic structural features were confirmed with a structure of the enzyme complexed with NAD^+ and 2, 3, 4, 5, 6-pentafluorobenzyl alcohol. From the crystal structures of the $NAD^+ \bullet$ *p*-bromobenzyl alcohol complex and the previously mentioned complex with ethanol and analog III it appears that the oxygen of the alcohol substrate coordinates with the Zn^{2+}, displacing the bound water. Binding of the chromophoric aldehyde 4-*trans*-(N, N-dimethylamino) cinnamaldehyde shifts the wavelength of maximum absorbance by 66 nm, suggesting that Zn^{2+} binds directly to the oxygen of this ligand, Resonance-enhanced Raman spectroscopy of the complex of dimethylamino-benzaldehyde with alcohol dehydrogenase also supports an intermediate in which the Zn^{2+} becomes coordinated directly with the substrate oxygen.

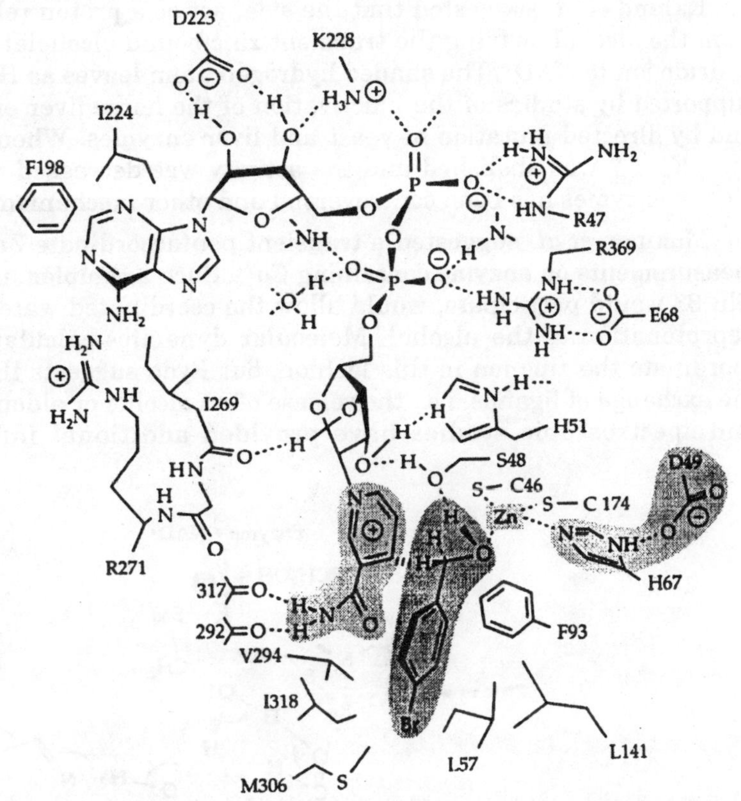

Fig. 10.5. Structure of the complex of horse liver alcohol dehydrogenase with NAD^+ and the slow substrate p-bromobenzyl alcohol. The zinc atom and the nicotinamide ring of the bound NAD^+ are shaded. Adjacent to them is the bound substrate.

Rapid scanning spectrophotometry of complexes with another chromophoric substrate, 3-hydroxy-4-nitrobenzyl alcohol, also suggested that the alcohol first formed a complex with an undissociated alcoholic –OH group and then lost a proton to form a zinc alcoholate complex which could react by hydride ion transfer step *c*.

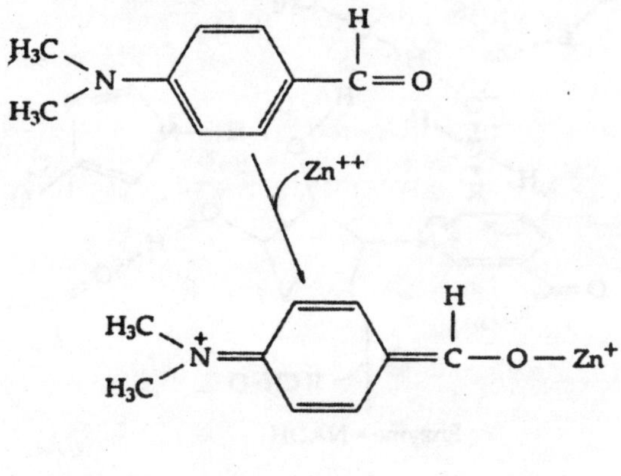

Eklund *et al.* suggested that the side, act as a proton relay system to remove the proton from the alcohol, leaving the transient zinc-bound alcoholate ion, which can then transfer a hydride ion to NAD^+, The shaded hydrogen atom leaves as H^+. The role of His 51 as a base is supported by studies of the inactivation of the horse liver enzyme by diethyl pyrocarbonate and by directed mutation of yeast and liver enzymes. When His 51 was substituted by Gin the pK_a of 7 was abolished and the activity was decreased ten-fold. The functioning of zinc ions in enzymes has been controversial and other mechanisms have been proposed.

Makinen *et al.* suggested a transient pentacoordinate Zn^{2+} complex on the basis of EPR measurements on enzyme containing Co^{2+}. Such a complex, in which the side chain of nearby Glu 68 would participate, would allow the coordinated water molecule to act as the base in deprotonation of the alcohol. Molecular dynamics calculations indicate that Glu 68 can coordinate the zinc ion in this fashion, but Ryde suggests that its function may be to assist the exchange of ligands, *i.e.*, the release of an alcohol or aldehyde product. A variety of kinetic and spectroscopic studies have provided additional information that makes alcohol

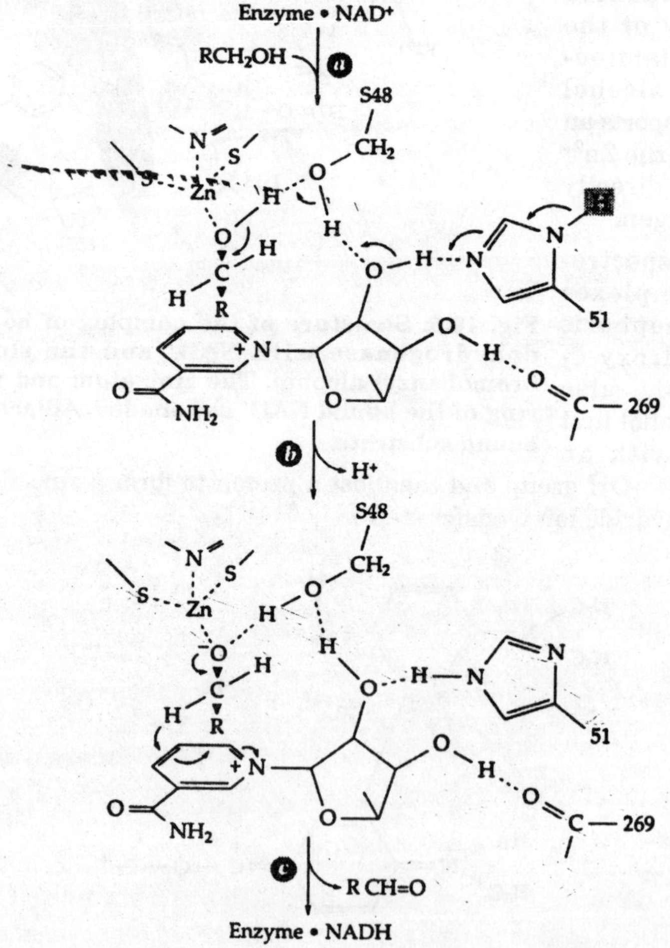

Enzyme • NAD^+

RCH_2OH (a)

S48 CH_2 N S S Zn H O H C H R H O H N N H 51 N O H O C 269 O NH_2 O

(b) H^+

S48 CH_2 N S S Zn H O H C H R H O H N N 51 N O H O C 269 O NH_2 O

(c) R CH=O

Enzyme • NADH

dehydrogenase one of the most investigated of all enzymes. Liver alcohol dehydrogenase is important to the metabolism of ethanol by drinkers. Human beings exhibit small individual differences in their rates of alcohol metabolism which may reflect the fact that there are several isoenzymes and a number of genetic variants whose distribution differs from one person to the next as well as among tissues. Inhibition of these isoenzymes by *uncompetitive* inhibitors, discussed, is a goal in treatment of poisoning by methanol or ethylene glycol.

Inhibition of the dehydrogenases slows the two-step oxidation of these substrates to toxic carboxylic acids. Yeast contains two cytosolic alcohol dehydrogenase isoenzymes. Alcohol dehydrogenase I, present in large amounts in cells undergoing fermentation, functions to reduce acetaldehyde in the fermentation process. Alcohol dehydrogenase II is synthesized by cells growing on such carbon sources as ethanol itself and needing to oxidize it to obtain energy. A third isoenzyme is present in mitochondria. An alcohol dehydrogenase isoenzyme of green plants is induced by anaerobic stress such as flooding. It permits ethanolic fermentation to provide energy temporarily to submerged roots and other tissues. Some bacteria contain an $NADP^+$-dependent, Zn^{2+}-containing alcohol dehydrogenase.

Other Alcohol Dehydrogenases and Aldo-keto Reductases

The oxidation of an alcohol to a carbonyl compound and the reverse reaction of reduction of a carbonyl group are found in so many metabolic pathways that numerous specialized dehydrogenases exist. A large group of "short-chain" dehydrogenases and reductases had at least 57 known members by l995. Their structures and functions are variable. Most appear to be single-domain proteins with a large β sheet, a nucleotide-binding pocket, and a conserved pair of residues: Tyr 152 and Lys 156. These may function in a manner similar to that of the His-Asp pair in lactate dehydrogenase. A possible role of a cysteine side chain has been suggested for another member of this group, 3-hydroxyisobutyrate dehydrogenase, an enzyme of valine catabolism. Dehydrogenases often act primarily to reduce a carbonyl compound rather than to dehydrogenate an alcohol. These enzymes may still be called dehydrogenases.

For example, in the lactic acid fermentation lactate is formed by reduction of pyruvate but we still call the enzyme lactate dehydrogenase. In our bodies this enzyme functions in both directions. However, some enzymes that act mainly in the direction of reduction are called reductases. An example is *aldose reductase*, a member of a family of *aldo-keto reductases* which have $(\alpha/\beta)_8$-barrel structures. The normal physiological function of aldose reductase is uncertain but it can cause a problem in diabetic persons by reducing glucose to sorbitol (glucitol), ribose to ribitol, etc.

The resulting sugar alcohols accumulate in the lens and are thought to promote cataract formation and may also be involved in the severe damage to retinas and kidneys that occurs in diabetes mellitus. Inhibitors of aldose reductase delay development of these complications in animals but the compounds tested are too toxic for human use. The active sites of aldo-keto reductases contain an essential tyrosine (Tyr 48) with a low pK_a value. The nearby His 110, Asp 43, and Lys 77 may also participate in catalysis. The kinetics are unusual. Both NAD^+ and NADH are bound tightly and the overall rate of reduction of a substrate is limited by the rate of dissociation of NAD^+ Citrate is a natural uncompetitive inhibitor of aldose reductase.

Dehydrogenation of Amino Acids and Amines

The dehydrogenation of an amine or the reverse reaction, the reduction of a Schiff base, is another important pyridine nucleotide-dependent process. *Glutamate dehydrogenase*, a large oligomeric protein whose subunits contain 450 or more residues, is the best known enzyme catalyzing this reaction. An intermediate Schiff base of 2-oxo-glutarate and NH_3 is a presumed intermediate. Similar reactions are catalyzed by dehydrogenases for alanine, leucine, phenylalanine, and other amino acids.

Glyceraldehyde-3-Phosphate Dehydrogenase and the Generation of ATP in Fermentation Reactions

The oxidation of an aldehyde to a carboxylic acid, a highly exergonic process, often proceeds through a thioester intermediate whose cleavage can then be coupled to synthesis of ATP. This sequence is of central importance to the energy metabolism of cells. The best known enzyme catalyzing the first step of this reaction sequence is glyceraldehyde 3-phosphate dehydrogenase which functions in the glycolytic sequence. It is present in both prokaryotes and eukaryotes as a tetramer of identical 36- to 37-kDa subunits. Three-dimensional structures have been determined for enzyme from several species, including lobster, *E. coli*, the thermophilic bacterium Bacillus *stearothermophilus,* and trypanosomes.

Recall that aldehydes are in equilibrium with their covalent hydrates. Dehydrogenation of such a hydrate yields an acid but such a mechanism offers no possibility of conserving the energy available from the reaction. However, during catalysis by glyceraldehyde-phosphate dehydrogenase the SH group of Cys 149, adds to the substrate carbonyl group to form an adduct, a thiohemiacetal. This adduct is oxidized by NAD^+ to a thioester, an S-acyl enzyme, which is then cleaved by the same enzyme through a displacement on carbon by an oxygen atom of P_i. The sulfhydryl group of the enzyme is released simultaneously and the product, the acyl phosphate 1, 3-*bisphosphoglycerate*, is formed. The imidazole group of His 176 may catalyze both steps *a* and b.

A separate enzyme then transfers the phospho group from the 1 position of 1, 3-bisphosphoglycerate to ADP to form ATP and 3-phosphoglycerate. The overall sequence the synthesis of one mole of ATP coupled to the oxidation of an aldehyde to a carboxylic acid and the conversion of NAD^+ to NADH.

$$R-CHO + H_2O \overset{a}{\rightleftharpoons} R-CH(OH)_2 \xrightarrow[b]{-2[H]} R-COOH$$

Covalent hydrate

In green plants and in some bacteria an NADP+-dependent cytoplasmic glyceraldehyde 3-phosphate dehydrogenase *does not* use inorganic phosphate to form an acyl phosphate intermediate but gives 3-phosphoglycerate with a free carboxylate. Because it doesn't couple ATP cleavage to the dehydrogenation, it drives the [NADPH]/[$NADP^+$] ratio to a high value favorable to biosynthetic processes.

Animal tissues also contain aldehyde dehydrogenases of a nonspecific type which are thought to act to remove toxic aldehydes from tissues. Like glyceraldehyde 3-phosphate, these enzymes form thioester intermediates but which are hydrolyzed rather than being converted to acyl phosphates. A mutation (E487K) in the mitochondrial enzyme occurs in about 50% of the Asian population. Although the structural alteration is not at the active site, the enzyme activity is low. Individuals carrying the mutation are healthy but have an aversion to alcohol, whose consumption causes an elevated blood level of acetaldehyde, facial flushing, dizziness, and other symptoms.

A similar effect is exerted by the drug disulfiram (Antabuse), which has been used to discourage drinking and whose metabolites are thought to inhibit aldehyde dehydrogenase.

Alcohol dehydrogenases also oxidize aldehydes, probably most often as the geminal diol forms, according. No ATP is formed. The same enzymes can catalyze the dismutation of aldehydes, with equal numbers of aldehyde molecules going to carboxylic acid and to the alcohol.

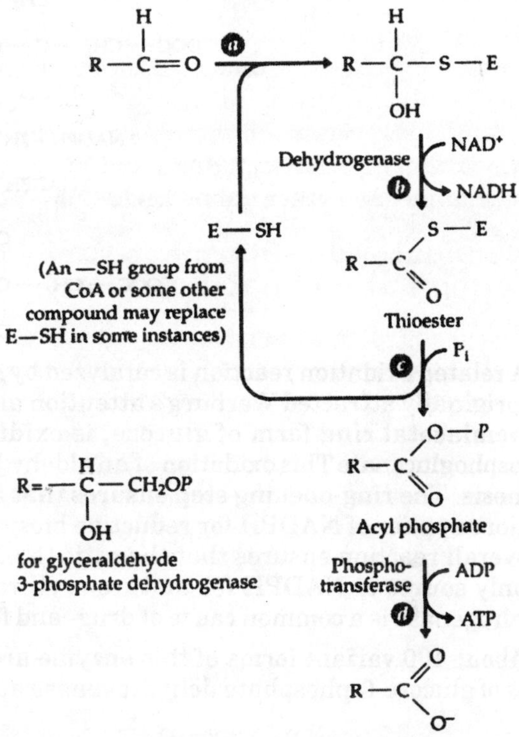

(An —SH group from CoA or some other compound may replace E—SH in some instances)

R= —C(H)(OH)— CH_2OP for glyceraldehyde 3-phosphate dehydrogenase

Fig. 10.6. Generation of ATP coupled to oxidation of an aldehyde to a carboxylic acid. The most important known example of this sequence is the oxidation of glyceraldehyde 3-phosphate to 3-phosphoglycerate. Other important sequences for "substrate-level" phosphorylation.

Reduction of Carboxyl Groups

The last two reaction essence the reverse of the sequence used for synthesis of a thioester such as a fatty acyl coenzyme A. Thus, the chemistry by which ATP is generated during glycolysis and that by which it is utilized in biosynthesis is nearly the same. Furthermore, a standard biochemical method for reduction of carboxyl groups to aldehyde groups is conversion, in an ATP-requiring process, to a thioester followed by reduction of the thioester. For example, the sequence reversed during gluconeogenesis.

The carboxyl group of the side chain of aspartate can be reduced in two steps to form the alcohol homoserine. The aldehyde generated by reduction of a thioester is not always released from the enzyme but may be converted on to the alcohol in a second reduction. This is the case for *3-hydroxy-3-methyl-glutaryl-CoA reductase* (HMG-CoA reductase), a large 887-residue protein that synthesizes mevalonate. This highly regulated enzyme controls the rate of synthesis of cholesterol and is a major target of drugs designed to block cholesterol synthesis. The structure of a smaller 428-residue bacterial enzyme is known. Aspartate 766 is a probable proton donor in both reduction steps and Glu 558 and His 865 may act as a catalytic pair that protonates the sulfur of coenzyme A.

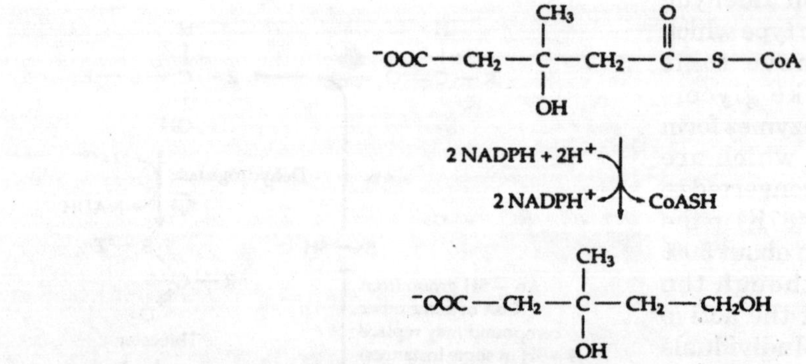

$$^{-}OOC-CH_2-\underset{OH}{\overset{CH_3}{C}}-CH_2-\overset{O}{\overset{\|}{C}}-S-CoA$$

$$\xrightarrow[2\,NADPH^+ \;\; CoASH]{2\,NADPH + 2H^+}$$

$$^{-}OOC-CH_2-\underset{OH}{\overset{CH_3}{C}}-CH_2-CH_2OH$$

A related oxidation reaction is catalyzed by *glucose-6-phosphate dehydrogenase*, the enzyme that originally attracted Warburg's attention and led to the discovery of $NADP^+$. The substrate, the hemiacetal ring form of glucose, is oxidized to a lactone which is then hydrolyzed to 6-phosphogluconate This oxidation of an aldehyde to a carboxylic acid is not linked directly to ATP synthesis. The ring-opening step ensures that the reaction goes to completion. This reaction is a major supplier of NADPH for reductive biosynthesis and the large Gibbs energy decrease for the overall reaction ensures that the ratio [NADPH]/[$NADP^+$] is kept high within cells. This is the only source of NADPH for mature erythrocytes and a deficiency of glucose 6-phosphate dehydrogenase is a common cause of drug- and food-induced hemolytic anemia in human beings.

About 400 variant forms of this enzyme are known. Like the sickle cell trait some mutant forms of glucose-6-phosphate dehydrogenase appear to confer resistance to malaria.

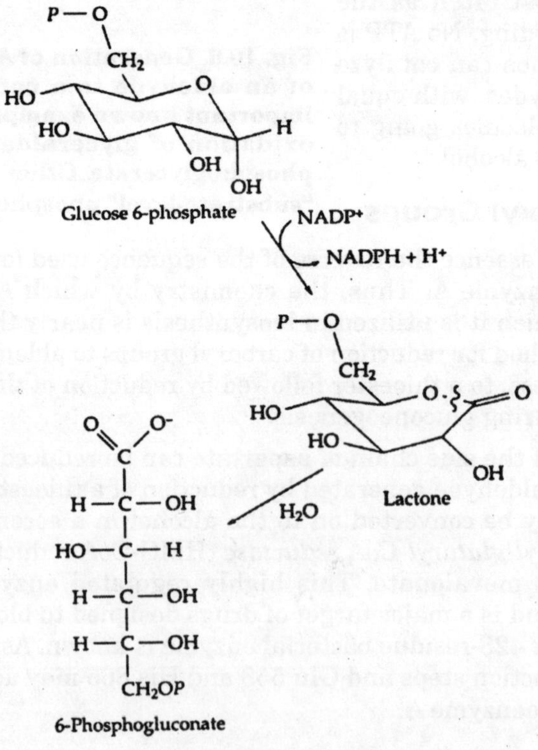

P—O, CH_2, O, HO, HO, H, OH, OH

Glucose 6-phosphate

$NADP^+$ → NADPH + H^+

P—O, CH_2, O, O, HO, HO, OH

Lactone

H_2O

$$\begin{array}{c} O \quad O^- \\ \backslash\!\!\backslash \;\; / \\ C \\ | \\ H-C-OH \\ | \\ HO-C-H \\ | \\ H-C-OH \\ | \\ H-C-OH \\ | \\ CH_2OP \end{array}$$

6-Phosphogluconate

Reduction of Carbon-Carbon Double Bonds

Neither NADP+ nor NAD^+ is a strong enough oxidant to carry out the dehydrogenation of an acyl-CoA. However, NADPH or NADH can participate in the opposite reaction. Thus, NADPH transfers a hydride ion to the β-carbon of an unsaturated acyl group during the biosynthesis of fatty acids and during elongation of shorter fatty acids. A discovery of medical importance is that isonicotinyl hydrazide (INH), the most widely used antituberculosis drug, forms an adduct (of an INH anion or radical) with NAD^+ of long-chain *enoylacyl carrier protein reductase* (enoyl-ACP reductase). This enzyme utilizes NADH in reduction of a C = C double bond during synthesis of mycolic acids. The same enzyme is blocked by *triclosan,* an antibacterial compound used widely in household products such as antiseptic soaps/toothpastes, cosmetics, fabrics, and toys.

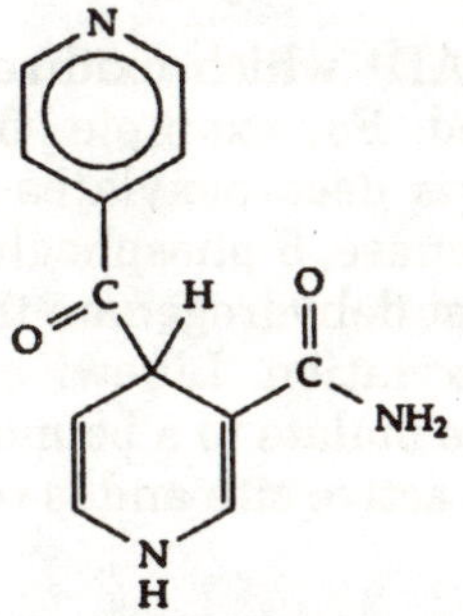

Adduct of isonicotinyl hydrazide anion with NAD^+

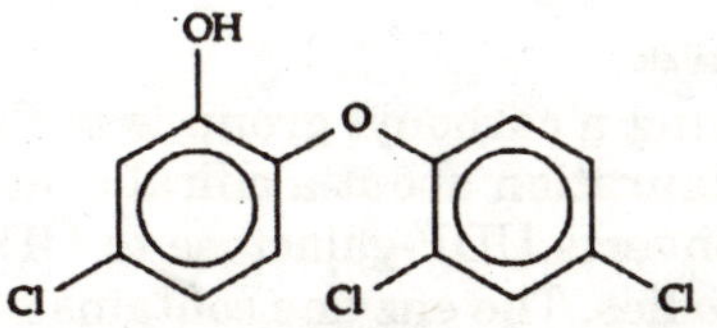

Triclosan

Less frequently NADPH is used to reduce an *isolated double bond*. An example is the hydrogenation of *desmosterol* by NADPH, the final one of the pathywas of biosynthesis of

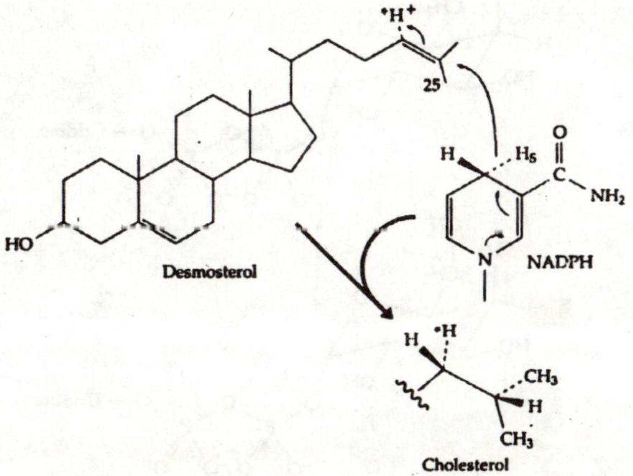

Desmosterol

NADPH

Cholesterol

cholesterol. In this and in other reactions of the same type, hydrogen transfer has been shown to be from the *pro-S* position in NADPH directly to C-25 of the sterol.

The proton introduced from the medium enters *trans* to the H^- ion from NADPH. The proton always adds to the more electron-rich terminus of the double bond, *i.e.*, it follows the Markovnikov rule. This result suggests that protonation of the double bond may precede H^- transfer. Additional pyridine nucleotide-dependent dehydrogenases include *glutathione reductase, dihydrofolate reductase*, isocitrate dehydrogenase, *sn*-glycerol-3-phosphate *dehydrogenase*, L-3-hydroxyacyl-CoA dehydrogenase, *retinol dehydrogenase,* and a bacterial quinone oxidoreductase. Some of these also contain a flavin coenzyme.

Transient Carbonyl Groups in Catalysis

Some enzymes contain bound NAD^+ which oxidizes a substrate alcohol to facilitate a reaction step and is then regenerated. For example, the *malolactic enzyme* found in some lactic acid bacteria and also in *Ascaris* decarboxylates L-malate to lactate. This reaction is similar to those of isocitrate dehydrogenase, 6-phosphogluconate dehydrogenase, and the malic enzyme, which utilize free NAD^+ to first dehydrogenate the substrate to a bound oxoacid whose β carbonyl group facilitates decarboxylation. Likewise, the bound NAD^+ of the malolactic enzyme apparently dehydrogenates the malate to a bound oxaloacetate which is decarboxylated to pyruvate. The latter remains in the active site and is reduced by the bound NADH to lactate which is released from the enzyme.

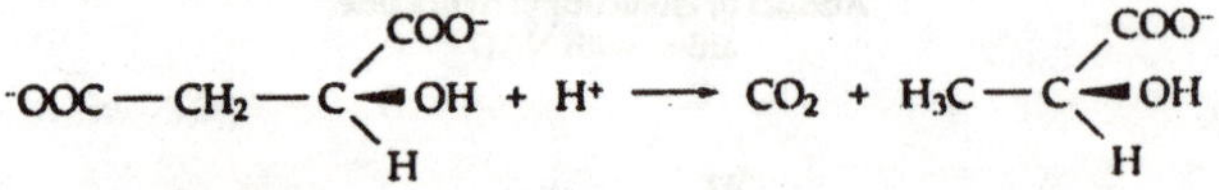

$$^-OOC-CH_2-C(COO^-)(H)-OH + H^+ \longrightarrow CO_2 + H_3C-C(COO^-)(H)-OH$$

L-Malate L-Lactate

Another reason for introducing a carbonyl group is to form a symmetric intermediate in a reaction that inverts the configuration about a chiral center. An example is *UDP-galactose 4-epimerase*, an enzyme that converts UDP-galactose to UDP-glucose and is essential in the metabolism of galactose in our bodies. The enzyme contains bound NAD^+ and forms a transient 4-oxo intermediate and bound NADH. Rotation of the intermediate allows nonstereo-specific reduction by the NADH, leading to epimerization.

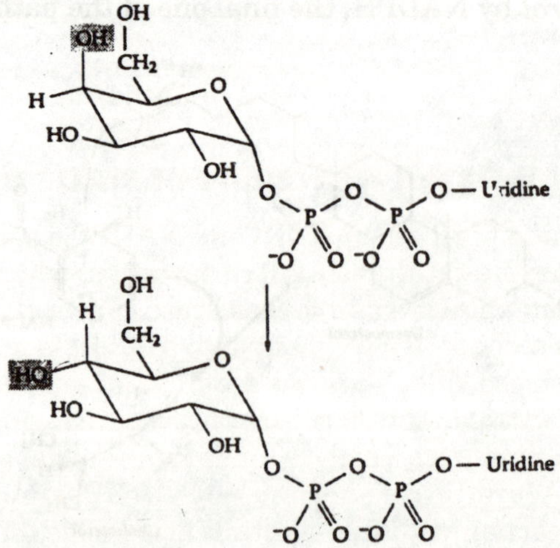

The enzyme is a member of the short-chain dehydrogenase group with a catalytic Tyr-Lys pair in the active site. Another way that formation of an oxo group can assist in epimerization of a sugar is through enolization with nonstereospecific return of a proton to the intermediate enediol. A third possible mechanism of epimerization is through aldol cleavage followed by aldol condensation, with inversion of configuration. In each case the initial creation of an oxo group by dehydrogenation is essential. Bound NAD^+ is also present in *S-adenosylhomocysteine hydrolase*, which catalyzes the irreversible reaction.

Transient oxidation at the 3 position of the ribose ring facilitates the reaction. The reader can doubtless deduce the function that has been established for the bound NAD^+ in this enzyme. However, the role of NAD in the *urocanase* reaction puzzling. This reaction, which is the second step in the catabolism of histidine, appears simple. However, there is no obvious mechanism and no obvious role for NAD^+. See Frey for a discussion.

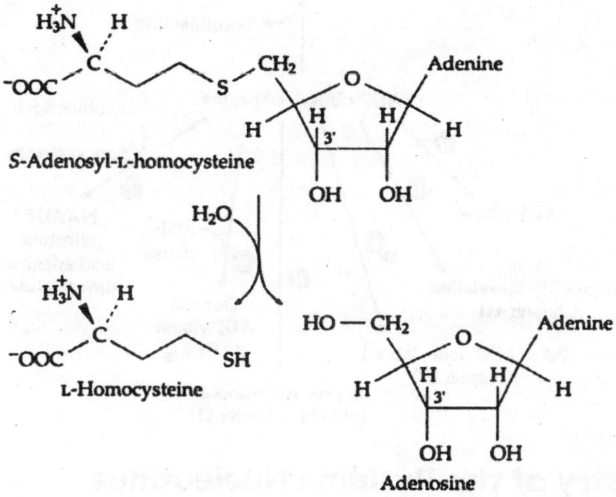

S-Adenosyl-L-homocysteine

H_2O

L-Homocysteine

Adenosine

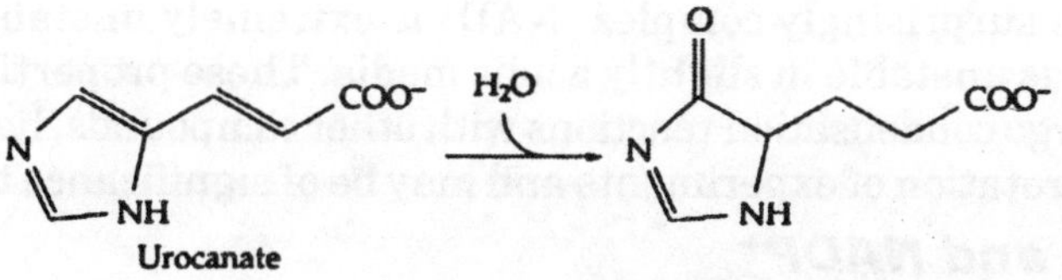

Urocanate

ADP Ribosylation and Related Reactions of NAD^+ and $NADP^+$

The linkage of nicotinamide to ribose in NAD^+ and $NADP^+$ is easily broken by nucleophilic attack on C-1 of ribose. In enzyme-catalyzed ADP ribosylation, which can be, discussed briefly. The nucleophilic group ^-Y from an ADP-ribosyltransferase carries the ADP-ribosyl group which can then be transferred by a second displacement onto a suitable nucleophilic acceptor group. Hydrolysis gives free ADP-ribose. Other known products of enzymatic action are indicated. Poly-(ADP ribosylation) is discussed. The structure of cyclic ADP-ribose (cADPR). The acceptor nucleophile is N-1 of the adenine ring which is made more nucleophilic by electron donation from the amino group. A similar reaction with ADP ribose produces a dimeric ADP-ribose $(ADPR)_2$ while reaction with free nicotinic acid yields, in an overall base exchange, nicotinic acid adenine dinucleotide (NAADP+). Some of these compounds, *e.g.*, cADPR, NAADP+, and $(ADPR)_2$, are involved in signaling with calcium ions.

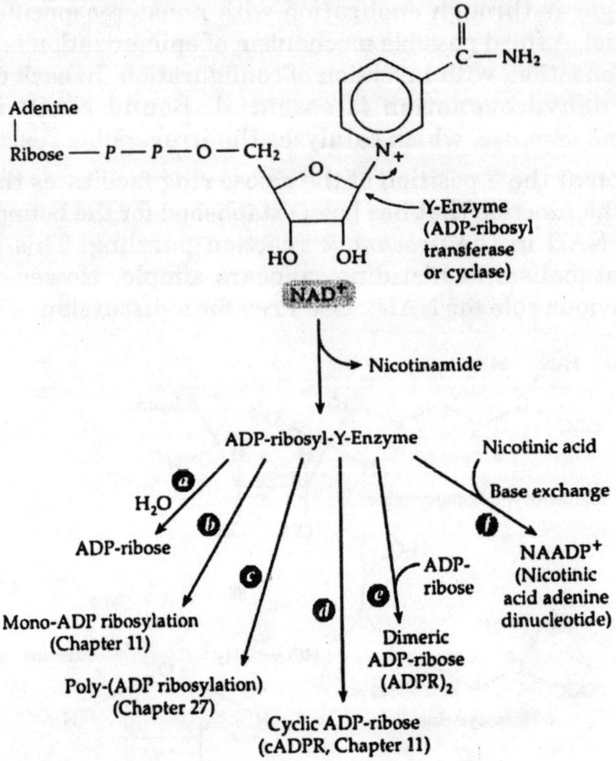

The Varied Chemistry of the Pyridine Nucleotides

Despite the apparent simplicity of their structures, the chemistry of the nicotinamide ring in NAD^+ and $NADP^+$ is surprisingly complex. NAD^+ is extremely unstable in basic solutions, whereas NADH is just as unstable in slightly acidic media. These properties, together with the ability of NAD^+ to undergo condensation reactions with other compounds, have sometimes caused serious errors in interpretation of experiments and may be of significance to biological function.

Addition to NAD^+ and $NADP^+$

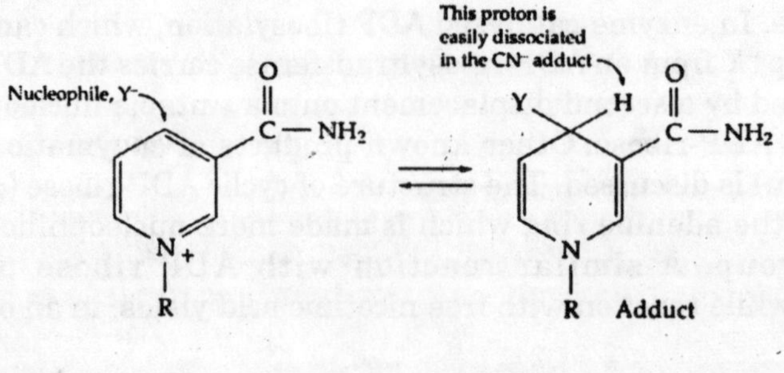

Many nucleophilic reagents add reversibly at the para (or 4) position to form adducts having structures resembling those of the reduced coenzymes. Formation of the cyanide adduct, whose absorption maximum is at 327 nm, has been used to introduce deuterium into the para position of the pyridine nucleotides. In the adduct, the hydrogen adjacent to the highly polarized C – N bond is easily dissociated as a proton. Thiolate ariions and bisulfite also add. Dithionite ion, $S_2O_4^{2-}$, can lose SO_2 and acquire a proton to form the sulfoxylate ion HSO_2^- which also adds to the 4 position of the NAD^+ ring. The resulting adduct is unstable and loses SO_2 to give NADH + H^+. Addition can also occur at the two ortho positions. The adducts of HO^- to the 4 position of NAD^+ are stable but those to the 2 position undergo ring opening in base-catalyzed reactions which are followed by further degradation.

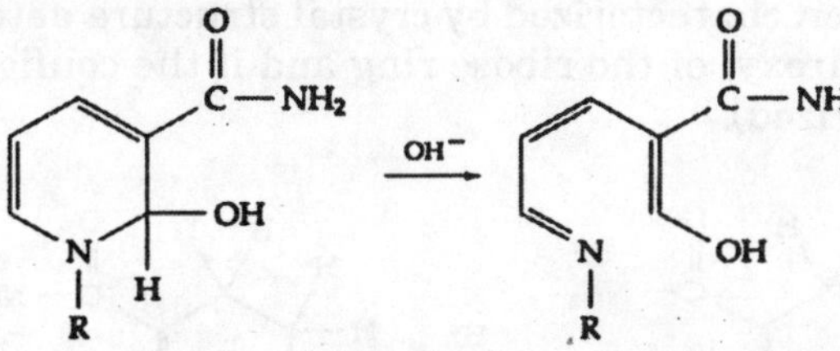

Another base-catalyzed reaction is the addition of enolate anions derived from ketones to the 4 position of the pyridine nucleotides. The adducts undergo ring closure and in the presence of oxygen are converted slowly to fluorescent materials. While forming the basis for a useful analytical method for determination of NAD^+, these reactions also have created a troublesome enzyme inhibitor from traces of acetone present in commercial NADH.

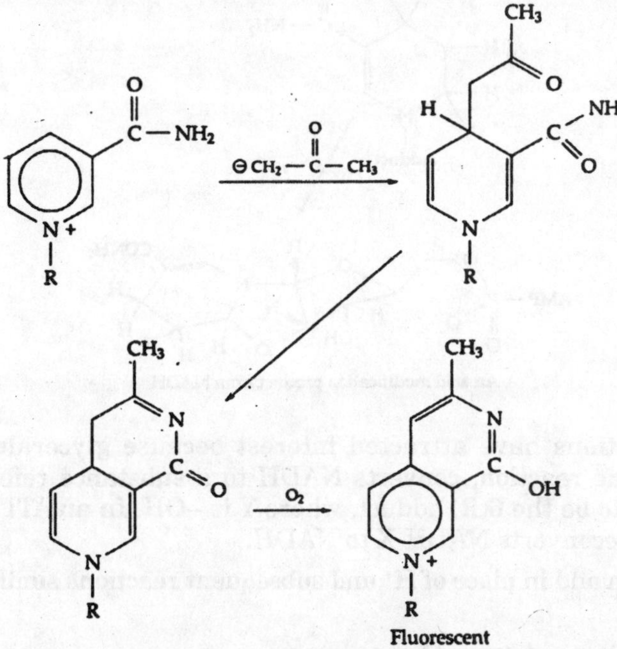

Fluorescent

The reactions occur nonenzymatically only under the influence of strong base but dehydrogenases often catalyze similar condensations relatively rapidly and reversibly. Pyruvate inhibits lactate dehydrogenase, 2-oxoglutarate inhibits glutamate dehydrogenase, and ketones inhibit a short-chain alcohol dehydrogenase in this manner.

Modification of NADH in Acid

Reduced pyridine nucleotides are destroyed rapidly in dilute HC1 and more slowly at pH 7 in reactions catalyzed by buffer acids. Apparently the reduced nicotinamide ring is first protonated at C-5, after which a nucleophile Y^- adds at the 6 position. The nucleophile may be OH^-, and the adduct may undergo further reactions. For example, water may add to the other double bond and the ring may open on either side of the nitrogen. The glycosidic linkage can be isomerized from P to a or can be hydrolytically cleaved. The early steps in the modification reaction are partially reversible, but the overall sequence is irreversible. One of the products, which has been characterized by crystal structure determination. It can arise if the group Y of the C-2′ hydroxyl of the ribose ring and if the configuration of the glyco-sidic linkage is inverted (anomerized).

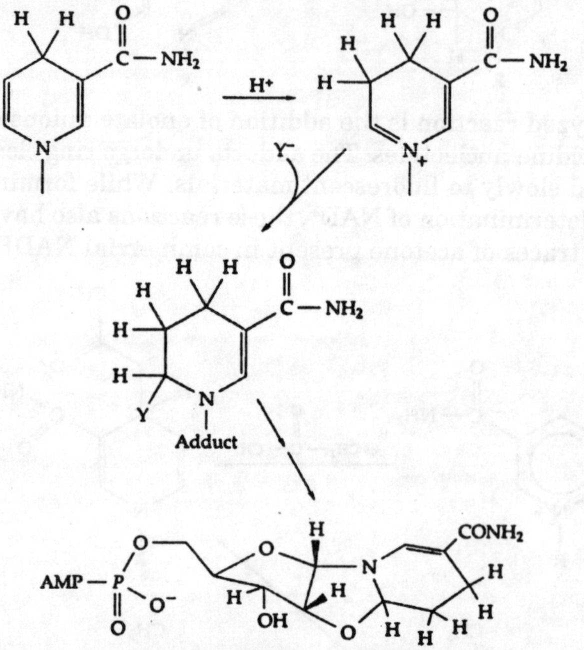

An acid modification product from NADH

The foregoing reactions have attracted interest because glyceraldehyde-3-phosphate dehydrogenase, in a side reaction, converts NADH to a substance referred to as NADH-X which has been shown to be the 6(R) adduct, where Y is –OH. In an ATP-dependent reaction an enzyme from yeast reconverts NADH-X to NADH.

Mercury (II) ions can add in place of H^+ and subsequent reactions similar to those promoted by acid can occur.

Other Reactions of Pyridine Nucleotides

Alka-line hexacyanoferrate (III) oxidizes NAD^+ and $NADP^+$ to 2-, 4-, and 6-pyridones. The 6-pyridone of N-methyl-nicotinamide is a well-known excretion product of nicotinic acid in mammals. Reoxidation of NADH and NADPH to NAD^+ and NADP+ can be accomplished with hexacyanoferrate (III), quinones, and riboflavin but not by H_2O_2 or O_2. However, O_2 does

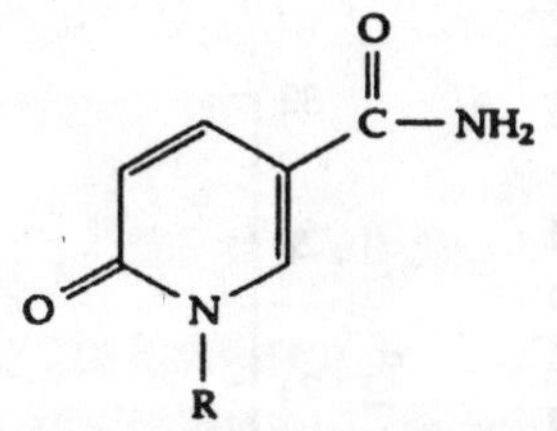

6-Pyridone

react at neutral pH with uptake of a proton to form a peroxide derivative of NADH. When heated in 0.1 N alkali at 100°C for 5 min, NAD^+ is hydrolyzed to nicotinamide and adenosine-cliphosphate-ribose.

Treatment of NAD^+ with nitrous acid deaminates the adenine ring. The resulting deamino NAD^+ as well as synthetic analogs containing the following groups in place of the carboxyamide have been used

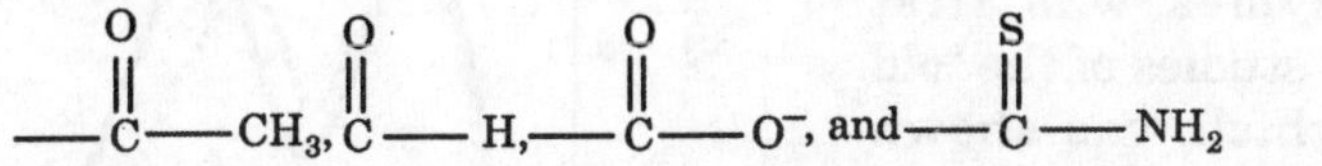

$$-\overset{O}{\overset{\|}{C}}-CH_3,\ \overset{O}{\overset{\|}{C}}-H,\ -\overset{O}{\overset{\|}{C}}-O^-,\ \text{and}\ \overset{S}{\overset{\|}{C}}-NH_2$$

widely in enzyme studies. In fact, almost every part of the coenzyme molecule has been varied systematically and the effects on the chemical and enzymatic properties have been, investigated. "Caged" NAD^+ and $NADP^+$ have also been made. These compounds do not react as substrates until they are released ("uncaged") by photolytic action of a laser beam.

THE FLAVIN COENZYMES

Flavin adenine diphosphate (FAD, flavin adenine dinucleotide) and *riboflavin 5'-monophosphate* (FMN, flavin mononucleotide), whose structures, are perhaps the most

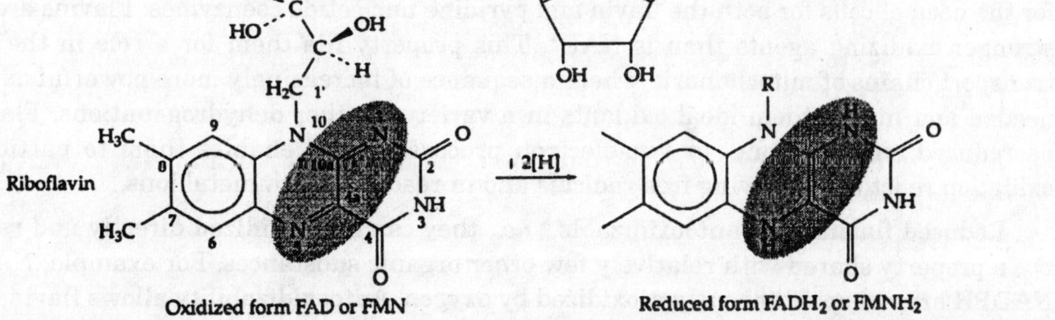

Riboflavin 5'-phosphate or FMN — AMP

Riboflavin

Oxidized form FAD or FMN — 2[H] → Reduced form $FADH_2$ or $FMNH_2$

Fig. 10.7. The flavin coenzymes flavin adenine dinucleotide (FAD) and riboflavin 5'-phosphate (FMN). Dotted lines enclose the region that is altered upon reduction.

versatile of all the oxidation coenzymes. The name flavin adenine dinucleotide is not entirely appropriate because the D-ribityl group is not linked to the riboflavin in a glycosidic linkage. Hemmerich suggested that FAD be called flavin adenine diphosphate. The attention of biochemists was first attracted to flavins as a result of their color and fluorescence. The study of spectral properties of flavins has been of importance in understanding these coenzymes. The biochemical role of the flavin coenzymes was first recognized through studies of the "old yellow enzyme" which was shown by Theorell to contain riboflavin 5′-phosphate.

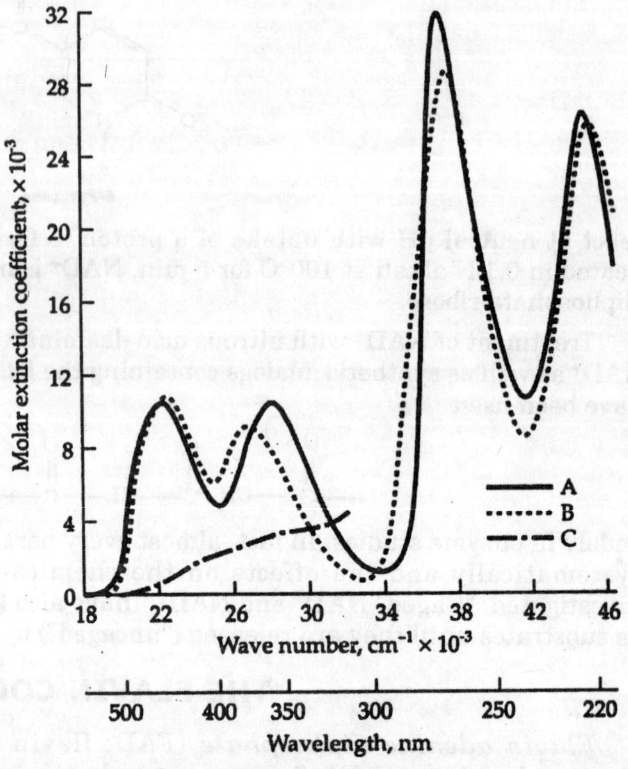

Fig. 10.8. Absorption spectrum of neutral, uncharged riboflavin (A), the riboflavin anion (B), and reduced to the dihydro form, the action of light in the presence of EDTA (C). A solution of 1.1 × 10^{-4} M riboflavin containing 0.01 M EDTA was placed 11.5 cm from a 40-W incandescent lamp for 30 min.

By 1938, FAD was recognized as the coenzyme of a different yellow protein, *D-amino acid oxidase* of kidney tissue. Like the pyridine nucleotides, the new flavin coenzymes were reduced by dithionite to nearly colorless dihydro forms revealing the chemical basis for their function as hydrogen carriers.

Flavins are also among the natural light receptors and display an interesting and much studied photochemistry. Flavins may function in some photoresponses of plants, and they serve as light emitters in bacterial bioluminescerice. Three facts account for the need of cells for both the flavin and pyridine nucleotide coenzymes: Flavins are usually stronger oxidizing agents than is NAD^+. This property fits them for a role in the electron transport chains of mitochondria where a sequence of increasingly more powerful oxidants is needed and makes them ideal oxidants in a variety of other dehydrogenations. Flavins can be reduced either by one- or two-electron processes. This enables them to participate in oxidation reactions involving free radicals and in reactions with metal ions.

Reduced flavins are "autooxidizable," *i.e.,* they can be reoxidized directly and rapidly by O_2, a property shared with relatively few other organic substances. For example, NADH and NADPH are not spontaneously reoxidized by oxygen. Autooxidizability allows flavins of some enzymes to pass electrons directly to O_2 and also provides a basis for the functioning of flavins in hydroxylation reactions.

Flavoproteins and their Reduction Potentials

Flavin coenzymes are usually bound tightly to proteins and cycle between reduced and oxidized states while attached to the same protein molecule. In a free unbound coenzyme the redox potential is determined by the structures of the oxidized and reduced forms of the couple. Both riboflavin and the pyridine nucleotides contain aromatic ring systems that are stabilized by resonance. Part of this resonance stabilization is lost upon reduction. The value of $E^{\circ\prime}$ depends in part upon the varying amounts of resonance in the oxidized and reduced forms. The structures of the coenzymes have apparently evolved to provide values of $E^{\circ\prime}$ appropriate for their biological functions. The relative strengths of binding of oxidized and reduced flavin coenzymes to a protein also have strong effects upon the reduction potential of the coenzyme. If the oxidized form is bound weakly, but the reduced form is bound tightly, a bound flavin will have a greater tendency to stay in the reduced form than it did when free.

The reduction potential $E^{\circ\prime}$ will be less negative than it is for the free flavindihydroflavin couple. On the other hand, if the oxidized form of the flavin is bound more tightly by the protein than is the reduced form, $E^{\circ\prime}$ will be more negative and the flavoenzyme will be a less powerful oxidizing agent. In fact, the values of $E^{\circ\prime}$ at pH 7 for flavoproteins span a remarkably wide range from –0.49 to +0.19 V. The state of protonation of the reduced flavin when bound to the enzyme will also have a major effect on the oxidation-reduction potential. For example, acyl-CoA dehydrogenases are thought to form an anionic species of reduced FAD ($FADH^-$) which is tightly bound to the protein. Every flavoprotein accepts electrons from the substrate that it oxidizes and passes these electrons on to another substrate, an oxidant. In the following sections we will consider for several enzymes how the electrons may get into the flavin from the oxidizable substrate and how they may flow out of the flavin into the final electron acceptor.

Typical Dehydrogenation Reactions Catalyzed by Flavoproteins

The functions of flavoprotein enzymes are numer-ous and diversified. A few of them are, and are classified there as follows: (A) oxidation of hemiacetals to lactones, (B) oxidation of alcohols to aldehydes or ketones, (C) oxidation of amines to imines, (D) oxidation of carbonyl compounds or carboxylic acids to α, β-unsaturated compounds, (E) oxidation of NADH and NADPH in electron transport chains, and (F) oxidation of dithiols to disulfides or the reverse reaction. Three-dimensional structures are known for enzymes of each of these types. Reactions of types A-C could equally well be catalyzed by pyridine nucleotide-requiring dehydrogenases. Recall that D-glucose-6-phosphate dehydro-genase uses NADP+ as the oxidant. The first product is the lactone which is hydrolyzed to 6-phosphogluconic acid. The similar reaction of free glucose is catalyzed by fungal *glucose oxidase*, a 580-residue FAD-containing enzyme.

A bacterial cholesterol oxidase has a similar structure. The important plant enzyme *glycolate oxidase* is a dimer containing riboflavin 5′-phosphate. It catalyzes a reaction of type B, which plays an important role in photorespiration. *Amino acid oxidases* (reaction type C) are well-known. The peroxisomal D-amino acid oxidase from kidney was the source from which Warburg first isolated FAD and has been the subject of much investigation of mechanism and structure. Many snake venoms contain an active 140-kDa L-amino-acid oxidase which contains FAD. Flavin-dependent *amine oxidases*, important in the human body, catalyze the related reaction with primary, secondary, or tertiary amines and in which a carboxyl group need not be present. Reduced flavin produced by all of these oxidases is reoxidized with molecular oxygen and hydrogen peroxide is the product. Nature has chosen to forego the use of an electron transport; giving up the possible gain of ATP in favor of simplicity and a more direct reaction with oxygen.

In some cases there is specific value to the organism in forming H_2O_2. In contrast to the flavin oxidases, flavin dehydro-genases pass electrons to carriers within electron transport chains and the flavin does *not* react with O_2. Examples include a bacterial *trimethylamine dehydrogenase* which contains an iron-sulfur cluster that serves as the immediate electron acceptor and yeast *flavocytochrome* b_2, *a* lactate dehydrogenase that passes electrons to a built-in heme group which can then pass the electrons to an external acceptor, another heme in cytochrome *c*. Like glycolate oxidase, these enzymes bind their flavin coenzyme at the ends of 8-stranded αβ barrels similar to that of triose phosphate isomerase. Flavocytochrome b_2 has an additional domain which carries the bound heme.

Two additional domains of trimethylamme dehydrogenase have a topology resembling that of the FAD- and NADH-binding domains of glutathione reductase. A bacterial *mandelate dehydrogenase* is structurally and mechanistically closely related to the glycolate oxidase family.

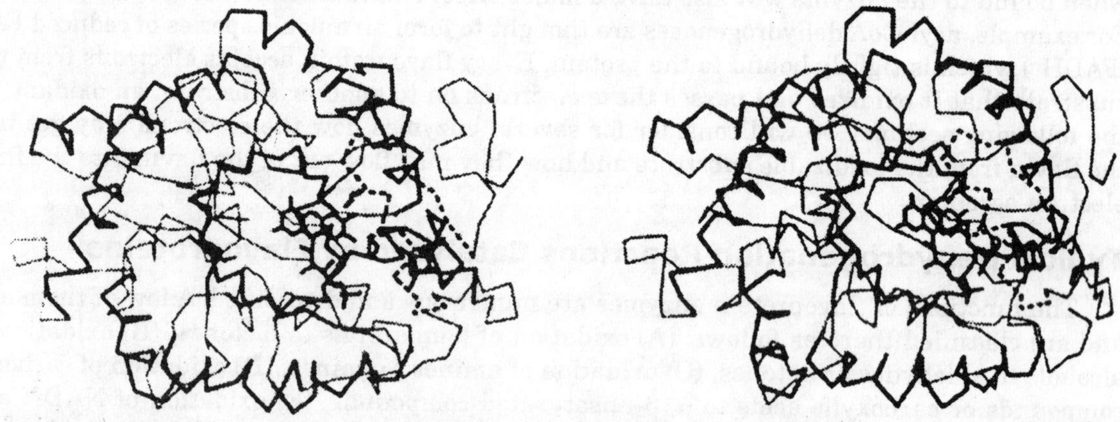

Fig. 10.9. Stereoscopic view of the large domain (residues 1-383) of trimethylamine dehydrogenase from a methylotrophic bacterium. The helices and β strands of the $(\alpha\beta)_8$ barrel are drawn in heavy lines as are the FMN (center) and the Fe_4S_4 iron-sulfur cluster at the lower right edge. The αβ loop to which it is bound is drawn with dashed lines. The 733-residue protein also contains two other structural domains.

Reaction type D of Table 10.2, the dehydrogenation of an acyl-coenzyme A (CoA), could not be accomplished by a pyridine nucleotide system because the reduction potential ($E°'$, pH 7 = –0.32 V) is inappropriate. The more powerfully oxidizing flavin system is needed. (However, the reverse reaction, hydrogenation of a C = C bond, is often carried out biologically, with a reduced pyridine nucleotide.) Dehydrogenation reactions of mis type are important in the energy metabolism of aerobic cells. For example, the first oxidative step in the β oxidation of fatty acids the αβ dehydrogenation of fatty acyl-CoA derivatives. The *pro-R* hydrogen atoms are removed from both the α- and β-carbon atoms to create the double bond.

Table 10.2. Some Dehydrogenation Reaction Catalyzed by Flavoproteins.

A $$\text{D-Glucose} \xrightarrow[\text{Glucose oxidase}]{-2[\text{H}]} \text{Gluconolactone} \xrightarrow{H_2O} \text{D-Gluconic acid}$$

(See Eq. 17-12 for structures in a closely related reaction.)

B $$HOOC{-}CH_2{-}OH \xrightarrow[\text{Glycolate oxidase}]{-2[\text{H}]} HOOC{-}CH{=}O$$

C $$R{-}CH(NH_2){-}COOH \xrightarrow{-2[\text{H}]} R{-}C({=}NH){-}COOH \xrightarrow{H_2O} R{-}C({=}O){-}COOH + NH_3$$

Amino acid oxidases

D $$R{-}CH_2{-}CH_2{-}C({=}O){-}S{-}CoA \xrightarrow[\text{Acyl-CoA dehydrogenases}]{-2[\text{H}]} R{-}CH{=}CH{-}C({=}O){-}S{-}CoA$$

E $$\text{NADH (or NADPH)} + H^+ \xrightarrow{-2[\text{H}]} NAD^+ \text{(NADP)}$$

F $$\text{(dithiol, HS, SH)}{-}C({=}O){-}NH{-}R \xrightarrow[\text{GSH}]{-2[\text{H}]} \text{(cyclic disulfide, S—S)}{-}C({=}O){-}NH{-}R$$

Dihydrolipoic acid amide

[a] These are shown as removal of two H atoms [H] and may occur by transfer of H^-, $H^+ + e^-$, or $2H^+ + 2e^-$. They represent reaction type 9C of the table inside the back cover.

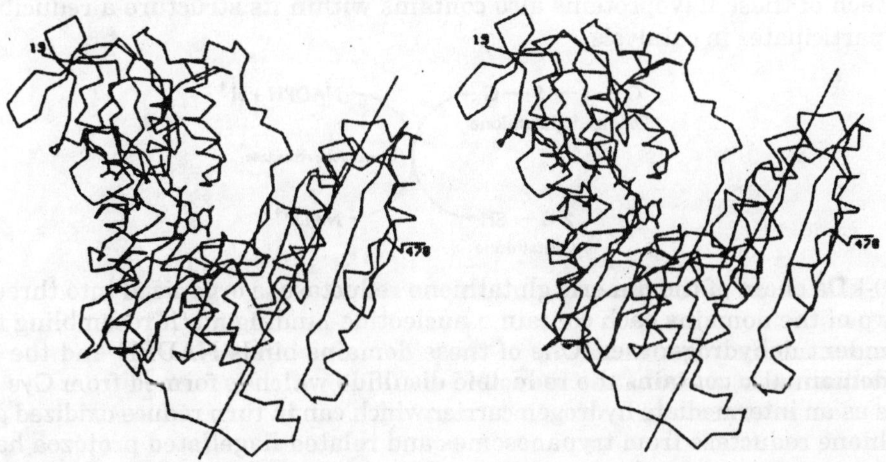

Fig. 10.10. The three-dimensional structure of glutathione reductase. Bound FAD is shown. NAD⁺ binds to a separate domain below the FAD. The two cysteine residues forming the reducible disulfide loop are indicated by dots.

Animal mitochondria contain several different *acyl-CoA dehydrogenases* with differing preferences for chain length or branching pattern. A related reaction that occurs in the citric acid cycle is dehydrogenation of succinate to fumarate by *succinate dehydrogenase*. The dehydrogenation also involves *trans* removal of one of the two hydrogens, one *pro-S* hydrogen and one *pro-R*. The enzyme has a large 621-residue flavoprotein subunit and a smaller 27-kDa iron-sulfur protein subunit.

$$^{-}OOC\text{–}CH_SH_R\text{–}CH_SH_R\text{–}COO^{-} \xrightarrow[-2[H]]{\text{Succinate dehydrogenase}} {}^{-}OOC\text{–}CH\text{=}CH\text{–}COO^{-}$$

Neither the acyl-CoA dehydrogenases nor succinate dehydrogenase react with O_2. Acyl-CoA dehydrogenases pass the electrons removed from substrates to another flavoprotein, a soluble electron transferring flavoprotein, which carries the electrons to an iron-sulfur protein embedded in the inner mitochondrial membrane where they enter the electron-transport chain. Succinate dehydrogenase as well as NADH dehydrogenase are embedded in the same membrane and also pass their electrons to iron-sulfur clusters and eventually to oxygen through the electron transport chain of the mitochondria.

Fumarate reductase has properties similar to those of succinate dehydrogenase but catalyzes the opposite reaction in "anaerobic respiration" as do similar reductases of bacteria and of some eukaryotes. *Dihydrolipoyl dehydrogenase* (lipoamide dehydrogenase), *glutathione reductase*, and human *thioredoxin reductase* belong to a subclass of flavoproteins that act on dithiols or disulfides. The reaction catalyzed by the first of these is illustrated. The other two enzymes usually promote the reverse type of reaction, the reduction of a disulfide to two SH groups by NADPH. Glutathione reductase splits its substrate into two halves while reduction of the small 12-kDa protein ***thioredoxin*** simply opens a loop in its peptide chain.

The reduction of lipoic acid opens the small disulfide-containing 5-membered ring in that molecule. Each of these flavoproteins also contains within its structure a reducible disulfide group that participates in catalysis.

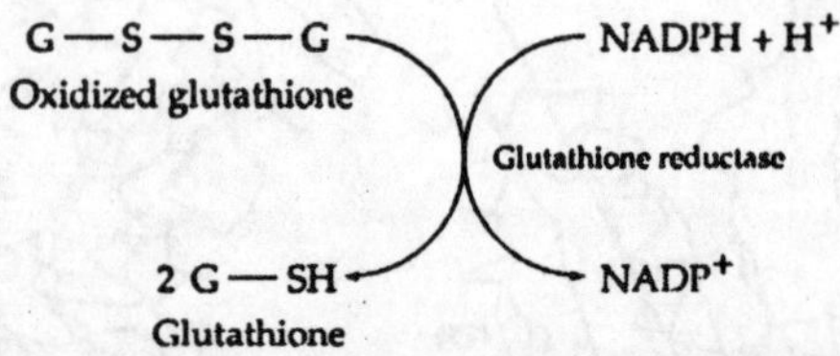

Each 50-kDa chain of the dimeric glutathione reductase is organized into three structural domains Two of the domains each contain a nucleotide-binding motif resembling those of the NAD$^+$-dependent dehydrogenases. One of these domains binds NADPH and the other FAD. The latter domain also contains the reducible disulfide which is formed from Cys 58 and Cys 63. It serves as an intermediate hydrogen carrier which can in turn reduce oxidized glutathione. A trypanothione reductase from trypanosomes and related flagellated protozoa has a similar structure and acts on trypanothione, which replaces glutathione in these organisms.

Because it is unique to trypanosomes, this enzyme is a target for design of drugs against these organisms which cause such terrible diseases as African sleeping sickness and Chagas disease. Another flavoprotein constructed on the glutathione reductase pattern is the bacterial plasmidencoded *mercuric reductase* which reduces the highly toxic Hg^{2+} to volatile elemental mercury, Hg^0. A reducible disulfide loop corresponding to that in glutathione reductase is present in this enzyme but there is also a second pair of cysteines nearby. All of these may participate in binding and reduction of Hg^{2+}.

More Flavoproteins

Flavoproteins function in virtually every area of metabolism and we have considered only a small fraction of the total number. Here are a few more. Flavin-dependent reductases use hydrogen atoms from NADH or NADPH to reduce many specific substances or classes of com-pounds. The FAD-containing ferredoxin: $NADP^+$ oxidoreductase catalyzes the reduction of free $NADP^+$ by reduced ferredoxin generated in the chloroplasts of green leaves. Similar enzymes, some of which utilize reduced flavodoxins, are found in bacteria. The FMN-containing subunit of NADH: ubiquinone oxidoreductase is an essential link in the mitochondrial electron transport chain for oxidation of NADH in plants and animals and for related processes in bacteria. *Glutamate synthase,* a key enzyme in the nitrogen metabolism of plants and microorganisms, uses electrons from NADPH to reduce 2-oxoglutarate to glutamate in a complex glutamine-dependent process. The enzyme contains both FMN and FAD and three different iron-sulfur clusters.

Flavin reductases use NADH or NADPH to reduce free riboflavin, FMN, or FAD needed for various purposes including emission of light by luminous bacteria. They provide electrons to many enzymes that react with O_2 such as the cytochromes P450 and nitric oxide synthase. An example is adrenodoxin reductase, which passes electrons from NADPH to cytochrome P450 via the small redox protein adrenodoxin. This system functions in steroid biosynthesis as is indicated. Other flavin-dependent reductases have protective functions catalyzing the reduction of ascorbic acid radicals, toxic quinones, and peroxides.

Modified Flavin Coenzymes

Mitochondrial succinate dehydrogenase, which catalyzes the reaction, contains a flavin prosthetic group that is covalently attached to a histidine side chain. This modified FAD was isolated and identified as 8α-($N^{\varepsilon 2}$-histidyl)-FAD. The same prosthetic group has also been found in several other dehydrogenases. It was the first identified member of a series of modified FAD or riboflavin 5′-phosphate derivatives that are attached by covalent bonds to the active sites of more than 20 different enzymes. These include 8α-($N^{\varepsilon 2}$-histidyl)-FMN, 8α-($N^{\delta 1}$-histidyl)-FAD, 8α-(*O*-tyrosyl-FAD), and 6-(*S*-cysteinyl)-riboflavin 5′-phosphate, which is found in trimethylamine dehydrogenase.

An 8-hydroxy analog of FAD (-OH in place of the 8-CH_3) has been isolated from a bacterial electron-transferring flavoprotein. Commercial FAD may contain some riboflavin 5′-pyrophosphate which activates some flavoproteins and inhibits others.

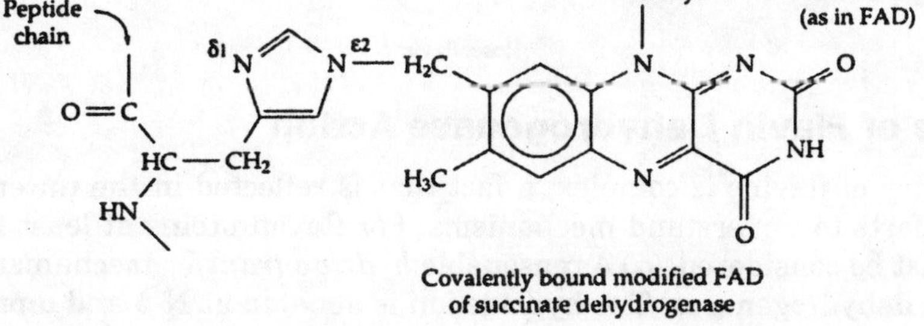

Covalently bound modified FAD of succinate dehydrogenase

Methanogenic bacteria contain a series of unique coenzymes among which is *coenzyme* F_{420}, a 5-deazaflavin substituted by H at position 7 and –OH at position 8 (8-hydroxy-7, 8-didemethyl-5-deazariboflavin).

Coenzyme F_{420} (oxidized form)

This unique redox catalyst links the oxidation of H_2 or of formate to the reduction of NADP and also serves as the reductant in the final step of methane biosynthesis. It resembles NAD^+ in having a redox potential of about –0.345 volts and the tendency to be only a two-electron donor. More recently free 8-hydroxy-7, 8-didemethyl-5-deazaribo-flavin has been identified as an essential light-absorbing chromophore in DNA photolyase of *Methanobacteriwn,* other bacteria, and eukaryotic algae. *Roseoflavin* is not a coenzyme but an antibiotic from *Streptomyces davmvensis.* Many synthetic flavins have been used in studies of mechanisms and for NMR and other forms of spectroscopy.

Roseoflavin

Mechanisms of Flavin Dehydrogenase Action

The chemistry of flavins is complex, a fact that is reflected in the uncertainity that has accompanied efforts to understand mechanisms. For flavoproteins at least four mechanistic possibilities must be considered. (*a*) A reasonable *hydride-transfer* mechanism can be written for flavoprotein dehydrogenases. The hydride ion is donated at N-5 and a proton is accepted at N-1. The oxidation of alcohols, amines, ketones, and reduced pyridine nucleotides can all be visualized in this way. Support for such a mechanism came from study of the nonenzymatic oxidation of NADH by flavins, a reaction that occurs at moderate speed in water at room temperature.

A variety of flavins and dihydropyridine derivatives have been studied, and the electronic effects observed for the reaction are compatible with the hydride ion mechanism. According to the mechanism, a hydride ion is transferred directly from a carbon atom in a substrate to the flavin. However, a labeled hydrogen atom transferred to N-5 or N-1 would immediately exchange with the medium, rapid exchange being characteristic of hydrogens attached to nitrogen.

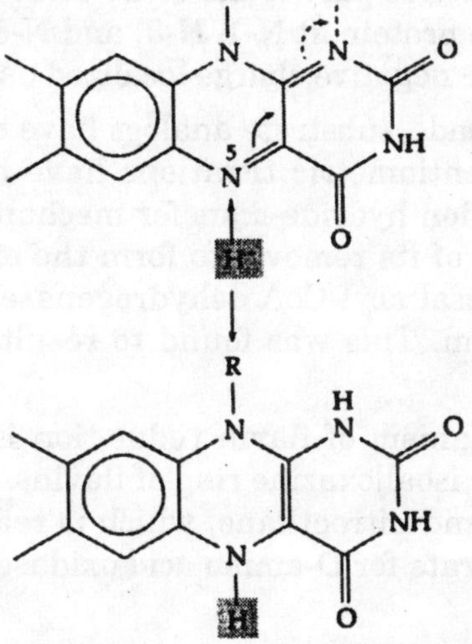

To avoid this problem, Brustlein and Bruice used a 5-*deazaflavin* to oxidize NADH nonenzymatically. When this reaction was carried out in ^2H_2O, no ^2H entered the product at C-5, indicating that a hydrogen atom had been transferred directly from NADH to the C-5 position. Similar direct transfer of hydrogen to C-5 of 5-deazariboflavin 5′-phosphate is catalyzed by flavoproteins such as N-methylglutamate synthase and acyl-CoA dehydrogenase.

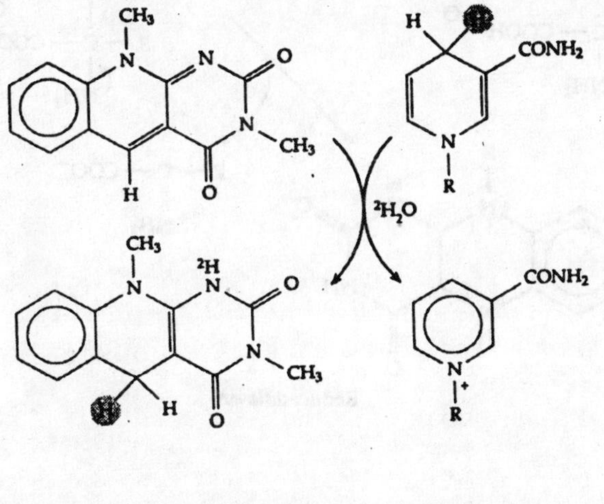

However, these experiments may not have established a mechanism for natural flavoprotein catalysis because the properties of 5-deazaflavins resemble those of NAD^+ more than of flavins. Their oxidation-reduction potentials are low, they do not form stable free radicals, and their reduced forms don't react readily with O_2. Nevertheless, for an acyl-CoA dehydrogenase the rate of reaction of the deazaflavin is almost as fast as that of natural FAD. For these enzymes a hydride ion transfer from the β CH is made easy by removal of the α-H of the acyl-CoA to form an enolate anion intermediate. The three-dimensional structure of the medium chain acyl-CoA dehydrogenases with bound substrates and inhibitors is known.

A conserved glutamate side chain is positioned to pull the *pro-R* proton from the a carbon to create the initial enolate anion. The *pro-R* P C-H lies by N-5 of the flavin ring seemingly ready to donate a hydride ion as. NMR spectroscopy has been carried out with ^{13}C or ^{15}N in each of the atoms of the redox active part of the FAD. The results show directly the effects of strong hydrogen bonding to the protein at N-1, N-3, and N-5 and also suggest mat the bound $FADH_2$ is really $FADH^-$ with the negative charge localized on N-1 by strong hydrogen bonding.

Many mutants have been made, substrate analogs have been tested, kinetic isotope effects have been measured, and potentiometric titrations have been done. All of the results are compatible with the enolate anion hydride-transfer mechanism. Questions about the acidity of the α-H and the mechanism of its removal to form the enolate anion are similar to those discussed. A peculiarity of several acyl-CoA dehydrogenases is a bright green color with an absorption maximum at 710 nm. This was found to result from tightly bound coenzyme A persulfide (CoA-S-S⁻).

(*b*) A second possible mechanism of flavin reduc-tion is suggested by the occurrence of addition reactions involving the isoalloxazine ring of flavins. Sulfite adds to flavins by forming an N-S bond at the 5 position and nitroethane, which is readily dissociated to the carbanion $H_3C-CH^--NO_2$ acts as a substrate for D-amino acid oxidase. This fact suggested a *carbanion*

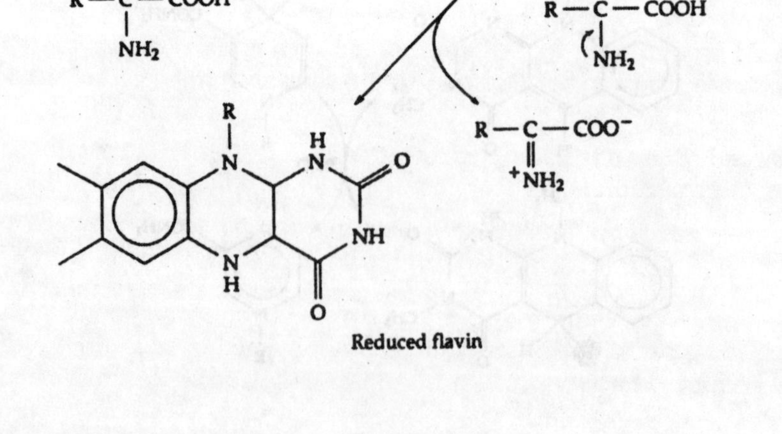

Reduced flavin

mechanism according to which normal D-amino acid substrates form carbanions by dissociation of the αH. Ionization would be facilitated by binding of the substrate carboxylate to an adjacent arginine side chain and the carbanion could react at N-5 of the flavin. Similar mechanisms have been suggested for other flavin enzymes.

(*c*) The adducts with nitroethane and other com-pounds pointed to reaction at N-5, but Hamilton suggested that a better position for addition of nucleophiles is carbon 4a, which together with N-5 forms a cyclic Schiff base. He argued that other electrophilic centers in the flavin molecule, such as carbons 2,4, and 10a, would be unreactive because of their involvement in amide or amidine-type resonance but an amine, alcohol, or other substrate could add to a flavin at position 4a. Cleavage of the newly formed C – O bond could then occur by movement of electrons from the alcohol part of the adduct into the flavin as indicated in step *b*. The products of this 4a *adduct mechanism* are the reduced flavin and an aldehyde,

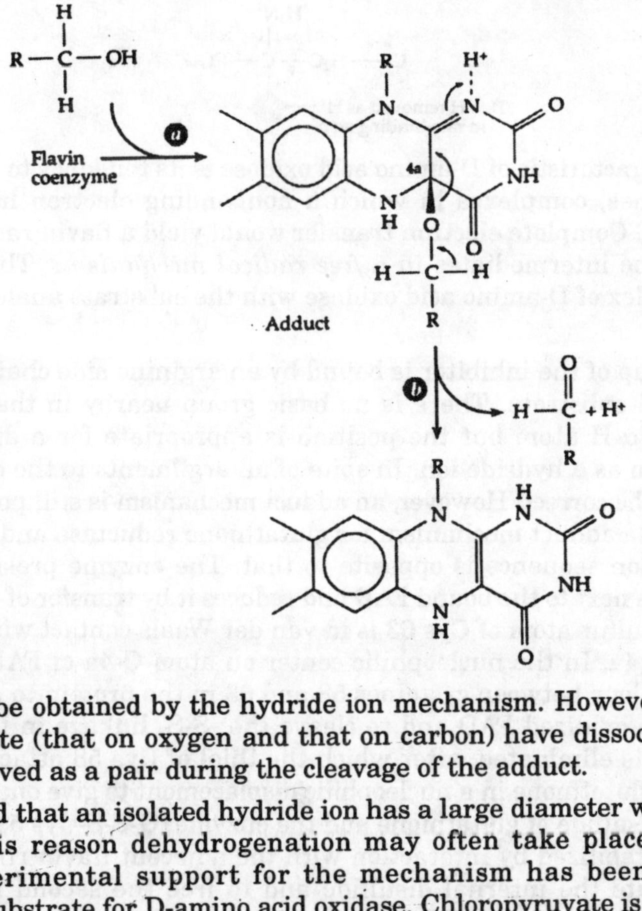

Flavin coenzyme

Adduct

the same as would be obtained by the hydride ion mechanism. However, both hydrogens in the original substrate (that on oxygen and that on carbon) have dissociated as protons, the electrons having moved as a pair during the cleavage of the adduct.

Hamilton argued that an isolated hydride ion has a large diameter while a proton is small and mobile; for this reason dehydrogenation may often take place by proton transfer mechanisms. Experimental support for the mechanism has been obtained using D-chloroalanine as a substrate for D-amino acid oxidase. Chloropyruvate is the expected product, but under anaerobic conditions pyruvate was formed. Kinetic data obtained with α-^{2}H and α-^{3}H substrates suggested a common intermediate for formation of both pyruvate and chloropyruvate. This intermediate could be an anion formed by loss of H^+ either from alanine or from a C-4a adduct. The anion could eliminate chloride ion as indicated by the dashed arrows in the following structure. This would lead to formation of pyruvate without reduction of the flavin. Alternatively, the electrons from the carbanion could flow into the flavin (green arrows), reducing. A similar mechanism has been suggested for other flavoenzymes. Objections to the carbanion mechanism are the expected very high pK_a for loss of the α-H to form the carbanion and the observed formation of only chloroalanine and no pyruvate in the reverse reaction of chloropyruvate, ammonia, and reduced flavoprotein.

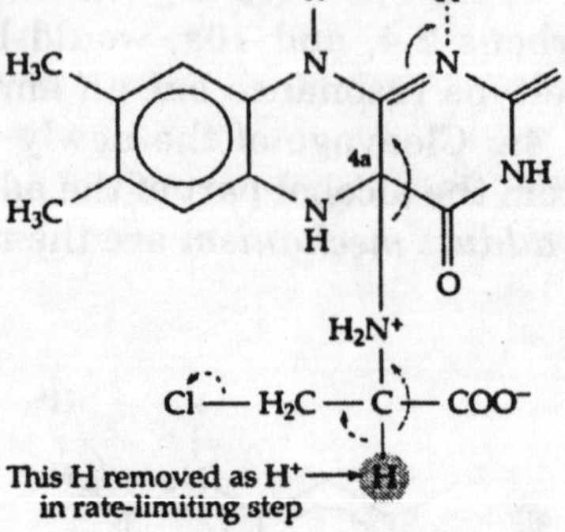

R H⁺ H₃C N N O 4a NH H₃C N H O H₂N⁺ Cl — H₂C — C — COO⁻

This H removed as H⁺ → H
in rate-limiting step

A long-known characteristic of D-amino acid oxidase is its tendency to form charge-transfer complexes with amines, complexes in which a nonbonding electron has been transferred partially to the flavin. Complete electron transfer would yield a flavin radical and a substrate radical which could be intermediates in *a free radical mechanism.* The three-dimensional structure of the complex of D-amino acid oxidase with the substrate analog benzoate has been determined.

The carboxyl group of the inhibitor is bound by an arginine side chain that probably also holds the amino acid substrate. There is no basic group nearby in the enzyme that could serve to remove the α-H atom but the position is appropriate for a direct transfer of the hydrogen to the flavin as a hydride ion. In spite of all arguments to the contrary the hydride ion mechanism could be correct! However, an adduct mechanism is still possible. Experimental evidence supports a 4a adduct mechanism for glutathione reductase and related enzymes. In this figure the reaction sequence is opposite to that. The enzyme presumably functions as follows. NADPH binds next to the bound FAD and reduces it by transfer of the 4-pro-S hydrogen of the NADPH. The sulfur atom of Cys 63 is in van der Waals contact with the bound FAD at or near carbon atom 4a. In the nucleophilic center on atom C-4a of $FADH_2$ attacks a sulfur atom of the disulfide loop between cysteines 58 and 63 in the protein to create a C-4a adduct of a thiolate ion with oxidized FAD and to cleave the -S-S- linkage in the loop. In the thiol group of cysteine 63 is eliminated, after which the thiol of Cys 58 attacks the nearer sulfur atom of the oxidized glutathione in a nucleophilic displacement to give one reduced glutathione (GSH) and a mixed disulfide of glutathione and the enzyme (G-S-S-Cys 58). The thiolate anion of Cys 63, which is stabilized by interaction with the adjacent flavin ring, then attacks this disulfide to regenerate the internal disulfide and to free the second molecule of reduced glutathione. The imidazole group of the nearby His 467 of the second subunit presumably participates in catalysis as may some other side chains.

The disulfide exchange reactions are similar to those discussed. A variation is observed for *E. coli* thioredoxin reductase. The reducible disulfide and the NADPH binding site are both on the same side of the flavin rather than on opposite sides. Mercuric reductase also uses NADPH as the reductant transferring the 4S hydrogen. The Hg^{2+} presumably binds to a sulfur atom of the reduced disulfide loop and there undergoes reduction. The observed geometry of the active site is correct for this mechanism.

Half-Reduced Flavins

A possible mechanism of flavin dehydrogenation consists of consecutive transfer of a hydrogen atom and of an electron with intermediate radicals being formed both on the flavin and on the substrate. Such a mechanism takes full advantage of one of the most characteristic properties of flavins, their ability to accept single electrons to form *semiquinone* radicals. If the oxidized form Fl of a flavin is mixed with the reduced form F1H, a single hydrogen atom is transferred from FlH_2 to Fl to form two •F1H radicals.

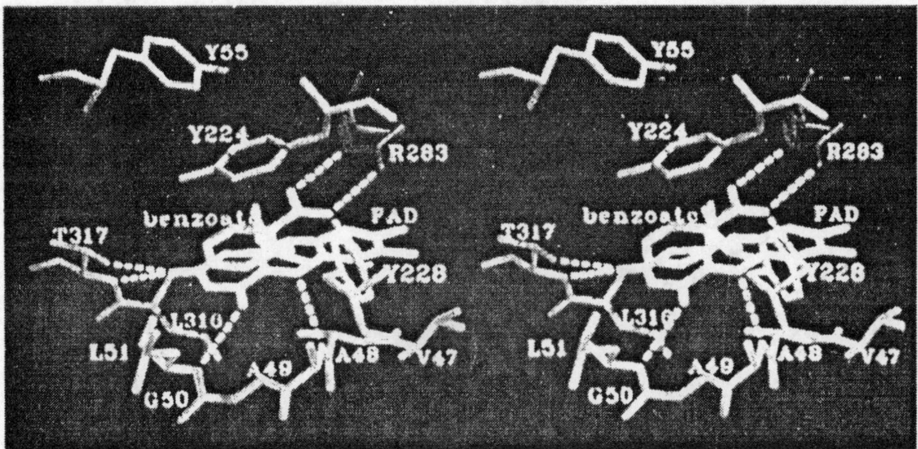

Fig. 10.11. Stereoscopic view of the benzoate ion in its complex with D-amino acid oxidase. A pair of hydrogen bonds binds the carboxylate of the ligand to the guanidinium group of R283. Several hydrogen bonds to the flavin ring of the FAD are also indicated.

$$Fl + FlH_2 \underset{}{\overset{K_f}{\rightleftharpoons}} 2\ {}^{\bullet}FlH \qquad ...(10.27)$$

The equilibrium represented by this equation is independent of pH, but because all three forms of the flavin have different pK_a values the apparent equilibrium constants relating total concentrations of oxidized, reduced, and radical forms vary with pH. The fraction of radicals present is greater at low pH and at high pH than at neutrality. For a 3-alkylated flavin the formation constant K_f has been estimated as 2.3×10^{-2} and for riboflavin as 1.5×10^{-2}. From these values and the pK_a values, it is possible to estimate the amount of radical present at any pH.

Neutral flavin radicals have a blue color (the wavelength of the absorption maximum, λ_{max}, is ~560 run) but either protonation at N-1 or dissociation of a proton from N-5 leads to red cation or anion radicals with λ_{max} at ~477 nm. Both blue and red radicals are observed in enzymes, with some enzymes favoring one and some the other. Hemmerich suggested that enzymes forming red radicals make a strong hydrogen bond to the proton in the 5 position of the flavin. This increases the basicity of N-1 leading to its protonation and formation of the red cation radicals.

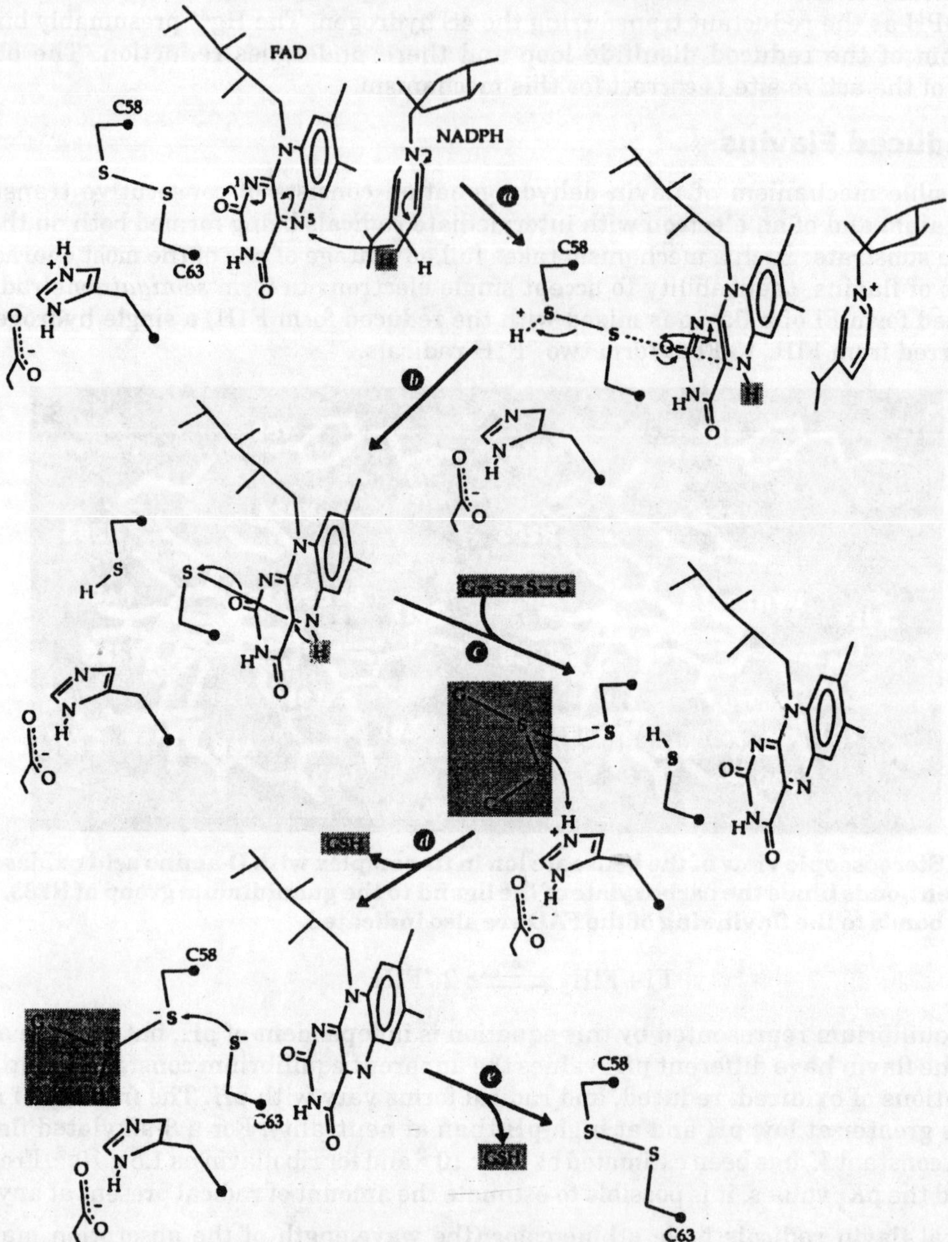

Fig. 10.12. Probable reaction mechanism for lipoamide dehydrogenase and glutathione reductase.

It is possible that many flavoprotein oxygenases and dehydrogenases react via free radicals. For example, instead of the mechanism, an electron could be transferred to the flavin, leaving a radical pair (at right). The crystallographic structure and modeling of the substrate complex

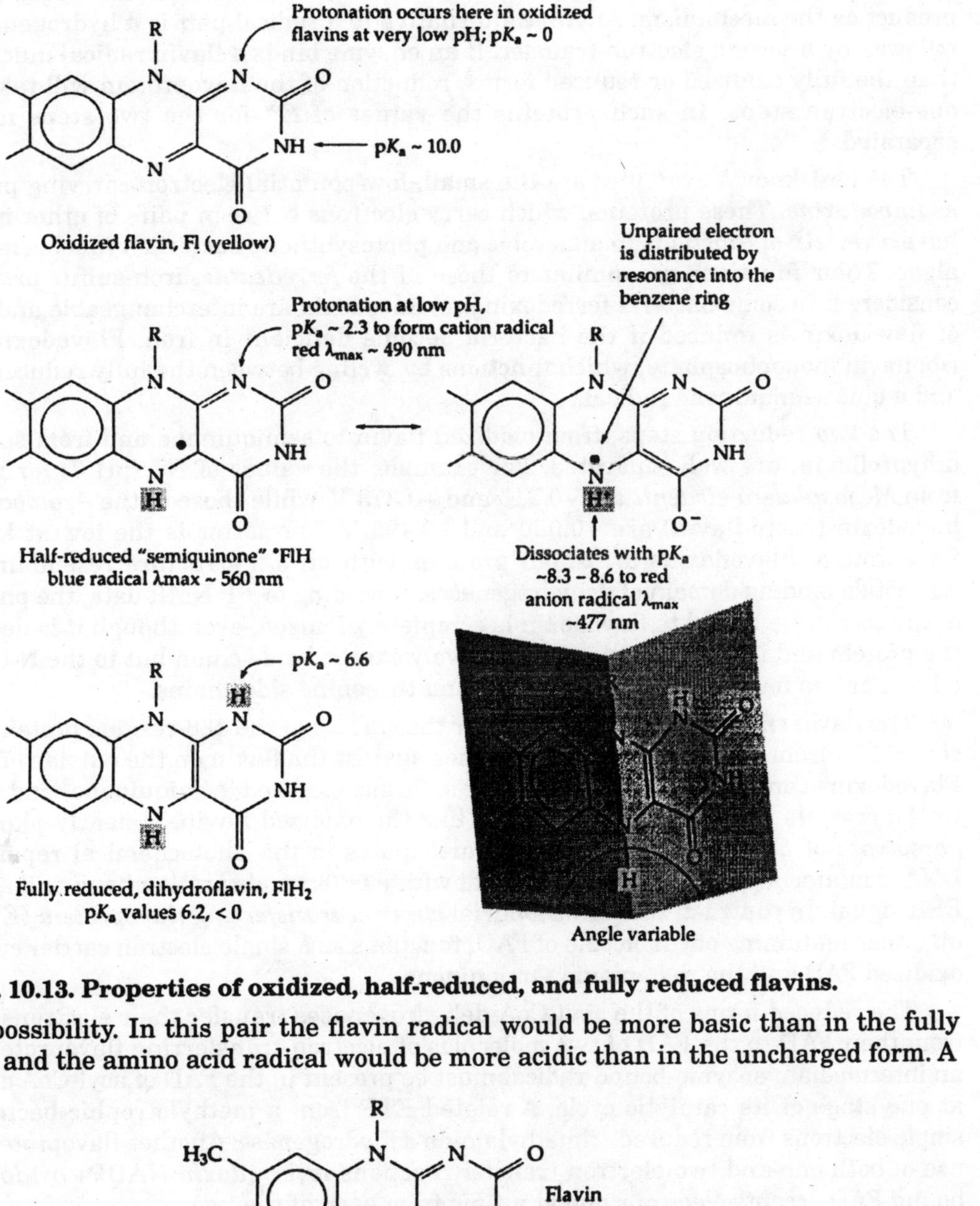

Protonation occurs here in oxidized flavins at very low pH; $pK_a \sim 0$

$pK_a \sim 10.0$

Oxidized flavin, Fl (yellow)

Unpaired electron is distributed by resonance into the benzene ring

Protonation at low pH, $pK_a \sim 2.3$ to form cation radical red $\lambda_{max} \sim 490$ nm

Half-reduced "semiquinone" $^{\bullet}$FlH blue radical λmax ~ 560 nm

Dissociates with pK_a ~8.3 – 8.6 to red anion radical λ_{max} ~477 nm

$pK_a \sim 6.6$

Fully reduced, dihydroflavin, FlH_2 pK_a values 6.2, < 0

Angle variable

Fig. 10.13. Properties of oxidized, half-reduced, and fully reduced flavins.

support this possibility. In this pair the flavin radical would be more basic than in the fully oxidized form and the amino acid radical would be more acidic than in the uncharged form. A

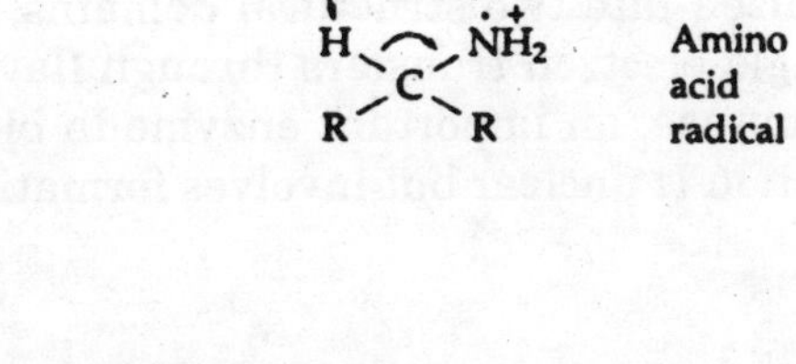

Flavin anion radical

Amino acid radical

proton transfer as indicated together with coupling of the radical pair would yield the same product as the mechanism. Another alternative to a radical pair is a hydrogen atom transfer followed by a second electron transfer. If an enzyme binds a flavin radical much more tightly than the fully oxidized or reduced forms, reduction of the flavoprotein will take place in two one-electron steps. In such proteins the values of $E^{\circ\prime}$ for the two steps may be widely separated.

The best known examples are the small, low-potential electron-carrying proteins known as *flavodoxins*. These proteins, which carry electrons between pairs of other redox proteins, have a variety of functions in anaerobic and photosynthetic bacteria, cyanobacteria, and green algae. Their functions are similar to those of the *ferredoxins*, iron-sulfur proteins that are considered. In some bacteria ferredoxin and flavodoxin are interchangeable and the synthesis of flavodoxin is induced if the bacteria become deficient in iron. Flavodoxins all contain riboflavin monophosphate, which functions by cycling between the fully reduced anionic form and a blue semiquinone radical.

The two reduction steps, from oxidized flavin to semiquinone and from semi-quinone to dihydroflavin, are well separated. For example, the values of $E^{\circ\prime}$ (pH 7) for the flavodoxin from *Megasphaera elsdenii* are –0.115 and –0.373 V, while those of the *Azotobacter vinlandii* flavodoxin (azoto-flavin) are +0.050 and –0.495 V. The latter is the lowest known for any flavoprotein. Flavodoxins are small proteins with an α/β structure resembling that of the nucleotide binding domain of dehydrogenases. According to ^{31}P NMR data, the phosphate group of the coenzyme bound to flavodoxin is completely ionized, even though it is deeply buried in the protein and is not bound to any positively charged side chain but to the N terminus of an a helix and to four –OH groups of serine and threonine side chains.

The flavin ring is partially buried near the surface of the 138-residue protein. An aromatic side chain, from tryptophan or tyrosine, lies against the flavin on the outside of the molecule. Flavodoxins can be crystallized in all three forms: oxidized, semiquinone, and fully reduced. In the crystals the flavin semi-quinone, like the oxidized flavin, is nearly planar. The *DNA photolyase* of *E. coli,* an enzyme that participates in the photochemical repair of damaged DNA, contains a blue neutral FAD radical with a 580-nm absorption band and an appropriate ESR signal. In contrast, the mitochondrial *electron-transferring flavoprotein* (ETF), a 57-kDa αβ dimer containing one molecule of FAD, functions as a single electron carrier cycling between oxidized FAD and the red anionic semiquinone.

The reduced forms of the acyl-CoA dehydrogenases transfer their electrons one at a time from their FAD to the FAD of two molecules of electron-transferring flavoprotein. Therefore, an intermediate enzyme-bound radical must be present in the FAD of acyl-CoA dehydrogenase at one stage of its catalytic cycle. A related ETF from a methylotrophic bacterium accepts single electrons from reduced trimethylamine dehydrogenase. Another flavoprotein that makes use of both one-and two-electron transfer reactions is *ferredoxin*-NADP+ *oxidoreductase.* Its bound FAD accepts electrons one at a time from each of the two

$$2Fd_{red} + NADP^{+} + H^{+} \rightarrow 2\,Fd_{ox}^{\ +} + NADPH \qquad ...(10.28)$$

reduced ferredoxins (Fd_{rcd}) in chloroplasts and then presumably transfers *a* hydride ion to the $NADP^{+}$. The enzyme is organized into two structural domains, one of which binds FAD and the other $NADP^{+}$. Similar single-electron transfers through flavoproteins also occur in many other enzymes. Chorismate mutase, an important enzyme in biosynthesis of aromatic rings, contains bound FMN. Its function is unclear but involves formation of a neutral flavin radical.

Metal Complexes of Flavins and Metalloflavoproteins

The presence of metal ions in many flavoproteins suggested a direct association of metal ions and flavins. Although oxidized flavins do not readily bind most metal ions, they form red complexes with Ag^+ and Cu^+ with a loss of a proton from N-3. Flavin semiquinone radicals also form strong red complexes with many metals. If the complexed metal ion can exist in more than one oxidation state, electron transfer between the flavin and *a* substrate could take place through the metal atom. However, *chelation by flavins in nature has not been observed.* Metalloflavoproteins probably function by having the metal centers close enough to the flavin for electron transfer to occur but not in direct contact.

This is the case for a bacterial trimethylamine dehydrogenase in which the FeS cluster is bound about 0.4 nm from the alloxazine ring of riboflavin 5′-phosphate. Some metalloflavoproteins contain heme groups. The previously mentioned *flavocytochrome* b_2 of yeast is a 230-kDa tetramer, one domain of which carries riboflavin phosphate and another heme. A flavocytochrome from the photosynthetic sulfur bacterium *Chromatium* (cytochrome *c*-552) is a complex of a 21-kDa cytochrome *c* and a 46-kDa flavoprotein containing 8α-(S-cysteinyl)-FAD. The 670-kDa *sulfite reductase* of *E. coli* has an $\alpha_8\beta_4$ subunit structure. The eight a chains bind four molecules of FAD and four of riboflavin phosphate, while the (3 chains bind three or four molecules of *siroheme* and also contain Fe_4S_4 clusters. Many nitrate and some nitrite reductases are flavoproteins which also contain Mo or Fe prosthetic groups. A group of *aldehyde oxidases and xanthine dehydrogenases* also contain molybdenum as well as iron. In every case the metal ions are bound independently of the flavin.

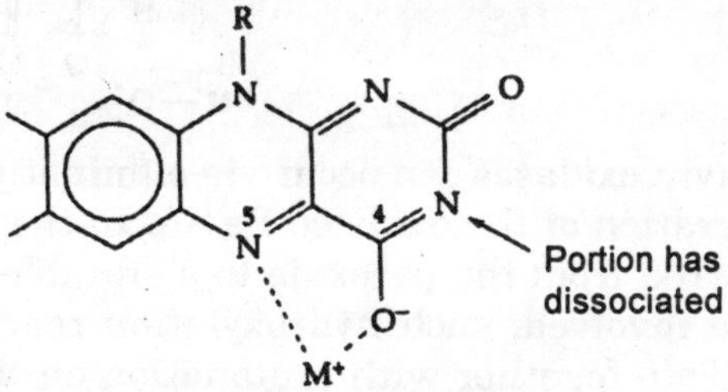

Portion has dissociated

Reactions of Reduced Flavins with Oxygen

Free dihydroriboflavin reacts nonenzymatically in seconds and reduced flavin oxygenases react even faster with molecular oxygen to form hydrogen peroxide.

$$FlH_2 + O_2 \rightarrow Fl + H_2O_2 \qquad (10.29)$$

The reaction is more complex than it appears. As soon as a small amount of oxidized flavin is formed, it reacts with reduced flavin to generate flavin radicals •F1H. The latter react rapidly with O_2 each donating an electron to form superoxide anion radicals $\bullet O_2$ which can then combine with flavin radicals.

$$\bullet FlH + O_2 \rightarrow \bullet O_2^- + H^+ + Fl \qquad (10.30a)$$

$$\bullet O_2 + \bullet FlH + H^+ \rightarrow Fl + H_2O_2 \qquad (10.30b)$$

During the corresponding reactions of reduced flavoproteins with O_2, intermediates have been detected. For example, spectrophotometric studies of the FAD-containing bacterial *p*-hydroxybenzoate hydroxylase revealed the consecutive appearance of three intermediate forms. The first, whose absorption maximum is at 380 – 390 nm, is thought to be an adduct at position 4a. That such a 4a peroxide really forms with the riboflavin phosphate of the light-emitting bacterial *luciferase* was demonstrated using coenzyme enriched with ^{13}C at position 4a. A large shift to lower frequency (from 104 to 83 ppm) accompanied formation of the transient adduct. Comparison with reference compounds showed that this change agreed with that predicted.

Other structures for O_2 adducts have also been considered, as has the possibility of rearrangements among these structures. Nevertheless, the products observed from many different flavoprotein reactions can be explained on the basis of a 4a peroxide.

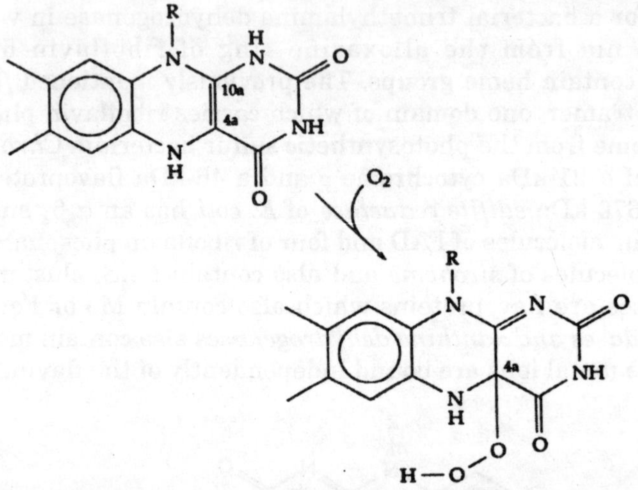

R
H
N
N
O
10a
4a
NH
N
H
O
O_2
R
N
N
O
4a
NH
N
H
O
O
H—O

Formation of H_2O_2 by flavin oxidases can occur via elimination of a peroxide anion HOO^- from the adduct with regeneration of the oxidized flavin. In the active site of a hydroxylase, an OH group can be transferred from the peroxide to a suitable substrate. Although radical mechanisms are likely to be involved, such hydroxylation reactions can also be viewed as transfer of OH^+ to the substrate together with protonation on the inner oxygen atom of the original peroxide to give a 4a –OH adduct. The latter is a covalent hydrate which can be converted to the oxidized flavin by elimination of H_2O. This hydrate is believed to be the third spectral intermediate identi-fied during the action of *p*-hydroxybenzoate hydroxylase.

LIPOIC ACID AND THE OXIDATIVE DECARBOXYLATION OF α-OXOACIDS

The isolation of lipoic acid in 1951 followed an earlier discovery that the ciliate protozoan *Tetrahymena gekii* required an unknown factor for growth. In independent experiments acetic acid was observed to promote rapid growth of *Lactobncillus casei,* but it could be replaced by an unknown "acetate replacing factor." Another lactic acid bacterium *Streptococcus faecalis* was unable to oxidize pyruvate without addition of "pyruvate oxidation factor." By 1949, all three unknown substances were recognized as identical. After working up the equivalent of 10 tons of water-soluble residue from liver, Lester Reed and his collaborators isolated 30 mg of a fat-soluble acidic material which was named lipoic acid (or 6-thioctic acid).

While *Tetmhymena* must have lipoic acid in its diet, we humans can make our own, and it is not considered a vitamin. Lipoic acid is present in tissues in extraordinarily small amounts. Its major function is to participate in the oxidative decarboxylation of α-oxoacids but it also plays an essential role in glycine catabolism in the human body as well as in plants. The structure is simple, and the functional group is clearly the cyclic disulfide which swings on the end of a long arm. Like biotin, which is also present in tissues in very small amounts, lipoic acid is bound in covalent amide linkage to lysine side chains in active sites of enzymes.

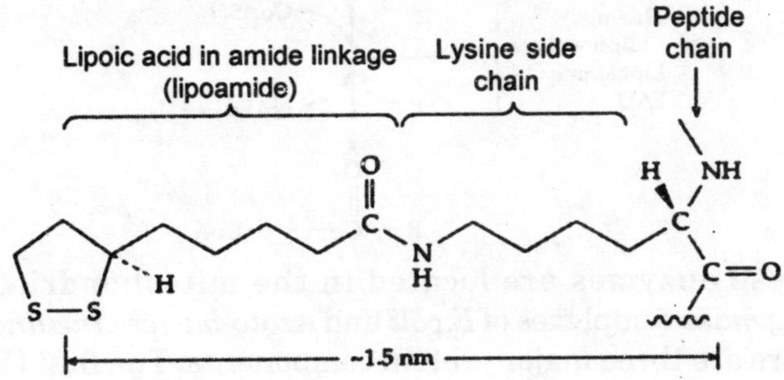

Lipoic acid in amide linkage (lipoamide) — Lysine side chain — Peptide chain — ~1.5 nm

Chemical Reactions of Lipoic Acid

The most striking chemical property of lipoic acid is the presence of ring strain of ~17 – 25 kJ moH in the cyclic disulfide. Because of this, thiol groups and cyanide ions react readily with oxidized lipoic acid to give mixed disulfides and isothiocyanates, respectively.

Another result of the ring strain is that the reduction potential $E^{\circ\prime}$ (pH 7, 25°C), is –0.30 V, almost the same as that of reduced NAD (–0.32V). Thus, reoxidation of reduced lipoic acid amide by NAD^+ is thermodynamically feasible. Yet another property attributed to the ring strain in lipoic acid is the presence of an absorption maximum at 333 nm.

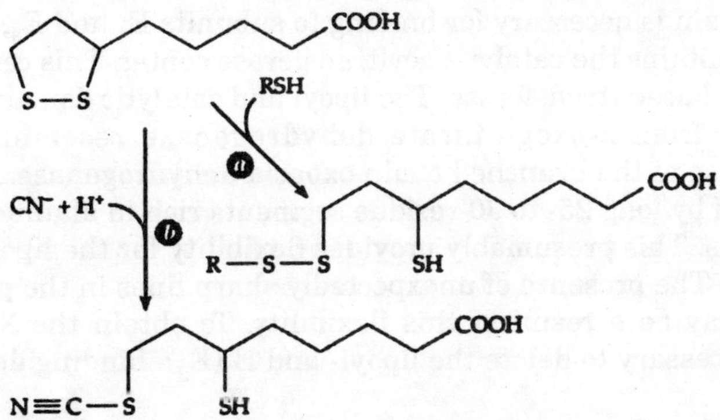

COOH; S—S; RSH (a); $CN^- + H^+$ (b); R—S—S, SH, COOH; N≡C—S, SH, COOH

Oxidative Decarboxylation of Alpha-Oxoacids

The oxidative cleavage of an α-oxoacid is a major step in the metabolism of carbohydrates and of amino acids and is also a step in the citric acid cycle. In many bacteria and in eukaryotes the process depends upon both thiamin diphosphate and lipoic acid. The oxoacid anion is cleaved to form CO_2 and the remaining acyl group is combined with coenzyme A NAD^+ serves as the oxidant. The reaction is catalyzed by a complex of enzymes whose molecular mass varies from

~4 to 10 × 10^6, depending on the species and exact substrate. Separate oxoacid dehydrogenase systems are known for pyruvate, 2-oxoglutarate, and the 2-oxoacids with branched side chains derived metabolically from leucine, isoleucine, and valine.

$$R-\overset{\displaystyle O}{\overset{\|}{C}}-COO^-$$

Thiamin diphosphate, Lipoamide, FAD → (NAD^+, CoASH in; CO_2, $NADH + H^+$ out)

$$R-\overset{\displaystyle O}{\overset{\|}{C}}-S-CoA$$

In eukaryotes these enzymes are located in the mitochondria. The *pyruvate* and *2-oxoglutarate dehydrogenase* complexes of *E.coli* and *Azoto-bacter vinelandii* have been studied most. In both cases there are three major protein components. The first (E_1) is a *decarboxylase* (also referred to as a dehydrogenase) for which the thiamin diphosphate is the dissociable cofactor. The second (E_2) is a lipoic acid amide-containing "core" enzyme which is a *dihydrolipoyl transacylase*. The third (E_3) is the flavoprotein *dihydrolipoyl dehydrogenase*, a member of the glutathione reductase family with a three-dimensional structure and catalytic mechanism similar to those of glutathione reductase. Electron microscopy of the core dihydrolipoyl transacylase from *E. coli* reveals a striking octahedral symmetry which has been confirmed by X-ray diffraction. The core from pyruvate dehydrogenase has a mass of ~2390 kDa and contains 24 identical 99.5-kDa E_2 subunits. The 2-oxoglutarate dehydrogenase from *E. coli* has a similar but slightly less symmetric structure. Each core subunit is composed of three domains.

A lipoyl group is bound in amide linkage to lysine 42 and protrudes from one end of the domain. A second domain is necessary for binding to subunits E_1 and E_3, while the third major 250-residue domain contains the catalytic acyltransferase center. This center closely resembles that of chloramphenicol acetyltransferase. The lipoyl and catalytic domains of the dihydrolipoyl succinyl-transferase from 2-oxoglutarate dehydrogenase resemble those of pyruvate dehydrogenase and also of the branched chain oxoacid dehydrogenase. The three domains of the proteins are joined by long 25- to 30-residue segments rich in alanine, proline, and ionized hydrophilic side chains. This presumably provides flexibility for the lipoyl groups which must move from site to site. The presence of unexpectedly sharp lines in the proton NMR spectrum of the core protein may be a result of this flexibility. To obtain the X-ray structure of the core protein it was necessary to delete the lipoyl- and $E_1(E_2)$- binding domains. The resulting 24-subunit structure.

It has been assumed for many years that 12 of the dimeric decarboxylase units (E_1) are bound to the 12 edges of the transacetylase cube, while six (50.6 × 2 kDa) flavoprotein (E_3) dimers bind on the six faces of the cube. The active centers of all three types of subunits are thought to come close together in the regions where the subunits touch, permitting the sequence of catalytic reactions indicated to take place. Eukaryotic as well as some bacterial pyruvate decarboxylases have a core of 60 subunits in an icosahedral array with 532 symmetry. This can be seen in the image reconstructions of the enzyme from *Saccharomyces cerevisiae*

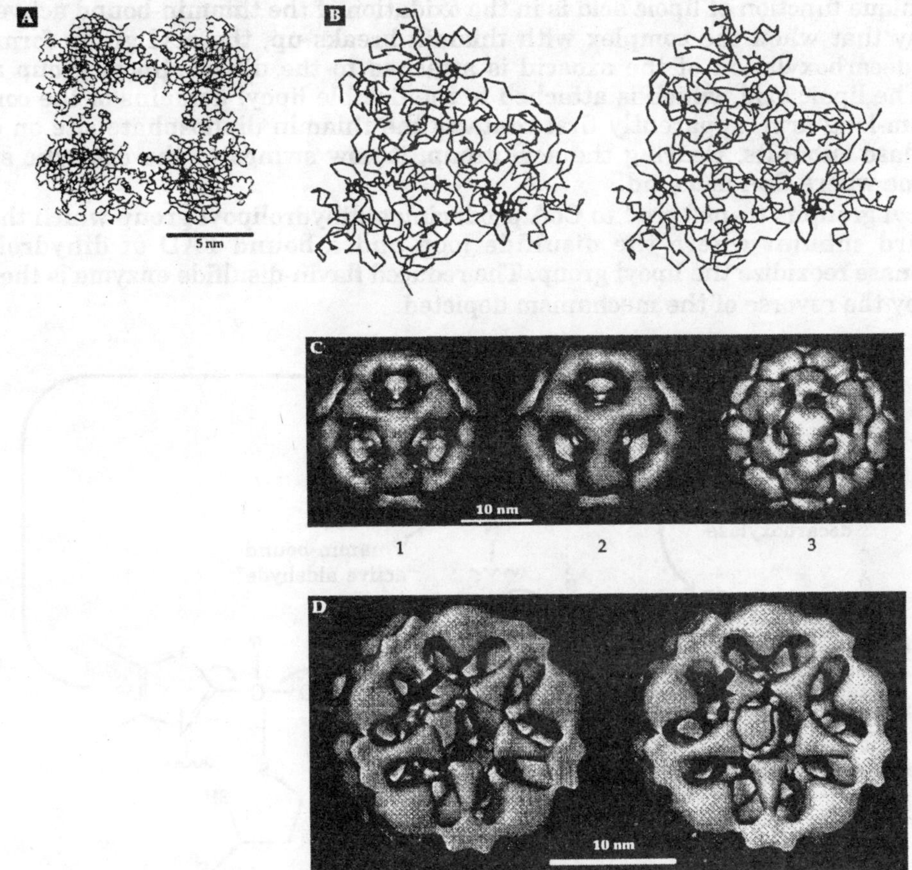

Fig. 10.14. (A) View of the complex down a fourfold axis through the face of the cube. (B) Stereoscopic view of three subunits of the 24-subunit cubic core of the *E. coli* pyruvate dehydrogenase complex. The view is down the threefold axis through opposite vertices of the cube. Bound lipoate and CoA molecules are visible in each subunit. (A) and (B) from Knapp *et al.* (C) Image of the 60-subunit core of yeast pyruvate dehydrogenase reconstructed from electron micrographs of frozenhydrated complexes. The five-, three-, and twofold symmetries can be seen. The view is along the threefold axis. 1. Truncated E_2; 2. Truncated E_2 plus binding proteins; 3. Truncated E_2 + binding protein + E_3. The reconstructions show that the E_3 and BP components are inside the truncated E_2 scaffold. The morphologies of E_3 and BP are not correctly represented since they are not bound with icosahedral symmetry. (D) Stereoscopic image of complex shown in B, 3 with the front half cut away to show the interior. The conical objects protruding in are the E_3 molecules. The arrow points to a binding site for attachment of the binding protein to the core.

and D. A surprising discovery is that in this yeast enzyme the E_3 units are not on the outside of the 5-fold symmetric faces but protrude into the inner cavity. Each of the 12 E_3 subunits is assisted in binding correctly to the E_2 core by a molecule of the 47-kDa E_3-*binding protein* (BP), also known as protein X. Absence of this protein is associated with congenital lactic acidosis.

The unique function of lipoic acid is in the oxidation of the thiamin-bound active aldehyde such a way that when the complex with thiamin breaks up, the acyl group formed by the oxidative decarboxylation of the oxoacid is attached to the dihydrolipoyl group at the S-8 position. The lipoic acid, which is attached to the flexible lipoyl domains of the core enzyme on a 1.5-nm-long arm, apparently first contacts the thiamin diphosphate site on one of the decarboxylase subunits. Bearing the acyl group, it now swings to the catalytic site on the core enzyme where CoA is bound.

The acyl group is transferred to CoA producing a dihydrolipoyl group which then swings to the third subunit where the disulfide loop and a bound FAD of dihydrolipoamide dehydrogenase reoxidize the lipoyl group. The reduced flavin-disulfide enzyme is then oxidized by NAD^+ by the reverse of the mechanism depicted.

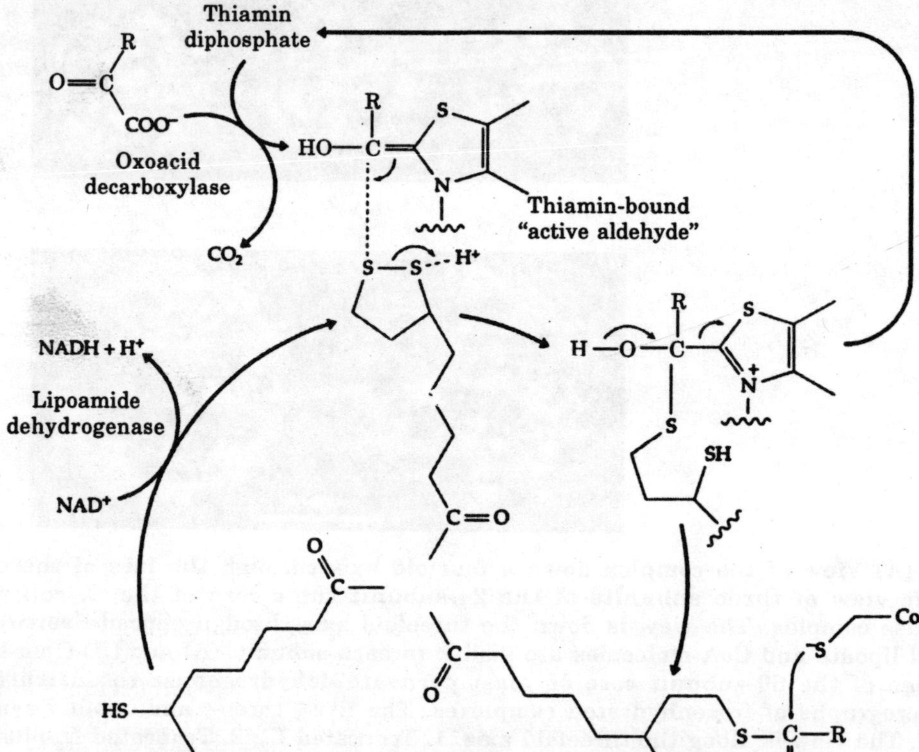

Fig. 10.15. Sequence of reactions catalyzed by α-oxoacid dehydrogenases. The substrate and product are shown in boxes, and the path of the oxidized oxoacid is traced by the heavy arrows. The lipoic acid "head" is shown rotating about the point of attachment to a core subunit. However, a whole flexible domain of the core is also thought to move.

Although the direct reaction of a lipoyl group with the thiamin-bound enamine (active aldehyde) is generally accepted, and is supported by recent studies, an alternative must be considered. Hexacyanoferrate (III) can replace NAD^+ as an oxidant for pyruvate dehydrogenase and is also able to oxidize nonenzymatically thiamin-bound active acetaldehyde to 2-acetylthiamin, *a* compound in which the acetyl group has a high group transfer potential. Thus, the lipoyl group could first oxidize the active aldehyde to a thiamin-bound acyl derivative, and then in the second step accept the acyl group by a nucleophilic displacement reaction.

This mechanism fails to explain the unique role of lipoic acid in oxidative decarboxylation. However, as we will see in the next section, oxidative decarboxylation does not always require lipoic acid and acetylthiamin is probably an intermediate whenever lipoamide is not utilized. Within many tissues the enzymatic activities of the pyruvate and branched chain oxoacid dehydrogenases complexes are controlled in part by a phosphorylation –dephosphorylation mechanism. Phosphorylation of the decarboxylase subunit by an ATP-dependent kinase produces an inactive phosphoenzyme. A phosphatase reactivates the dehydrogenase to complete the regulatory cycle.

The regulation is apparently accomplished, in part, by controlling the affinity of the protein for thiamin diphosphate. The lipoamide dehydrogenase component of all three dehydrogenase complexes appears to be the same.

Other Functions of Lipoic Acid

In addition to its role in the oxidative decarboxylation of 2-oxoacids, lipoic acid functions in the human body as part of the essential mitochondrial glycinecleavage system described. It may also participate in bacterial glycine reductase and other enzyme systems. Dihydrolipoamide dehydrogenase binds to G4-DNA structures of telomeres and may have a biological role in DNA-binding. Lipoic acid is being utilized as a nutritional supplement and appears to help in maintaining cellular levels of glutathione.

Additional Mechanisms of Oxidative Decarboxylation

Pyruvate is a metabolite of central importance and a variety of mechanisms exist for its cleavage. The pyruvate dehydrogenase complex is adequate for strict aerobes and is very important in aerobic bacteria and in facultative anaerobes such as *E. coli.* Lactic acid bacteria, which lack cytochromes and other heme proteins, are able to carry out limited oxidation with flavoproteins such as *pyruvate* oxidase. However, *strict* anaerobes have to avoid accumulation of reduced pyridine nucleotides and use a nonoxidative cleavage of pyruvate by *pyruvate formatelyase*. The pyruvate dehydrogenase complex of *our* bodies generates NADH which can be oxidized in mitochondria to provide energy to our cells. However, NADH is a weak reductant. Some cells, such as those of nitrogen-fixing bacteria, require more powerful reductants such as reduced ferredoxin and utilize *pyruvate: ferredoxin oxidoreductase. Escherichia coli,* a facultative anaerobe, is adaptable and makes use of all of these types of pyruvate cleavage.

Pyruvate oxidase

The soluble flavoprotein pyruvate oxidase, which was discussed briefly, acts together with a membrane-bound electron transport system to convert pyruvate to acetyl phosphate and CO_2. Thiamin diphosphate is needed by this enzyme but lipoic acid is not. The flavin probably dehydrogenates the thiamin-bound intermediate to 2-acetylthiamin. The electron acceptor is the bound FAD and the reaction may occur in two steps as shown with a thiamin diphosphate

radical intermediate. Reaction with inorganic phosphate generates the energy storage metabolite *acetyl phosphate*.

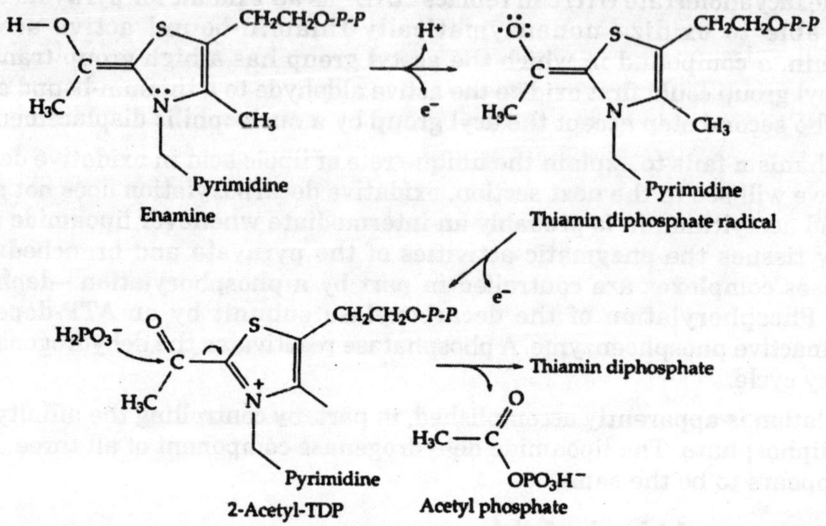

Enamine

Thiamin diphosphate radical

2-Acetyl-TDP

Acetyl phosphate

Pyruvate:ferredoxin Oxidoreductase

Within clostridia and other strict anaerobes this enzyme catalyzes *reversible* decarboxylation of pyruvate. The oxidant used by clostridia is the low-potential iron-sulfur ferredoxin. Clostridial ferredoxins contain two Fe-S clusters and are therefore two-electron oxidants. Ferredoxin substitutes for NAD^+, but the Gibbs energy decrease is much less (–16.9 vs –34.9 kJ/mol. for oxidation by NAD^+).

$$\text{Pyruvate}^- + \text{CoA} \longrightarrow \text{Acetyl-CoA} + CO_2$$

Ferredoxin (ox) → Ferrodoxin (red)

$$DG' \text{ (pH7)} = -16.9 \text{ kJ mol}^{-1}$$

The enzyme does not require lipoic acid. It seems likely that a thiamin-bound enamine is oxidized by an iron-sulfide center in the oxidoreductase to 2-acetyl-thiamin which then reacts with CoA. A free radical intermediate has been detected and the proposed sequence for oxidation of the enamine intermediate is that but with the Fe-S center as the electron acceptor. Like pyruvate oxidase, this enzyme transfers the acetyl group from acetylthiamin to coenzyme A.

Cleavage of the resulting acetyl-CoA is used to generate ATP. An indolepyruvate: ferredoxin oxidoreductase has similar properties. In methanogenic bacteria the low-potential 5-deazaflavin coenzyme F_{420} serves as the reductant in a reversal. A similar enzyme, *2-oxo-glutarate synthase*, apparently functions in synthesis of 2-oxoglutarate from succinyl-CoA and CO_2 by photosynthetic bacteria. Either reduced ferredoxin or, in *Azotobacter*, reduced flavodoxin is generated in nitrogen-fixing bacteria by cleavage of pyruvate and is used in the N_2 fixation process.

Oxidative Decarboxylation by Hydrogen Peroxide

The nonenzymatic oxidative decarboxylation of a-oxoacids by H_2O_2 is well-known. The first step is the formation of an adduct, an organic peroxide, which breaks up as indicated. An

enzyme-catalyzed version of this reaction is promoted by *lactate monooxygenase*, a 360-kDa octameric flavoprotein obtained from *Mycobacterium smegmatis* and a member of the glycolate oxidase family. Under anaerobic conditions, the enzyme produces pyruvate by a simple dehydrogenation. However, the pyruvate dissociates slowly and in the presence of oxygen it forms acetic acid, with one of the oxygen atoms of the carboxyl group coming from O_2. Hydrogen peroxide is the usual product formed from oxygen by flavoprotein oxidases, and it seems likely that with lactate monooxygenase the hydrogen peroxide formed immediately oxidizes the pyruvate according. The α-oxoacids formed by amino acid oxidases *in vitro* are also oxidized by accumulating hydrogen peroxide. However, if catalase is present it destroys the H_2O_2 and allows the oxoacid to accumulate.

H_2O_2 ; $H_2O + CO_2$

Pyruvate Formate-lyase Reaction

Anaerobic cleavage of pyruvate to acetyl-CoA and formate is essential to the energy economy of many cells, including those of *E. coli.* No external oxidant is needed, and the reaction does not require lipoic acid.

The reaction has sometimes been called the "phosphoroclastic reaction" because the product acetyl-CoA usually reacts further with inorganic phosphate to form acetyl phosphate. The latter can then transfer its phospho group to ADP to form ATP.

$CH_3-CO-COO^- + CoA-SH \rightarrow CH_3-CO-S-CoA + H-COO^-$

$\Delta G'$ (pH 7) = −12.9 kJ mol^{-1}

The mechanism of the cleavage of the pyruvate not obvious. Thiamin diphosphate is not involved, and free CO_2 is not formed. The first identified intermediate is an acetyl-enzyme containing a thioester linkage to a cysteine side chain. This is cleaved by reaction with CoA-SH to give the final product. A clue came when it was found by Knappe and coworkers that the active enzyme, which is rapidly inactivated by oxygen, contains a long-lived free radical.

Under anaerobic conditions cells convert the inactive form E_i to the active form E_a by an enzymatic reaction with S-adenosylmethionine and reduced flavodoxin Fd(red). A deactivase reverses the process.

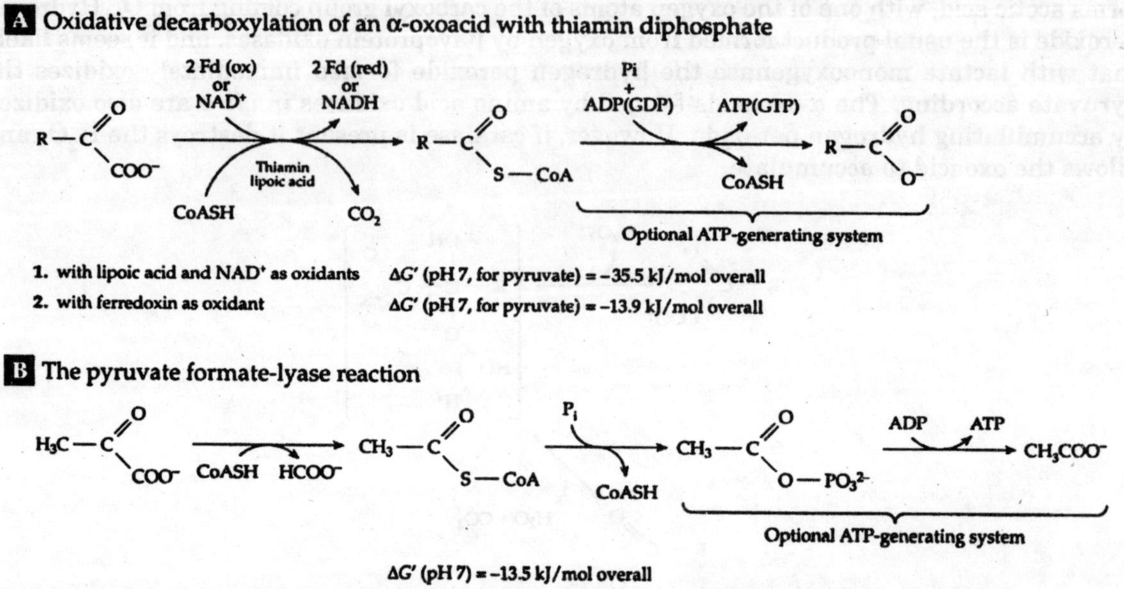

A Oxidative decarboxylation of an α-oxoacid with thiamin diphosphate

1. with lipoic acid and NAD⁺ as oxidants $\Delta G'$ (pH 7, for pyruvate) = −35.5 kJ/mol overall

2. with ferredoxin as oxidant $\Delta G'$ (pH 7, for pyruvate) = −13.9 kJ/mol overall

B The pyruvate formate-lyase reaction

$\Delta G'$ (pH 7) = −13.5 kJ/mol overall

Fig. 10.16. Two systems for oxidative decarboxylation of α-oxoacids and for "substrate-level" phosphorylation. The value of $\Delta G'$ = +34.5 kJ mol^{-1} was used for the synthesis of ATP^{4-} from ADP^{3-} and HPO_4^{2-} in computing the values of $\Delta G'$ given.

The activating enzyme, which is allosterically activated by pyruvate, is an iron-sulfur (Fe_4S_4) protein. Formation of the observed radical may proceed via a 5′-deoxyadenosyl radical

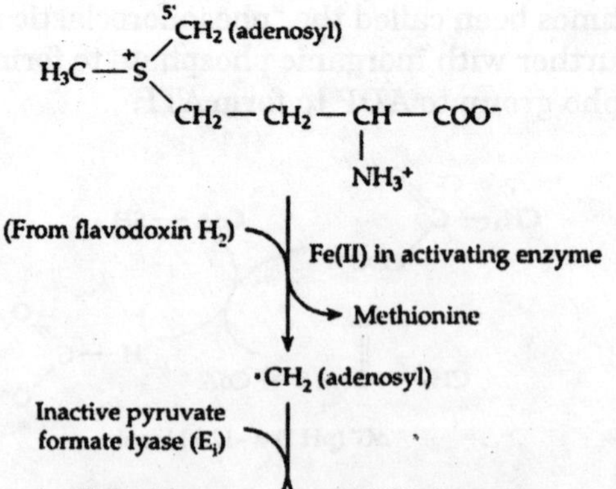

as has been proposed for lysine 2, 3-aminomutase. The activation reaction also resembles the free radical-dependent reactions of vitamin B_{12} which are discussed. When subjected to O_2 of air at 25°C pyruvate formatelyase is destroyed with a half-life of ~10s. The peptide chain is cleaved and sequence analysis and mass spectrometry of the resulting fragments show that the specific sequence Ser-Gly-Tyr at positions 733 – 755 is cut with formation of an oxalyl group on the N-terminal tyrosine of one fragment. Various ^{13}C-containing amino acids were supplied to growing cells of *E. coli* and were incorporated into the proteins of the bacteria. Pyruvate formatelyase containing ^{13}C in carbon-2 of glycine gave an EPR spectrum with hyperfine splitting arising from coupling of the unpaired electron of the radical with the adjacent ^{13}C nucleus. This experiment, together with the results described, suggested that the radical is derived from Gly 734.

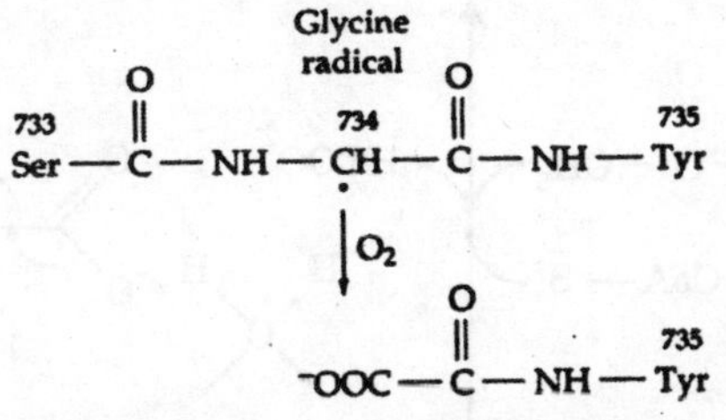

Glycine radical

733 Ser—C(=O)—NH—734 ĊH—C(=O)—NH—735 Tyr

↓ O_2

$^-$OOC—C(=O)—NH—735 Tyr

The activating enzyme will also generate radicals from short peptides such as Arg-Val-Ser-Gly-Tyr-Ala-Val, which corresponds to residues 731 – 737 of the pyruvate formate-lyase active site. If Gly 734 is replaced by L-alanine, no radical is formed, but radical *is* formed if D-alanine is in this position. This suggests that the *pro-S* proton of Gly 734 is removed by the activating enzyme as illustrated. It has been sug-gested that the peptide exists in the β-bend conforma-tion shown.

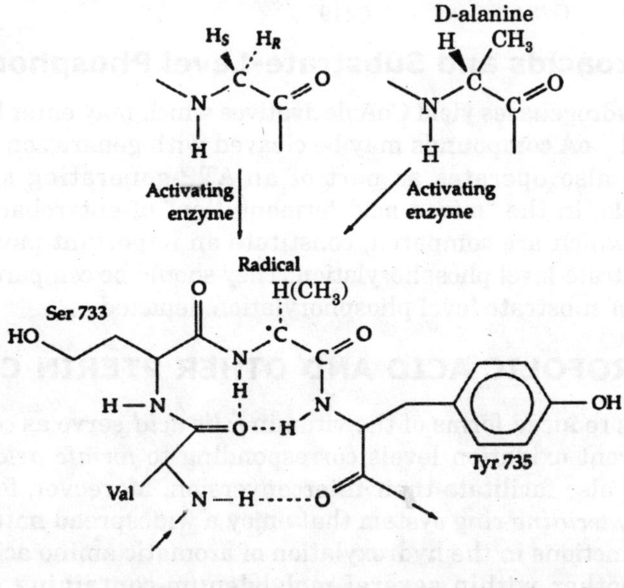

H_S H_R

D-alanine H CH_3

Activating enzyme

Activating enzyme

Radical

$H(CH_3)$

Ser 733

HO

H—N

Val

N—H

OH

Tyr 735

How does the enzyme work? The α-CH proton of the glycyl radical of the original *pro-R* proton undergoes an unexpected exchange with the deuterium of ^2H_2O. The exchange is catalyzed by the thiol group of Cys 419, suggesting that it is close to Gly 734 in the three-dimensional structure. The mutants C419S and C418S are inactive but still allow formation of the Gly 734 radical. The mechanism has been proposed.

Cleavage of α-Oxoacids and Substrate-Level Phosphorylation

The α-oxoacid dehydrogenases yield CoA derivatives which may enter biosynthetic reactions. Alternatively, the acyl-CoA compounds may be cleaved with generation of ATP. The pyruvate formatelyase system also operates as part of an ATP-generating system for anaerobic organisms, for example, in the "mixed acid fermentation" of enterobacteria such as *E. coli*. These two reactions, which are compared, constitute an important pair of processes both of which accomplish substrate-level phosphorylation. They should be compared with the previously considered examples of substrate level phosphorylation depicted.

TETRAHYDROFOLIC ACID AND OTHER PTERIN COENZYMES

In most organisms reduced forms of the vitamin *folic acid* serve as *carriers for one-carbon groups* at three different oxidation levels corresponding to *formic acid, formaldehyde*, and the *methyl group* and also facilitate their interconversion. Moreover, folic acid is just one of the derivatives of the *pteridine* ring system that enjoy a widespread natural distribution. One of these derivatives functions in the hydroxylation of aromatic amino acids and in nitric oxide synthase and yet another within several molybdenum-containing enzymes. Pteridines also provide coloring to insect wings and eyes and to the skins of amphibians and fish. They appear to act as protective filters in insect eyes and may function as light receptors. An example is a folic acid derivative found in some forms of the DNA photorepair enzyme *DNA photolyase*.

Structure of Pterins

Because of the prevalence of its derivatives, 2-amino-4-hydroxypteridine has been given the trivial name *pterin*. Its structure resembles closely that of guanine, the 5-membered ring of the latter having been expanded to a 6-membered ring. In fact, pterins are derived biosynthetically from guanine. The two-ring system of pterin is also related structurally and biosynthetically to that of riboflavin.

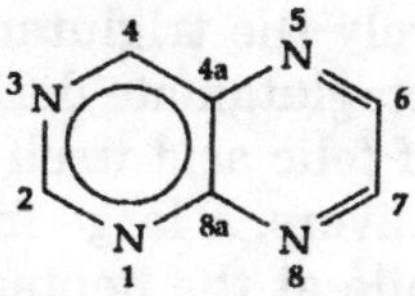

Pteridine

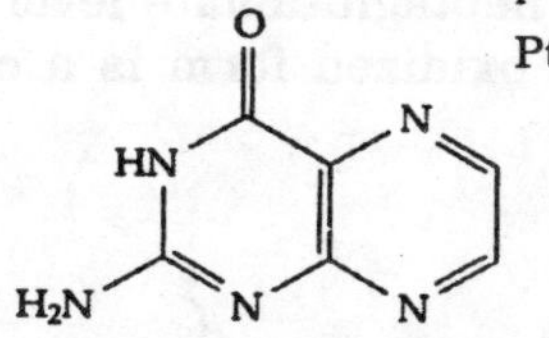

Pterin: 2-amino-4-hydroxypteridine 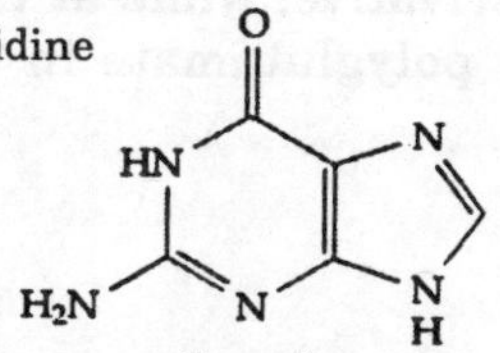 Guanine

Interest in pteridines began with Frederick G. Hopkins, who in 1891 started his investigation of the yellow and white pigments of butterflies.

Almost 50 years and a million butterflies later, the structures of the two pigments, *xanthopterin* and *leucopterin*, were established. These pigments are produced in such quantities as to suggest that their synthesis may be a means of deposition of nitrogenous wastes in dry form. Among the simple pterins isolated from the eyes of *Drosophila* is *sepiapterin*, in which the pyrazine ring has been reduced in the 7, 8 position and a short side chain is present at position 6. Reduction of the carbonyl group of sepiapterin with $NaBH_4$ followed by air oxidation produces *biopterin*, the most widely distributed of the pterin compounds. First isolated from human urine, biopterin is present in liver and other tissues where it functions in a reduced form as a *hydroxylation coenzyme*. It is also present in nitric oxide synthase.

Other functions in oxidative reactions, in regulation of electron transport, and in photosynthesis have been proposed. *Neopterin*, found in honeybee larvae, resembles biopterin but has a *D-erythro* configuration in the side chain. The red eye pigments of *Drosophila*, called *drosopterins*, are complex dimeric pterins containing fused 7-membered rings. In the pineal gland, as well as in the retina of the eye, light-sensitive pterins may be photochemically cleaved to generate such products as *6-formylpterin*, a compound that could serve as a metabolic regulator. Another pterin acts as a chemical attractant for aggregation of the ameboid cells of *Dictyostelium lacteum*. *Molybdopterin* is a component of several Mo-containing enzymes discussed in and methanopterin is utilized by methanogenic bacteria.

Folate Coenzymes

The coenzymes responsible for carrying single-carbon units in most organisms are derivatives of 5, 6, 7, 8-*tetrahydrofolic acid* (abbreviated H_4PteGlu, H_4folate or THF). However, in methanogenic bacteria the tetrahydro derivative of the structurally unique methanopterin is the corresponding single-carbon carrier. Naturally occurring tetrahydrofolates contain a chiral center of the S configuration. They exhibit negative optical rotation at 589 nm. The folate coenzymes are present in extremely low concentrations and the reduced ring is readily oxidized by air. In addition to the single L-glutamate unit present in tetrahydrofolic acid, the coenzymes occur to the greatest extent as conjugates called *folyl polyglutamates* in which one to eight or more additional molecules of L-glutamic acid have been combined via amide linkages. The first two of the extra glutamates are always joined through the g (side chain) carboxyl groups but in *E. coli* the rest are joined through their a carboxyls. The distribution of the polyglutamates varies from one organism to the next.

Some bacteria contain exclusively the triglutamate derivatives, while in others almost exclusively the tetraglutamate or octaglutamate derivatives predominate. The serum of many species contains only derivatives of folic acid itself but pteroylpentaglutamate is the major folate derivative present in rodent livers. A large fraction of these pentaglutamates consists of the 5-methylTHF derivative, while at the heptaglutamate level most consists of the free THF derivative. Folyl polyglutamate in its oxidized form is a component of some DNA photolyases.

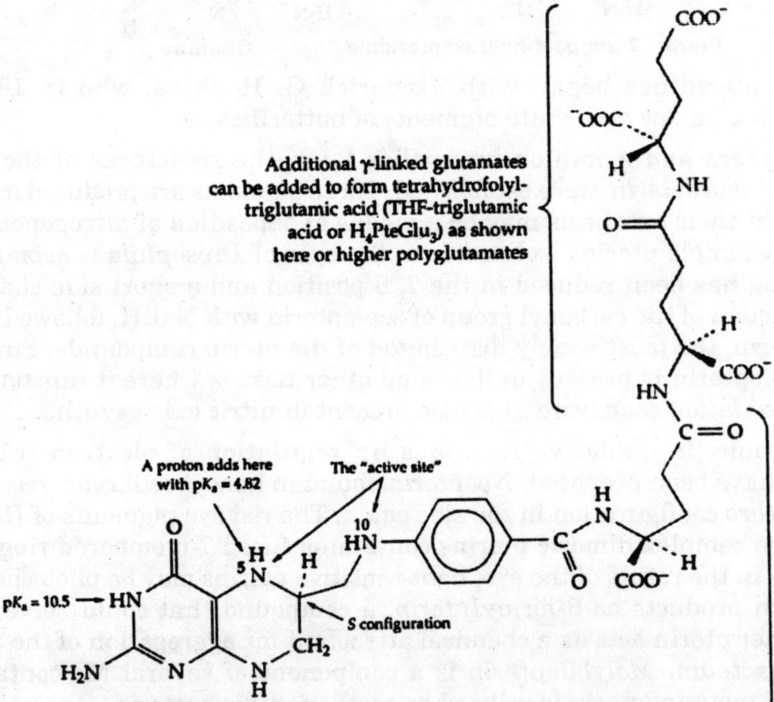

Additional γ-linked glutamates can be added to form tetrahydrofolyl-triglutamic acid (THF-triglutamic acid or $H_4PteGlu_3$) as shown here or higher polyglutamates

A proton adds here with $pK_a = 4.82$

The "active site"

$pK_a = 10.5$

S configuration

Tetrahydrofolic acid (THF) or tetrahydropteroylglutamic acid ($H_4PteGlu$)

Dihydrofolate Reductase

Folic acid and its polyglutamyl derivatives can be reduced to the THF coenzymes in two stages: the first step is a slow reduction with NADPH to 7, 8-dihydrofolate. The same enzyme that catalyzes this reaction rapidly reduces the dihydrofolates to tetrahydrofolates. Again,

NADPH is the reducing agent and the enzyme has been given the name *dihydrofolate reductase*. An unusual fact is that bacteriophage T4 not only carries a gene for its own dihydrofolate reductase *but* also the enzyme is a structural unit in the phage baseplate.

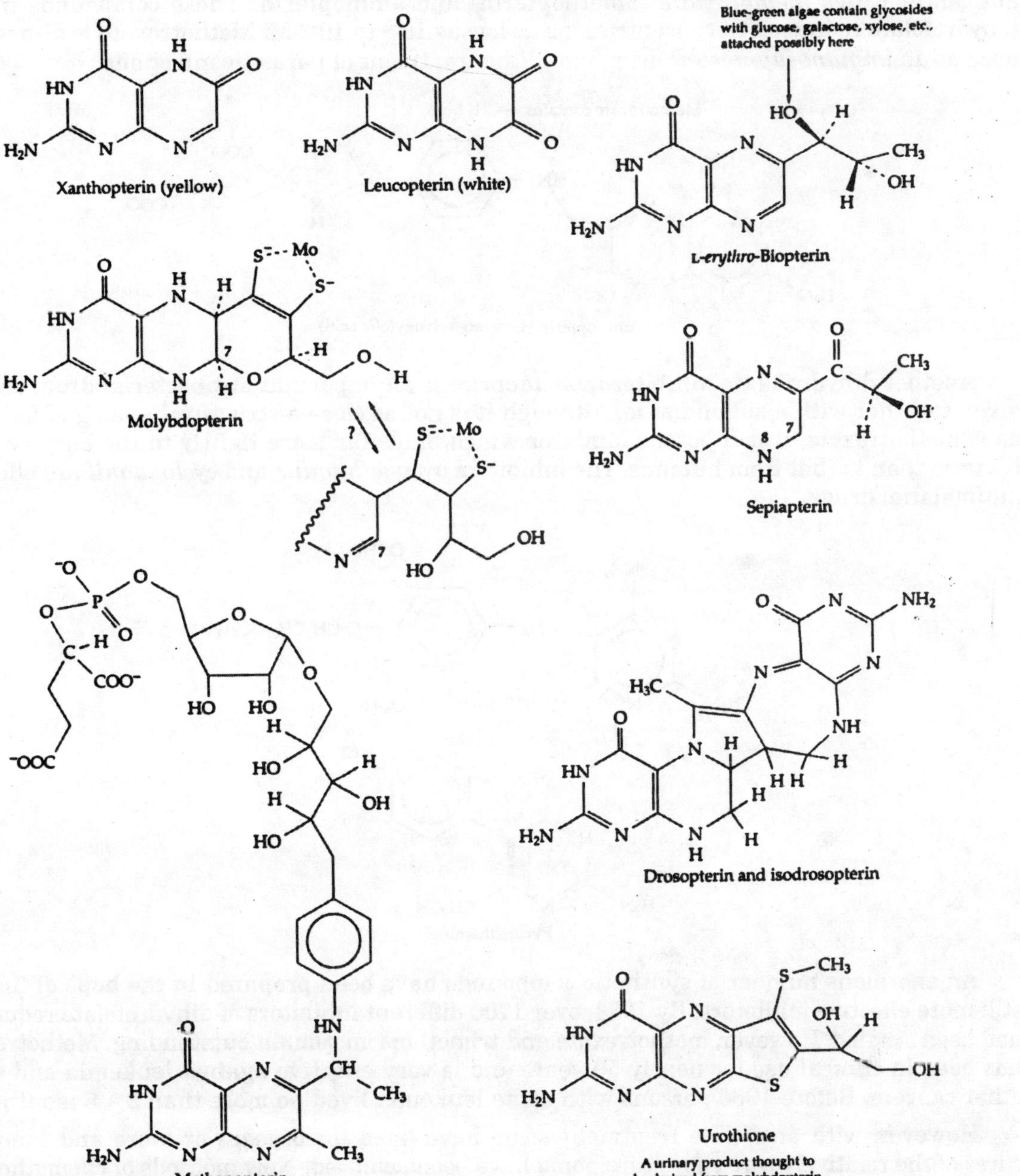

Fig. 10.17. Structures of several biologically important terins.

Inhibitors

Aside from its role in providing reduced folate coenzymes for cells, this enzyme has attracted a great deal of attention because it appears to be a site of action of the important anticancer drugs *methotrexate* (amethopterin) and aminopterin. These compounds inhibit dihydrofolate reductase in concentrations as low as 10^{-8} to 10^{-9} M. Methotrexate is also widely used as an *immunosuppresant* drug and in the treatment of parasitic infections.

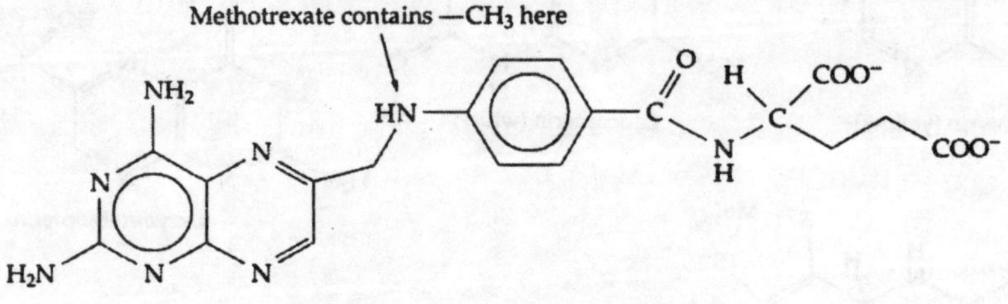

Methotrexate contains $—CH_3$ here

Aminopterin (4-amino-4-deoxyfolic acid)

Another dihydrofolate inhibitor *trimethoprim* is an important antibacterial drug, usually given together with a sulfonamide. Although it is not as close a structural analog of folk acid as is methotrexate, it is a potent inhibitor which binds far more tightly to the enzyme from bacteria than to that from humans. The inhibitors *pyrimethamine* and *cycloguanil* are effective antimalarial drugs.

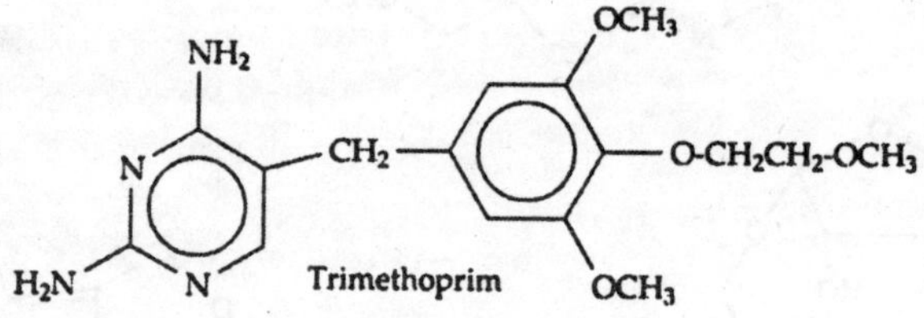

Trimethoprim

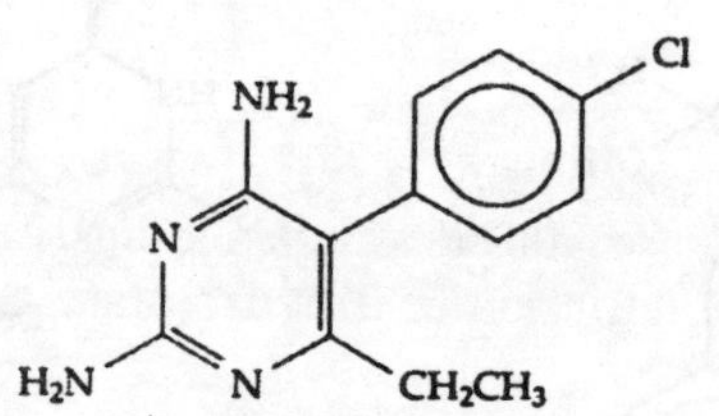

Pyrimethamine

An enormous number of synthetic compounds have been prepared in the hope of finding still more effective inhibitors. By 1984, over 1700 different inhibitors of dihydrofolate reductase had been studied. However, methotrexate and trimethoprim remain outstanding. Methotrexate has been in clinical use for nearly 50 years and is very effective against leukemia and some other cancers. Before 1960 persons with acute leukemia lived no more than 3 – 6 months.

However, with antifolate treatment some have lived for 5 years or more and complete cures of the relatively rare choriocarcinoma have been achieved. New methods of chemotherapy use the antifolates in combination with other drugs. Folate coenzymes are required in the biosynthesis of both purines and thymine. Consequently, rapidly growing cancer cells have a high requirement for activity of this enzyme. However, since all cells require the enzyme the antifolates are toxic and cannot be used for prolonged therapy. An even more serious problem is the development of resistance to the drug by cancer cells, often through "amplification" of the dihydrofolate reductase gene as discussed. Cells may also become resistant to methotrexate and other antifolates as a result of mutations that prevent efficient uptake of the drug, cause increased action of effux pumps in the cell, reduce the affinity of dihydrofolate reductase for the drug, or interfere with conversion of the drug to polyglutamate derivatives which are better inhibitors than free methotrexate.

Cell surface *folate receptors*, present in large numbers in some tumor cells, can be utilized to bring suitably designed antifolate drugs or even unrelated cytotoxic compounds into tumor cells. Some tumor cells are more active than normal cells in generating folyl polyglutamates, contributing to the effectiveness of methotrexate. Resistance of *E. coli* cells to trimethoprin may result from acquisition by the bacteria of a new form of dihydrofolate reductase carried by a plasmid.

Structure and Mechanism

Dihydrofolate reductase from *E. coli* is a small 159-residue protein with a central parallel stranded sheet, while that from higher animals is 20% larger. The three-dimensional structures of the enzymes from *Lactobacillus,* chicken, mouse, and human are closely similar. The NADP binds at the C-terminal ends of p strands as in other dehydrogenases. In the reduced nicotinamide end of $NADP^+$ is seen next to a molecule of bound dihydrofolate. The side chain carboxylate of Asp 27 makes a pair of hydrogen bonds to the pterin ring as shown on the right-hand side. As can be seen from, the nicotinamide ring of $NADP^+$ is correctly positioned to have donated a hydride ion to C-6 to form THF with the 6*S* configuration. Notice that the NH hydrogen atom at the 3′ position of the pterin makes a hydrogen bond to Asp 27.

However, X-ray studies have shown that the pteridine ring of methotrexate is turned 180° so that its 4-NH_2 group forms hydrogen bonds to backbone carbonyls of Ala 97 and Leu 4 at the edge of the central pleated sheet while a prtonated N-1 interacts with the side chain carboxylate of Asp. 27. Because of its significance in cancer therapy and its small size, dihydrofolate reductase is one of the most studied of all enzymes. Numerous. NMR studies and investigations of catalytic mechanism and of other properties have been created. For example, substitution of Asp 27 (Asp 26 in *L. casei*) of the *E. coli* enzyme by Asn reduced the specific catalytic activity to 1/300 that of native enzyme indicating that this residue is important for catalysis. The value of k_{cat} for reduction of dihydrofolate is highest at low pH and varies around a pK_a of ~6.5.

One interpretation is that this pK_a belongs to the Asp 27 carboxyl group and another is that it belongs to an N-5 protonated species of the coenzyme. As we have it is often impossible to assign a pK_a to a single group because there may be a mixture of interacting tautomeric species. Despite the enormous amount of study, we still don't know quite how the proton gets to N-5 in this reaction. Only one possibility is illustrated. Asp 27 is protonated, the pterin ring is enolized, and a buried water molecule serves to relay a proton to N-5. The X-ray crystallographic studies on the *E. coli* enzyme show that after the binding of substrates a conformational change closes a lid over the active site. This excludes water from N-5 but it may permit intermittent access, allowing transfer of a proton from a water molecule bound to O-4. During the reduction of folate to dihydrofolate a different mechanism of proton donation must be followed to allow protonation at C-7.

Single-Carbon Compounds and Groups in Metabolism

Some single-carbon compounds and groups important in metabolism order (from left to right) of increasing oxidation state of the carbon atom. Groups at three different oxidation levels, corresponding to *formic acid*, *formaldehyde*, and the *methyl group*, are carried by tetrahydrofolic acid coenzymes. While the most completely reduced compound, methane, cannot exist in a combined form, its biosynthesis depends upon reduced methanopterin, as does that of carbon monoxide. Summarizes the metabolic interrelationships of the compounds and groups in this table.

Serine as a C1 Donor

For many organisms, from *E. coli* to higher animals, *serine* is a major precursor of C-1 units. The β-carbon of serine is removed as formaldehyde through direct transfer to tetrahydrofolate with formation of methylene-THF and glycine. This is a stereospecific transfer in which the *pro-S* hydrogen on C-3 of serine enters the *pro-S* position also in methylene-THF. The glycine formed in this reaction can, in turn, yield another single-carbon unit by loss of CO_2 under the influence of the THF and the PLP-requiring glycine cleavage system which is discussed in the next paragraph. Free formaldehyde in a low concentration can also combine with THF to form methylene-THF.

H_2CO THF HO CH₂ H 10 N 5 N H_2C H :N HO⁻ +N H⁺ H H C N N

The Glycine Decarboxylase-synthetase System

Glycine is cleaved reversibly within mitochondria of plants and animals and also by bacteria to CO_2, NH_3 and a methylene group which is carried by tetrahydrofolic acid. Four proteins are required. The P-protein consists of two identical 100-kDa subunits, each containing a molecule of PLP. This protein is a *glycine decarboxylase* which, however, replaces the lost CO_2 by an electrophilic sulfur of lipoate rather than by *a* proton. Serine hydroxymethyltransferase can also catalyze this step of the sequence. The lipoate is bound to a second protein, the H-protein. A third protein, the T-protein, carries bound tetrahydrofolate which displaces the aminomethyl group from the dihydrolipoate and converts it to N^5, N^{10}-methylene tetrahydrofolate with release of ammonia. The dihydrolipoate is then reoxidized by NAD^+ and dihydrolipamide dehydrogenase. Glycine can be oxidized completely in liver mitochondria with the methylene group of methylene-THF being converted to CO_2 through reaction. The glycine cleavage system is reversible and is used by some organisms to synthesize glycine.

Table 10.3. Single-Carbon Compounds in Order of Oxidation State of Carbon

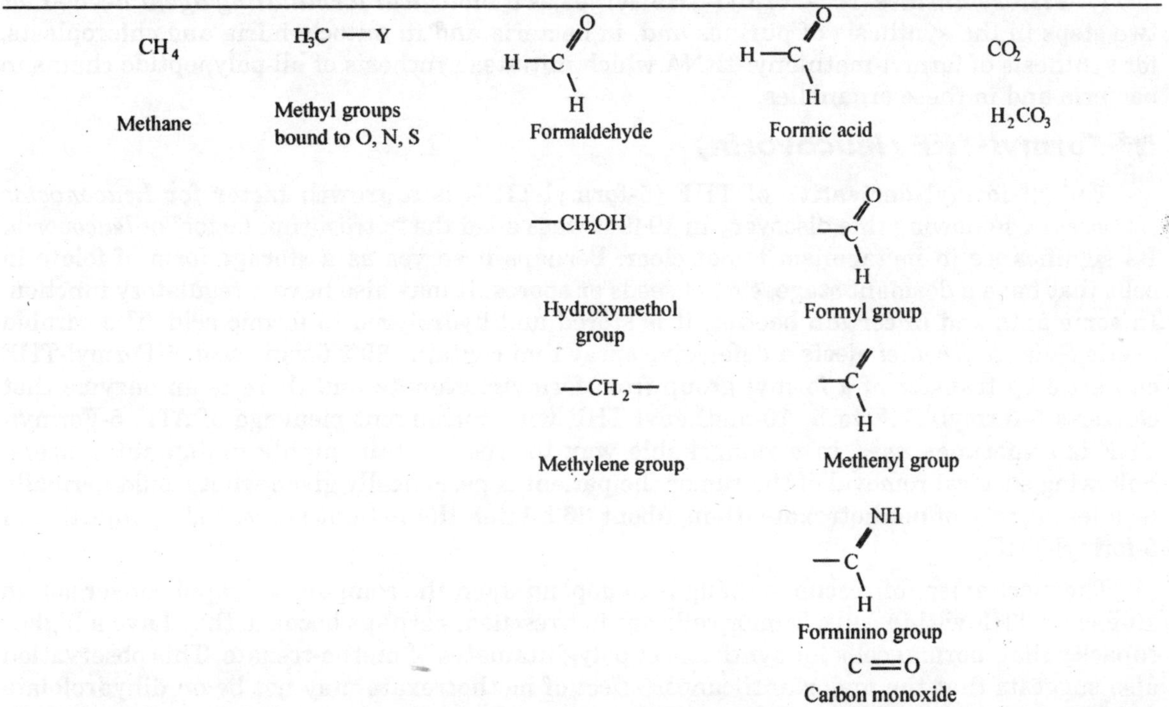

Methane	Methyl groups bound to O, N, S	Formaldehyde	Formic acid	
CH_4	$H_3C—Y$	H—C(=O)—H	H—C(=O)—H	CO_2
Methane	Methyl groups bound to O, N, S	Formaldehyde	Formic acid	H_2CO_3
		—CH_2OH Hydroxymethol group	—C(=O)—H Formyl group	
		—CH_2— Methylene group	—C(=)—H Methenyl group	
			—C(=NH)—H Forminino group	
			C=O Carbon monoxide	

Whether it arises from the hydroxymethyl group of serine or from glycine, the single-carbon unit of methylene-THF (which is at the formaldehyde level of oxidation) can either be oxidized further to 5, 10-methenyl-THF and 10-formyl-THF or it can be reduced to methyl-THF. The reactions of folate metabolism occurs in both cytoplasm and mitochondria at rates that are affected by the length of the polyglutamate chain. The many complexities of these pathways are not fully understood.

Starting with Formate or Carbon Dioxide

Most organisms can, to some extent, also utilize formate as a source of single-carbon units. Human beings have a very limited ability to metabolize formate and the accumulation of formic acid in the body following ingestion of methanol is often fatal. However, many bacteria are able to subsist on formate as a sole carbon source. Archaea may generate the formate by reduction of CO_2. Utilization of formate begins with formation of 10-formyl-THF, lower left corner). 10-Formyl-THF can be reduced to methylene-THF and the single-carbon unit can be transferred to glycine to form serine. In some bacteria three separate enzymes are required to convert formate to methylene-THF: 10-formyl-THF synthetase, presumably via formyl phosphate), methenyl-THF cyclohydrolase, and methylene-THF dehydrogenase. NADH is used by the acetogens but in E. *coli* and in most higher organisms NADPH is the reductant. In some bacteria and in some tumor cells, two of the three enzymes are present as a bifunctional enzyme.

In most eukaryotes all three of the enzymatic activities needed for converting formate to methylene-THF are present in a single large (200-kDa) dimeric trifunctional protein called *formyl-THF synthetase.* N^{10}-Formyl-THF serves as *a biological formulating agent* needed for two steps in the synthesis of purines and, in bacteria and in mitochondria and chloroplasts, for synthesis of formyl-methionyl-tRNA which initiates synthesis of all polypeptide chains in bacteria and in these organelles.

N^5-Formyl-THF (leucovorin)

The N^5-formyl derivative of THF (5-formyl-THF) is a growth factor for *Leuconostoc citrovorum.* Following this discovery in 1949 it was called the "citrovorum factor" or *leucovorin.* Its significance in metabolism is not clear. Perhaps it serves as a storage form of folate in cells that have a dormant stage, *e.g.,* of seeds or spores. It may also have a regulatory function. In some ants and in certain beetles, it is stored and hydrolyzed to formic acid. The carabid beetle *Galerita lecontei* ejects a defen-sive spray that contains 80% formic acid. 5-Formyl-THF can arise by transfer of a formyl group from formylglutamate and there is an enzyme that converts 5-formyl-THF to 5, 10-methenyl-THF with concurrent cleavage of ATP. 5-Formyl-THF is sometimes used in a remarkable way to treat certain highly malignant cancers. Following surgical removal of the tumor the patient is periodically given what would normally be a lethal dose of methotrexate, then, about 36 h later, the patient is rescued by injection of 5-formyl-THF.

The mechanism of rescue is thought to depend upon the compound's rapid conversion to 10-formyl-THF within cells. Tumor cells are not rescued, perhaps because they have a higher capacity than normal cells for synthesis of polyglutamates of metho-trexate. This observation also suggests that the major anti-cancer effect of methotrexate may not be on dihydrofolate reductase but on enzymes of formyl group transfer that are inhibited by polyglutamate derivatives of methotrexate.

Catabolism of Histidine and Purines

Another source of single-carbon units in metabolism is the degradation of histidine which occurs both in bacteria and in animals via *formiminoglutamate*. The latter transfers the –CH = NH group to THF forming 5-form-imino-THF, which is in turn converted. to 5, 10-methenyl-THF and ammonia. In bacteria that ferment purines, *formiminoglycine* is an intermediate. Again, the forniimino group is transferred to THF and deaminated, the eventual product being 10-formyl-THF. In these organisms, the enzyme 10-formyl-THF synthase also has a very high activity. It probably operates in reverse as a mechanism for synthesis of ATP in this type of fermentation. In other organisms excess 10-formyl-THF may be oxidized to CO_2 via an NADP+-dependent dehydrogenase, providing a mechanism for detoxifying formic acid. In some organisms excess 10-formyl-THF may simply be hydrolyzed with release of formate.

Thymtidylate Synthase

Methylene-THF serves as the direct precursor of the 5-methyl group of thymine as well as of the hydroxymethyl groups of *hydroxymethylcytosine* and of *2-oxopantoate,* an intermediate in the formation of pantothenate and coenzyme A. The latter is a simple hydroxymethyl transfer reaction that is related to an aldol condensation and which may proceed through an imine of the kind.

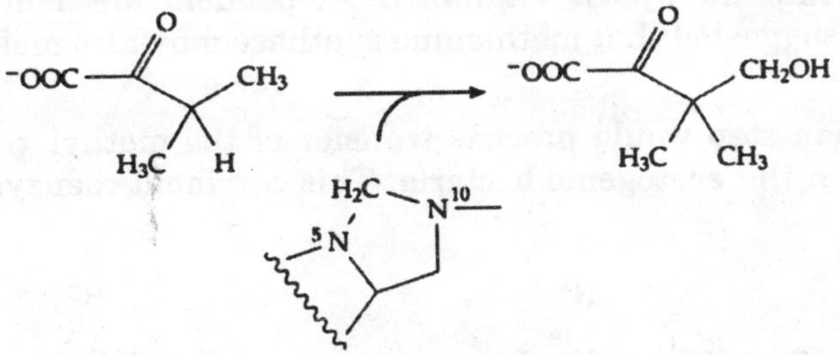

$N^{5,10}$-Methylene-THF

During thymine formation the coenzyme is oxidized to dihydrofolate, which must be reduced by dihydrofolate reductase to complete the catalytic cycle. A possible mechanistic sequence for *thymidylate synthase*, an enzyme of known three-dimensional structure. In the first step (*a*) a thiolate anion, from the side chain of Cys 198 of the 316-residue *Lactobacillus* enzyme, adds to the 5 position of the substrate 2′-deoxyuridine monophosphate (dUMP). As a consequence the 6 position becomes a nucleophilic center which can combine with methylene-THF. After tautomerization this adduct eliminates THF to give a 5-methylene derivative of the dUMP. The latter immediately oxidizes the THF by a hydride ion transfer to form thymidylate and dihydrofolate. In protozoa thymidylate synthase and dihydro-folate reductase exist as a single bifunctional protein.

Consequently, in methotrexate-resistant strains of *Leishmania* (and also in cancer cells) both enzyme activities are increased equally by gene amplification. This presents a serious problem in the treatment of protozoal parasitic diseases for which few suitable drugs are known. Cells of *E. coli,* when infected with T-even bacterio-phage, convert dCMP to 5-hydroxymethyl-dCMP and suitably infected cells of *Bacillus subtilis* form 5-hydroxymethyl-dUMP. These products can be formed by attack of OH^- on quinonoid intermediates of the type postulated for thymidylate synthetase. A related reaction is the posttranscriptional conversion of a single uracil residue in tRNA mole-cules of some bacteria to a thymine ring (a *ribothymidylic acid* residue). In this case, $FADH_2$ is used as a reducing agent to convert the initial adduct the ribothymidylic acid and THF.

Synthesis of Methyl Groups

The reduction of methylene-THF to 5-methyl-THF within all living organisms from bacteria to higher plants and animals provides the methyl groups needed in biosynthesis. These are required for formation of methionine and, from it, S-adenosylmethionine. The latter is used to modify proteins, nucleic acids, and other biochemicals through methylation of specific groups.

In the methanogens, reduction of a corresponding methylene derivative of methanopterin gives rise to methane. Mammalian methylene-THF reductase is a FAD-containing flavoprotein that utilizes NADPH for the reduction to 5-rnethyl-THF. Matthews suggested that the mechanism of this reaction involves an internal oxidation-reduction reaction that generates a 5-methyl-quinonoid dihydro-THF. Methylene-THF reductase of acetogenic bacteria is also a flavoprotein but it contains Fe-S centers as well. The 237-kDa $\alpha_4\beta_4$ oligomer contains two molecules of FAD and four to six of both Fe and S^{2-} ions. The methyl group of methyl-THF is incorporated into methionine by the vitamin B_{12}-dependent methionine synthase which is discussed. Matthews suggested that methionine synthase may also make use of the 5-methyl-quinonoid-THF.

An initial reduction step would precede transfer of the methyl group. Methyl-THF is a precursor to acetate in the acetogenic bacteria. This corrinoid coenzyme-dependent process is also considered.

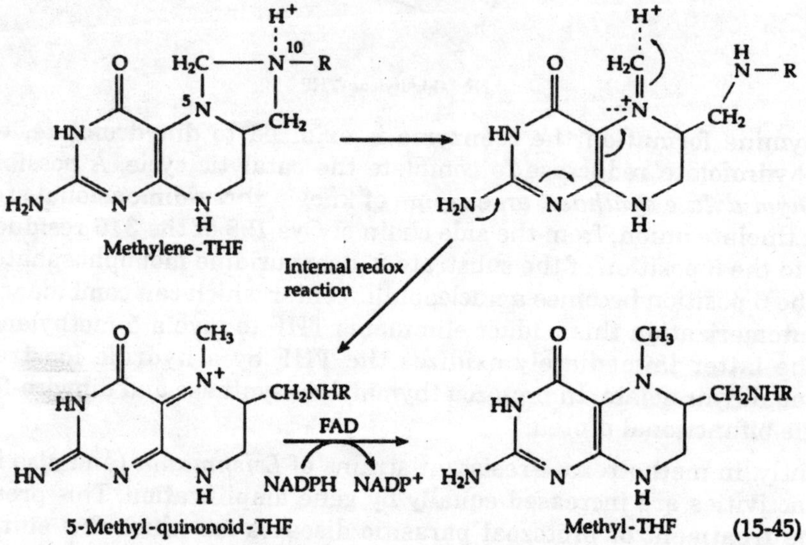

Methylene-THF

Internal redox reaction

FAD

NADPH $NADP^+$

5-Methyl-quinonoid-THF

Methyl-THF (15-45)

SPECIALIZED COENZYMES OF METHANOGENIC BACTERIA

Methane-producing bacteria obtain energy by reducing CO_2 with molecular hydrogen:

$$CO_2 + 4H_2 \rightarrow CH_4 + 2H_2O$$

$$\Delta G° = -131 \text{ kJ/mol} \qquad \text{...(10.46)}$$

Some species are also able to utilize formate or formal-dehyde as reducing agents. These compounds are oxidized to CO_2, the reducing equivalents formed being used to reduce CO_2 to methane. Carbon monoxide can also be converted to CO_2.

$$CO + H_2O \rightarrow CO2 + H_2$$

$$\Delta G° = -20 \text{ KJ/ mol} \qquad \text{...(10.47)}$$

By using the combined reactions the bacteria can subsist on CO alone.

$$4CO + 2\,H_2O \rightarrow 3CO_2 + CH_4$$

$$\Delta G° = -211 \text{kJ/mol of methane formed} \qquad \text{...(10.48)}$$

Some species reduce methanol to methane via.

$$CH_3OH + H_2 \rightarrow CH_4 + H_2O$$

$$\Delta G° = -113 \text{ kJ/ mol} \qquad ....(10.49)$$

$$4CH_3OH \rightarrow 3CH_4 + CO_2 + H_2O$$

$$\Delta G° = -107 \text{kJ/mol of methane formed} \qquad ...(10.50)$$

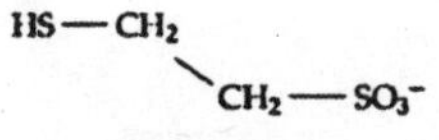

Coenzyme M

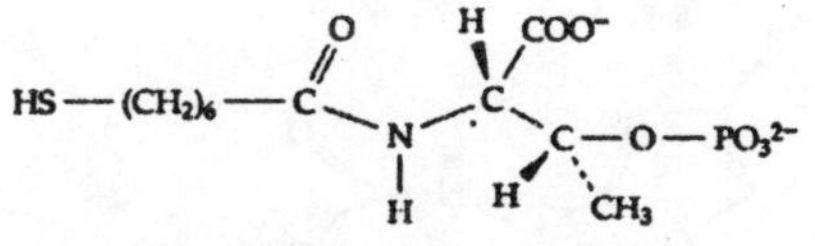

7-Mercaptoheptanoylthreonine phosphate

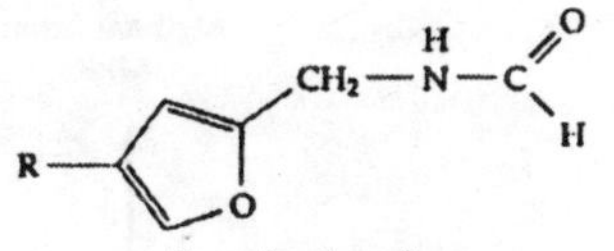

Formylmethanofuran

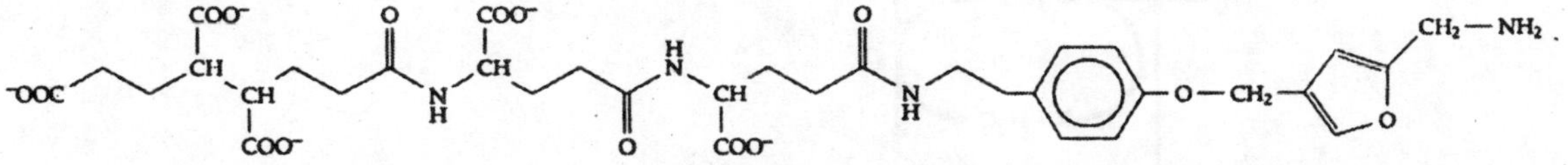

Methanofuran

To accomplish these reactions a surprising variety of specialized cofactors are needed. The first of these, *coenzyme* M, 2-mercaptoethane sulfonate, was discovered in 1974. It is the simplest known coenzyme. Later, the previously described 5-*deazaflavin* F, a *nickel tetrapyrrole* F_{430}, *methanopterin* the "carbon dioxide reduction factor" called *methanofuran*, and 7-*mercaptoheptanoylthreonine phosphate* were also identified. A sketch of the metabolic pathways followed in methane formation. In the first step (a) the amino group of methanofuran is thought to add to CO_2 to form a carbarnate which is reduced to formylmethanofuran by H_2 and an intermediate carrier H_2X in step *b*.

The formyl group is then transferred to tetrahydromethanopterin (H_4MPT) and is cyclized and reduced in two stages in steps *d, e,* and/. The reductant is the deazaflavin F_{420} and the reactions parallel those for conversion of formyl-THF to methyl-THF. The methyl group of methyl-H_4MPT is then transferred to the sulfur atom of the thiolate anion of coenzyme M, from which it is reduced off as CH_4. This is a complex process requiring the nickel-containing F_{430}, FAD, and 7-mercaptoheptanoylthreo-nine phosphate (HS-HTP). The HS-HTP may be the 2-electron donor for the reduction.

The mixed-disulfide CoM-S-S-HTP appears to be an intermediate and also an allosteric effector for the first step, the reduction of CO_2 to formylmethanofuran. A complex of proteins that is unstable in oxygen is also needed. The principal component of the methyl reductase binds two molecules of the nickel-containing F_{430}. It also binds two moles of CoM-SH noncovalently. The binding process requires ATP as well as the presence of other proteins, notably those of the hydrogenase system. Coenzyme M, previously found only in methanogenic archaeobacteria, has recently been discovered in both gram-negative and gram-positive alkeneoxidizing eubacteria. It seems to function in cleavage of epoxy rings and as a carrier of hydroxyalkyl groups.

QUINONES, HYDROQUINONES, AND TOCOPHEROLS

Pyrroloquinoline Quinone (PQQ)

Bacteria that oxidize methane or methanol (*methylotrophs*) employ a periplasmic methanol dehydrogenase that contains as a bound coenzyme, the pyrroloquinoline quinone designated PQQ or methoxatin. This fluorescent *ortho*-quinone is released when the protein is denatured. Some other bacterial alcohol dehydrogenases and glucose dehydrogenases contain the same cofactor.

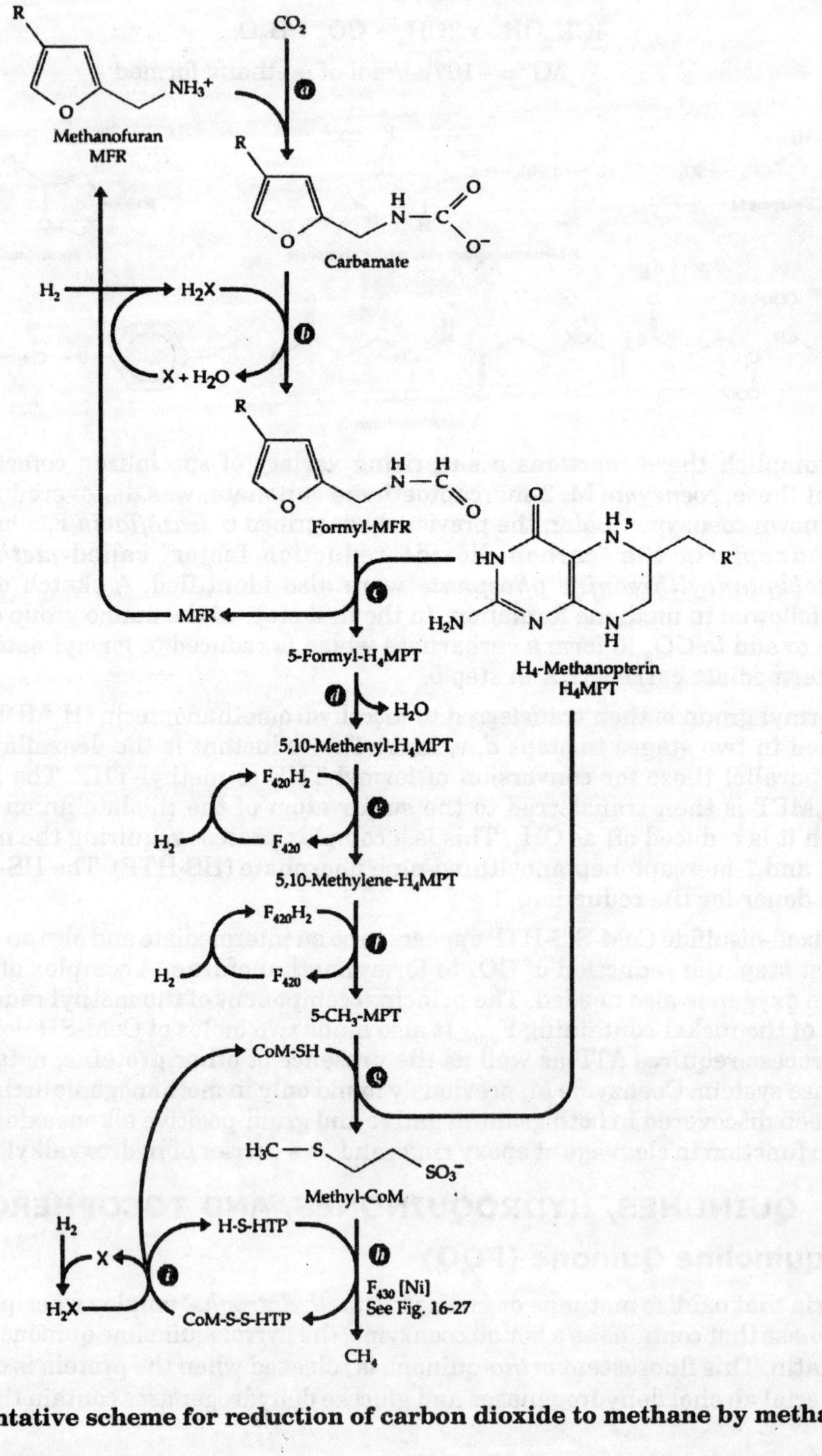

CO_2

(a)

Methanofuran MFR

Carbamate

H_2 → H_2X

(b)

$X + H_2O$

Formyl-MFR

(c)

MFR

5-Formyl-H_4MPT

H_4-Methanopterin H_4MPT

(d) H_2O

5,10-Methenyl-H_4MPT

$F_{420}H_2$

(e)

H_2 F_{420}

5,10-Methylene-H_4MPT

$F_{420}H_2$

(f)

H_2 F_{420}

5-CH_3-MPT

CoM-SH

(g)

$H_3C—S$ SO_3^-

Methyl-CoM

H_2

H-S-HTP

X

(i)

(h)

H_2X

CoM-S-S-HTP

F_{430} [Ni] See Fig. 16-27

CH_4

Fig. 10.18. Tentative scheme for reduction of carbon dioxide to methane by methanogens.

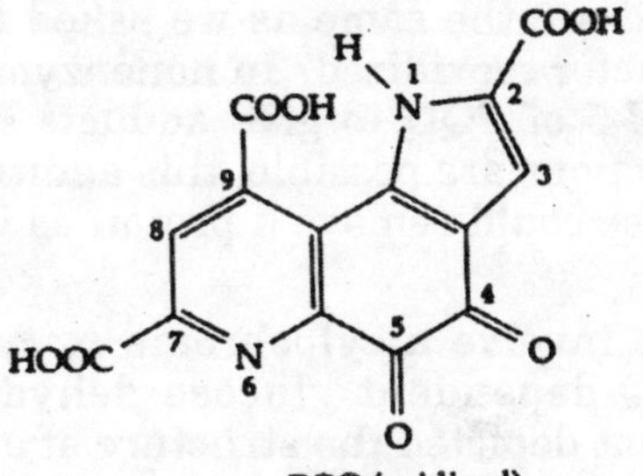

PQQ (oxidized)

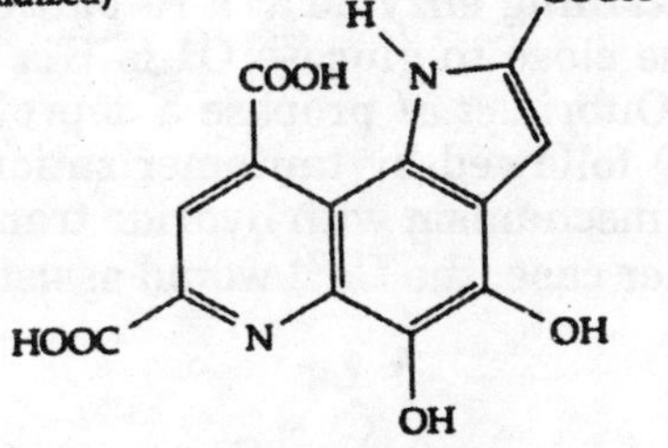

PQQH_2 (reduced)

PQQ-containing methanol dehydrogenase from *Methylophilus* is a dimer of a 640-residue protein consisting of two disulfide-linked peptide chains with one noncovalently bound PQQ. The large subunit contains a β propellor, which is similar to that in the protein G. The PQQ binding site lies above the propellor and is formed by a series of loops. The coenzyme interacts with several polar protein side chains and with a bound calcium ion.

Bacterial PQQ-dependent glucose dehydrogenases have also been studied intensively. The 450-residue soluble enzyme from *Acinetobacter calcoaceticus* is partially homologous to the methanol dehydrogenase. PQQ and the other quinone prosthetic groups described here all function in reactions that would be possible for pyridine nucleotide or flavin coenzymes. All

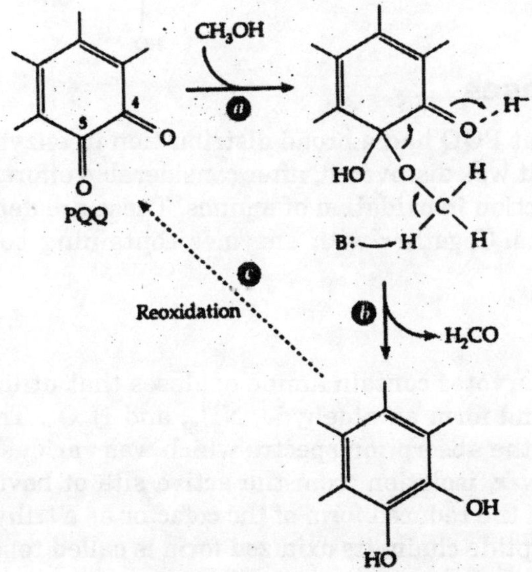

CH_3OH

PQQ

Reoxidation

H_2CO

of them, like the flavins, can exist in oxidized, half-reduced semiquinone and fully reduced dihydro forms. The questions to be asked are the same as we asked for flavins. How do the substrates react? How is the reduced cofactor reoxidized? In nonenzymatic reactions alcohols, amines, and enolate anions all add at C-5 of PQQ to give adducts such as that shown for methanol. Although many additional reactions are possible, this addition is a reasonable first step in the mechanism. An enzymatic base could remove a proton as is indicated in step *b* to give $PQQH_2$.

The pathway for reoxidation might involve a cytochrome *b*, cytochrome *c*, or bound ubiquinone. Although the soluble PQQ-dependent glucose dehydrogenase forms, with methylhydrazine, an adduct similar to that depicted the structure of a glucose complex of the $PPQH_2$- and Ca^{2+}-containing enzyme at a resolution of 0.2 ran suggests a direct hydride ion transfer. The only base close to glucose C1 is His 144. The C1-H proton lies directly above the $PQQH_2$ C5 atom. Oubrie *et al* propose a deprotonation of the C1-OH by His 144 and H^- transfer to C5 of PQQ followed by tautomerization. Theoretical calculations by Zheng and Bruice favor a simpler mechanism with hydride transfer to the oxygen atom of the C4 carbonyl (green arrows). In either case, the Ca^{2+} would assist by polarizing the C5 carbonyl group.

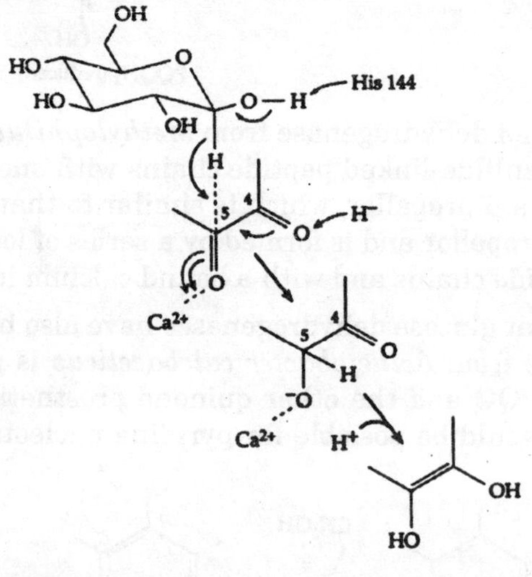

Copper Amine Oxidases

At first it appeared that PQQ had a broad distribu-tion in enzymes, including eukaryotic amine oxidases. However, it was discovered, after considerable effort, that there are additional quinone cofactors that function in oxidation of amines. These are derivatives of tyrosyl groups of specific enzyme proteins. Together with enzymes containing bound PQQ they are often called *quinoproteins*.

Topaquinone (TPQ)

Both bacteria and eukaryotes contain amine oxidases that utilize bound copper ions and O_2 as electron acceptors and form an aldehyde, NH_3, and H_2O_2. The presence of an organic cofactor was suggested by the absorption spectra which was variously attributed to pyridoxal phosphate or PQQ. However, isolation from the active site of bovine serum amine oxidase established the structure of the reduced form of the cofactor as a trihydroxyphenylalanyl group covalently linked in the peptide chain. Its oxidized form is called topaquinone:

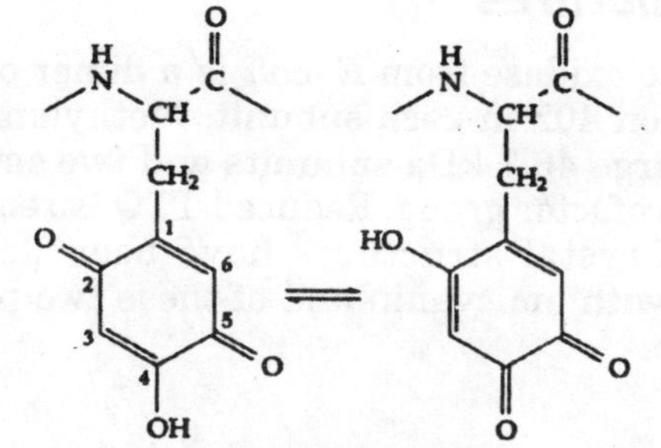

Topaquinone (TPQ)

The same cofactor is present in an amine oxidase from *E. coli* and in other bacterial fungal, plant, and mammalian amine oxidases.

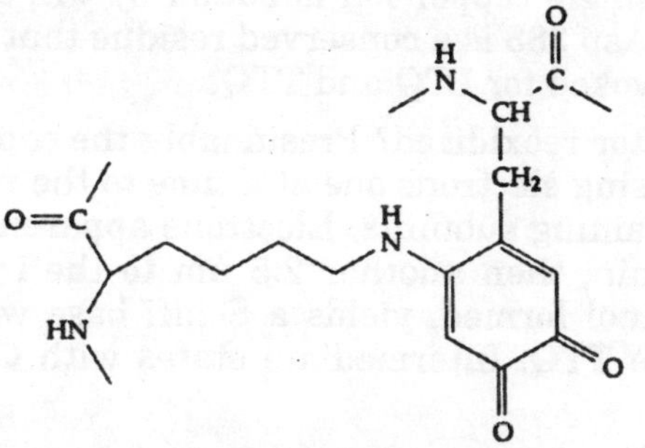

Lysine tyrosylquinone (LTQ)

Lysine Tyrosylquinone *(LTQ)*

Another copper amine oxidase, ***lysyl oxidase***, which oxidizes side chains of lysine in collagen and elastin contains a cofactor that has been identified as having a lysyl group of a different segment of the protein in place of the –OH in the 2 position of topaquinone. Lysyl oxidase plays an essential role in the crosslinking of collagen and elastin.

Tryptophan Tryptophanylquinone *(TTQ)*

This recently discovered quinone cofactor is similar to the lysyl tyrosylquinone but is formed from two tryptopharvyl side chains. It has been found in *methyl-amine dehydrogenase* from methylotrophic grain-negative bacteria and also in a bacterial aromatic amine dehydrogenase.

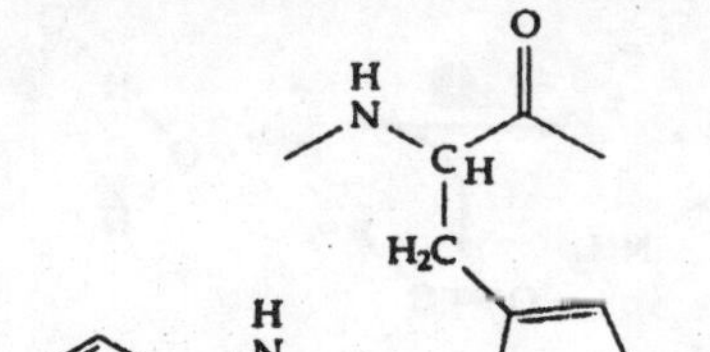

Tryptophan tryptophylquinone (TTQ)

Three-dimensional Structures

The TPQ-containing amine oxidase from *E. coli* is a dimer of 727-residvie subunits with one molecule of TPQ at position 402 in each subunit. Methylamine dehydrogenase is also a large dimeric protein of two large 46.7-kDa subunits and two small 15.5-kDa subunits. Each large subunit contains a TTQ cofactor group. Reduced TTQ is reoxidized by the 12.5-kDa blue copper protein *amicyanin*. Crystal structures have been determined for complexes of methylamine dehydrogenase with amicyanin and of these two proteins with a third protein, a small bacterial cytochrome *c*.

Mechanisms

Studies of model reactions and of electronic, Raman, ESR, and NMR spectra and kinetics have contributed to an understanding of these enzymes. For these copper amine oxidases the experimental evidence suggests an aminotransferase mechanism. The structure of the *E.coli* oxidase shows that a single copper ion is bound by three histidine imidazoles and is located adjacent to the TPQ. Asp 383 is a conserved residue that may be the catalytic base. A similar mechanism can be invoked for LTQ and TTQ.

How is the reduced cofactor reoxidized? Presumably the copper ion adjacent to the TPQ functions in this process, passing electrons one at a time to the next carrier in a chain. There is no copper in the TTQ-containing subunits. Electrons apparently must jump about 1.6 nm to the copper ion of amicyanin, then another 2.5 nm to the iron ion of the cytochrome *c*. Reoxidation of the aminoquinol formed, yields a Schiff base whose hydrolysis will release ammonia and regenerate the TTQ. Intermediate states with Cu^+ and a TTQ semiquinone radical have been observed.

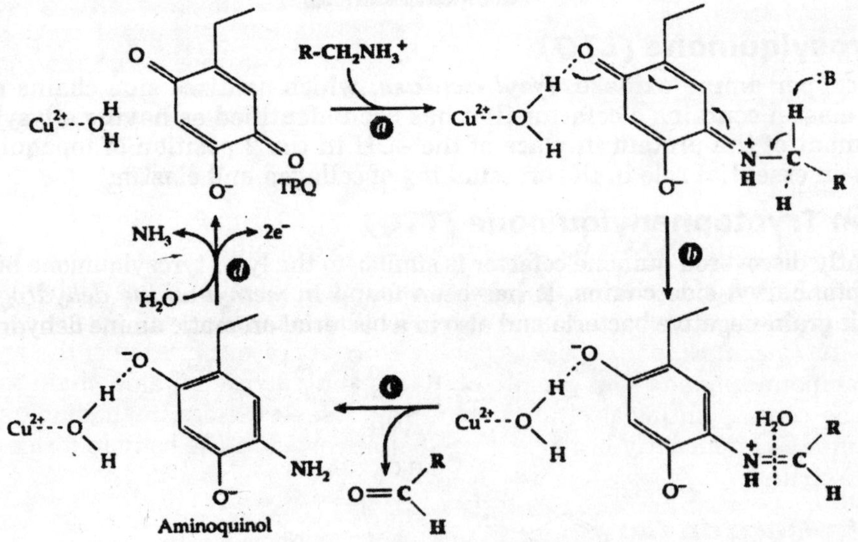

TPQ

$R\text{-}CH_2NH_3^+$

Cu^{2+}

NH_3 $2e^-$ H_2O

Aminoquinol

Ubiquinones, Plastoquinones, Tocopherols, and Vitamin K

In 1955, R. A. Morton and associates in Liverpool announced the isolation of a quinone which they named ubiquinone for its ubiquitous occurrence. It was characterized as a derivative of benzoquinone attached to an unsaturated polyprenyl (isoprenoid) side chain. In fact, there is a family of ubiquinones: that from bacteria typically contains six prenyl units in its side chain, while most ubiquinones from mammalian mitochondria contain ten. Ubiquinone was also isolated by F. L. Crane and associates using isooctane extraction of mitochondria. These workers proposed that the new quinone, which they called *coenzyme Q*, might participate in electron transport.

This function has been fully established. Both the name ubiquinone and the abbreviation Q are in general use. A subscript indicates the number of prenyl units, *e.g.*, Q_{10} Ubiquinones can be reversibly reduced to the hydroquinone forms, providing a basis for their function in electron transport within mitochondria and chloroplasts. While vitamin vitamin are dietary essentials for humans, ubiquinone is apparently not. Animals are able to make ubiquinones in quantities adequate to meet the need for this essential component of mitochondria. However, an extra dietary supplement may sometimes be of value.

A reduced level of ubiquinone has been reported in gum tissues of patients with periodontal disease. Ubiquinones are being tested as possible protectants against heart damage caused by lack of adequate oxygen or by drugs. A closely related series of *plastoquinones* occur in chloroplasts. In these compounds the two methoxyl groups of ubiquinone are replaced by methyl groups. The most abundant of these compounds, plastoquinone A, contains nine prenyl units. Addition of the hydroxyl group of the reduced ubiquinone or plastoquinone to the adjacent double bond leads to a chroman-6-ol structure. The compounds derived from ubiquinone in this way are called *ubichromanols*. The corresponding ubichromenol has been isolated from human kidney. *Plastochromanols* are derived from plastoquinone. That from plastoquinone A, first isolated from tobacco, is also known as solanochromene. A closely related and important family of chromanols are the *tocopherols* or vitamins.

Tocopherols are plant products found primarily in plant oils and are essential to proper nutrition of humans and other animals. α-Tocopherol is the most abundant form of the vitamin E family; smaller amounts of the β, δ, and γ forms occur, as do a series of *tocotrienols* which contain unsaturated isoprenoid units. The configuration of α-tocopherol is 2R, 4′R, 8′R as indicated. When α-tocopherol is oxidized, *e.g.*, with ferric chloride, the ring can be opened by hydrolysis to give *tocopherolquinones*, which can in turn be reduced to tocopherolhydroquinones. Large amounts of the tocopherol-quinones have been found in chloroplasts. Another important family of quinones, related in structure to those already discussed, are the *vitamins*. These occur naturally as two families.

The vitamins K_1 (*phylloquinones*) have only one double bond in the side chain and that is in the prenyl unit closest to the ring. This suggests again the possibility of chromanol formation. In the vitamin K_2 (*menaquinone*) series, a double bond is present in each of the prenyl units. A synthetic compound *menadione* completely lacks the polyprenyl side chain and bears a hydrogen in the corresponding position on the ring. Nevertheless, menadione serves as a synthetic vitamin K, apparently because it can be converted in the body to forms containing polyprenyl side chains.

Quinones as Electron Carriers

Ubiquinones function as electron transport agents within the inner mitochondrial membranes and also within the reaction centers of the photosynthetic membranes of bacteria. The plastoquinones also function in electron transport within these membranes. Within some mycobacteria vitamin K apparently participates in electron transport chains in the same way. Some bacteria contain both menaquinones and ubiquinones.

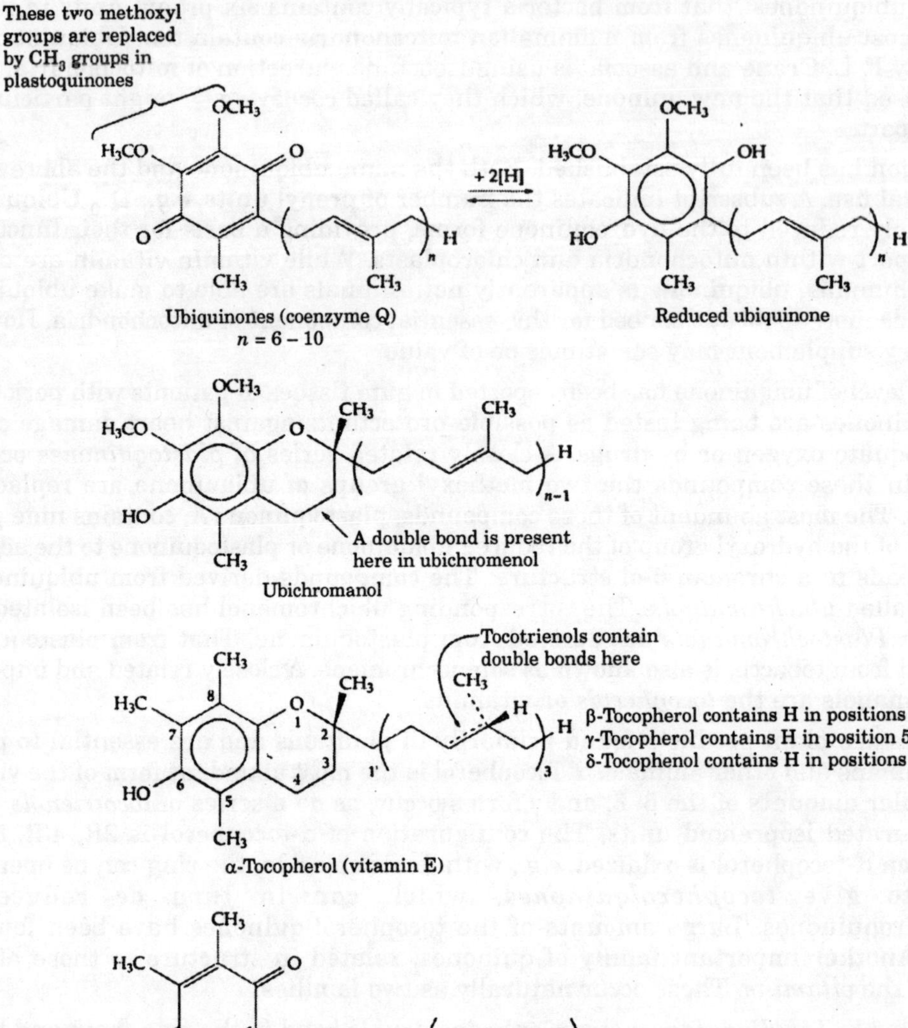

These two methoxyl groups are replaced by CH_3 groups in plastoquinones

OCH_3 H_3CO O O CH_3 CH_3 H n

+ 2[H]

OCH_3 H_3CO OH HO CH_3 CH_3 H n

Ubiquinones (coenzyme Q)
$n = 6 - 10$

Reduced ubiquinone

OCH_3 H_3CO HO CH_3 O CH_3 CH_3 H n-1

A double bond is present here in ubichromenol

Ubichromanol

Tocotrienols contain double bonds here

CH_3 H_3C 8 7 6 5 HO CH_3 O 1 2 3 4 CH_3 1′ CH_3 H H 3

β-Tocopherol contains H in positions 7
γ-Tocopherol contains H in position 5
δ-Tocophenol contains H in positions 5 and 7

α-Tocopherol (vitamin E)

CH_3 H_3C O O CH_3 HO CH_3 CH_3 H 3

Tocopherolquinone of chloroplasts

O O CH_3 CH_3 H n

$n = 4.5$, In the phylloquinone (vitamin K_1) series only one double bond is present in the innermost unit of the isoprenoid side chain.

Vitamins K

Fig. 10.19. Structures of the isoprenoid quinines and vitamin E.

The vitamin E derivative α-*tocopherolquinone* can also serve as an electron carrier, being reversibly reduced to the hydroquinone form α-tocopherolquinol. Such a function has been proposed for the anaerobic rumen bacterium *Butyrovibrio fibrisolvens*. When a single hydrogen atom is removed from a hydroquinone or from a chromanol such as a tocopherol, a free radical is formed. Phenols substituted in the 2, 4, and 6 positions give especially stable radicals.

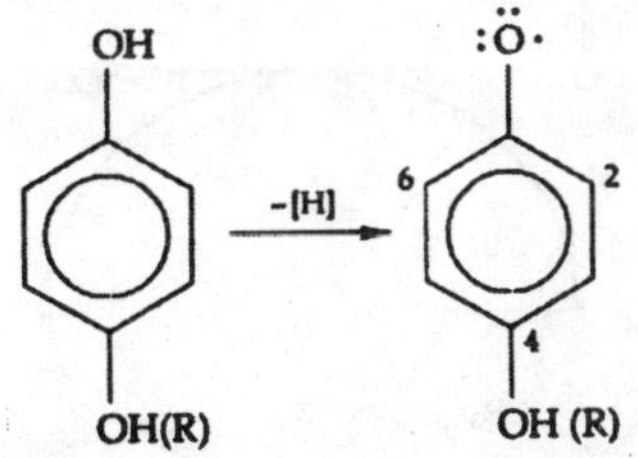

OH → -[H] → :Ö·
OH(R) OH (R)

Both the presence of methyl substituents in the tocopherols and their chromanol structures increase the ability of these compounds to form relatively stable radicals. This ability is doubtless probably important also in the function of ubiquinones and plastoquinones. Ubiquinone radicals (semiquinones) are probably intermediates in mitochondrial electron transport and radicals amounting to as much as 40% of the total ubiquinone in the NADH-ubiquinone reductase of heart mitochondria have been detected by EPR measurements. The equilibria governing semiquinone formation from quinones are similar to those for the flavin semiquinones which were discussed.

Two consecutive one-electron redox steps can be defined. Their redox potentials will vary with pH because of a pK_a for the semiquinone in the pH 4.5 – 6.5 region. For ubiquinone this pK_a is about 4.9 in water and 6.45 in methanol. A pK_a of over 13 in the hydroquinone form will have little effect on redox potentials near pH 7. The potential for the one-electron reaction $Q + e^- \rightarrow Q^-$ is evaluated most readily. For this reaction $E^{\circ\prime}$ (pH 7) is –0.074, –0.13, –0.17, and –0.23 for 2, 3-dimethylbenzoquinone, plastoquinone, ubiquinone, and phylloquinone, respectively. Why does this entire family of compounds have the long polyprenyl side chains? A simple answer is that they serve to anchor the compounds in the lipid portion of the cell membranes where they function. In the case of ubiquinones both the oxidized and the reduced forms may move freely through the lipid phase shuttling electrons between carriers.

Vitamin K and γ-Carboxyglutamate Formation

In higher animals the only known function of vitamin K is in the synthesis of γ-carboxyglutamate (Gla)-containing proteins, several of which are needed in blood clotting. Following the discovery of γ-carboxyglutamate, it was shown that liver microsomes were able to incorporate ^{14}C-containing bicarbonate or CO_2 into the Gla of prothrombin and could also generate Gla in certain simple peptides such as Phe-Leu-Glu-Val. Three enzymes are required. All are probably bound to the microsomal membranes. An NADPH-dependent reductase reduces vitamin K quinone to its hydroquinone form. Conversion of Glu residues to Gla residues requires this reduced vitamin K as well as O_2 and CO_2. During the carboxylation reaction the reduced vitamin K is converted into vitamin K 2,3-epoxide.

The mechanism is uncertain but a peroxide intermediate such as that probably involved. This could be used to generate a hydroxide ion adjacent to the *pro-S* -H of the glutamate side

chain of the substrate. This hydrogen could be abstracted by the OH^- to form H_2O and a carbanion which would be stabilized by the adjacent carboxyl group.

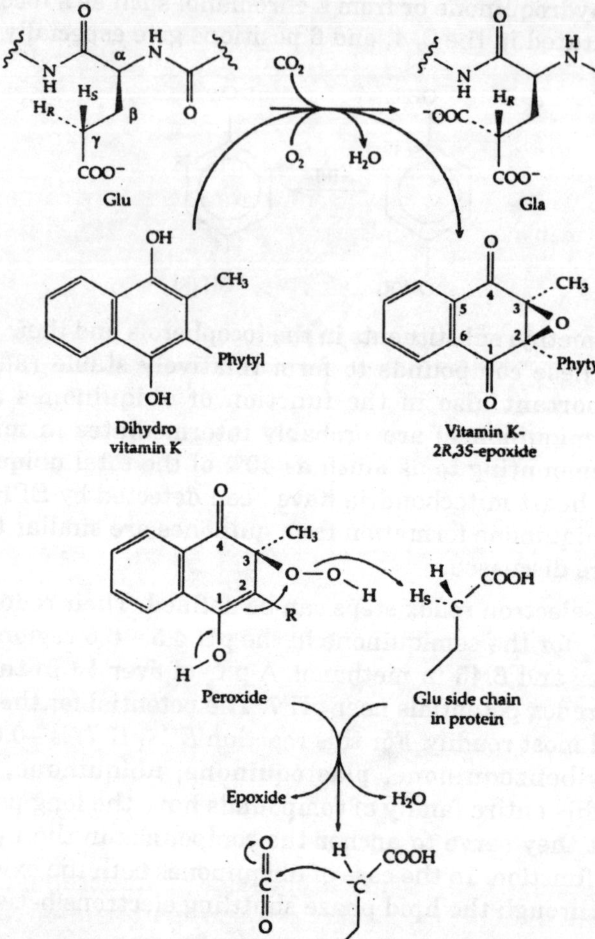

Glu
Gla
CO_2
O_2
H_2O
Dihydro vitamin K
Vitamin K-2R,3S-epoxide
Peroxide
Glu side chain in protein
Epoxide
H_2O

Dowd and coworkers raised doubts that a hydroxide ion released in the active site in this manner is a strong enough base to generate the anion. They hypothesized a *"base strength amplification" mechanism* that begins with a peroxide formed at C-4 followed by ring closure to form a dioxetane and rearrangement to the following hypothetical strong base.

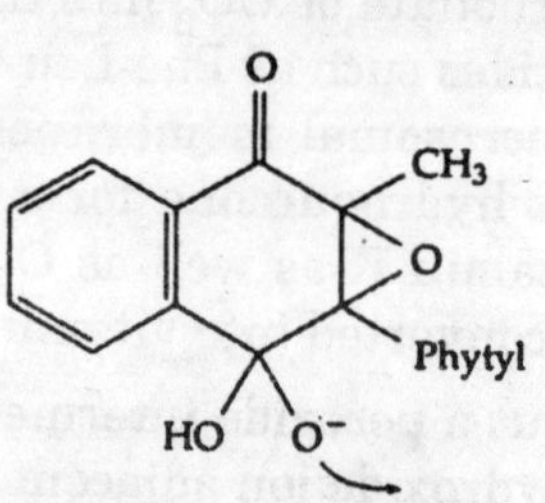

Hypothetical strong base

The proposal was supported by model experiments and also by observation of some incorporation of both atoms of $^{18}O_2$ into vitamin K epoxide. However, theoretical calculations support the simpler mechanism. The fact that L-threo-γ-fluoroglutamate residues, in which the fluorine atom is in the position corresponding to the *pro-S* hydrogen, are not carboxylated but that an *erythro*-γ-fluoroglutamate is carboxylated suggested the indicated stereospecificity. This was confirmed by observation of a kinetic isotope effect when ^2H is present in the *pro-S* position.

Addition of the carbanion to CO_2 would generate the Gla residue. The two glutamates in the following sequence, in which X may be various amino acids, are carboxylated if the protein also carries a suitable N-terminal signal sequence: EXXXEXC. If a suitable glutamyl peptide is not available for carboxylation the postulated peroxide intermediate is still converted slowly to the 2, 3-epoxide of vitamin K.

A third enzyme is required to reduce the epoxide to vitamin K. The biological reductant is uncertain but dithiols such as dithiothreitol serve in the laboratory. Protonation of an intermediate enolate anion would give 3-hydroxy-2, 3-dihydrovitamin K, an observed side reaction product.

$$\text{Vitamin K epoxide} + 2[H] \rightarrow H_2O + \text{Vitamin K} \qquad ...(10.57)$$

This reaction is of interest because of its specific inhibition by such coumarin derivatives as Warfarin:

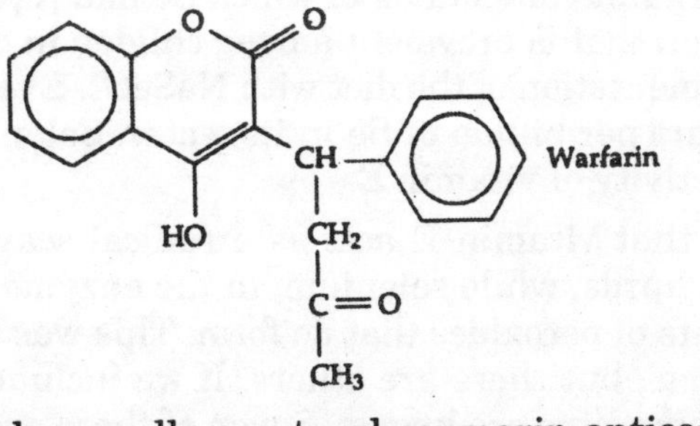

O O HO CH CH₂ C=O CH₃ Warfarin

This synthetic compound, as well as natural coumarin anticoagulants, inhibits both the vitamin K reductase and the epoxide reductase. The matter is of considerable practical importance because of the spread of warfarin-resistant rats in Europe and the United States.

One resistance mutation has altered the vitamin K epoxide reductase so that it is much less susceptible to inhibition by war-farin. While glutamate residues in peptides of appropriate sequence are carboxylated by the vitamin K-dependent system, aspartate peptides scarcely react. Beta-carboxyaspartate is present in protein C of the blood anticoagulant system and in various other proteins containing EGF homology domains but the mechanism of its formation is unknown.

Tocopherols (Vitamin E) as Antioxidants

The major function of the tocopherols is thought to be the protection of phospholipids of cell membranes against oxidative attack by free radicals and organic peroxides. Peroxidation of lipids, which is described, can lead to rapid development of rancidity in fats and oils. However, the presence of a small amount of tocopherol inhibits this decomposition, presumably by trapping the intermediate radicals in the form of the more stable tocopherol radicals which may dimerize or react with other radicals to terminate the chain. Even though only one molecule of tocopherol is present for a thousand molecules of phospholipid, it is enough to protect membranes. That vitamin E does function in this way is supported by the observation that much of the tocopherol requirement of some species can be replaced by N, N′-diphenyl-*p*-phenylenediamine, a synthetic antioxidant.

Three generations of rats have been raised on a tocopherol-free diet containing this synthetic antioxidant. However, not all of the deficiency symptoms are prevented. The antioxidant role of α-tocopherol in membranes is generally accepted. It is thought to be critical to defense against oxidative injury and to help the body combat the development of tumors and to slow aging. Gamma-tocopherol may be more reactive than α-tocopherol in removing radicals created by NO and other nitrogen oxides. Its actions are strongly linked to those of ascorbic acid and selenium. Ascorbate may reduce tocopherol semiquinone radicals, while selenium acts to enhance breakdown of peroxides as described in the next section.

SELENIUM-CONTAINING ENZYMES

In 1957, Schwartz and associates showed that the toxic element selenium was also a nutritional factor essential for prevention of the death of liver cells in rats. Liver necrosis would be prevented by as little as 0.1 ppm of selenium in the diet. Similar amounts of selenium were shown to prevent a muscular dystrophy called "white muscle disease" in cattle and sheep grazing on selenium-deficient soil. Sodium selenite and other inorganic selenium compounds were more effective than organic compounds in which Se had replaced sulfur. *Keshan disease*, an often fatal heart condition that is prevalent among childen in Se-deficient regions of China, can be prevented by supplementation of the diet with $NaSeO_3$. Even the little crustacean "water flea" *Daphniu* needs 0.1 part per billion of Se in its water. Selenium has long been known to enhance the antioxidant activity of vitamin E.

Recent work suggests that vitamin E acts as a radical scavenger, preventing excessive peroxidation of membrane liprds, while selenium, in the enzyme *glutathione peroxidase*, acts to destroy the small amounts of peroxides that do form. This was the first established function of selenium in human beings, but there are others. If we include proteins from animals and bacteria, at least ten selenoproteins are known. Seven of these are enzymes and most catalyze redox processes. The active sites most often contain *selenocysteine*, whose selenol side chain is more acidic (pK_a ~5.2) than that of cysteine and exists as $-CH_2 - Se^-$ at neutral pH. Glutathione peroxidases catalyze the reductive decomposition of H_2O_2 or of organic peroxides by glutathione (G-SH) according.

At least three isoenzyme forms have been identified in mammals: a cellular form, a plasma form, and a form with a preference for organic peroxides derived from phospholipids. A related selenoprotein has been found in a human poxvirus. Selenocysteine is present at a position near the N terminus of an a helix, in the ~180-residue polypeptides. A possible reaction mechanism involves attack by the selenol on the peroxide to give a selenic acid intermediate which is reduced by glutathione in two nucleophilic displacement.

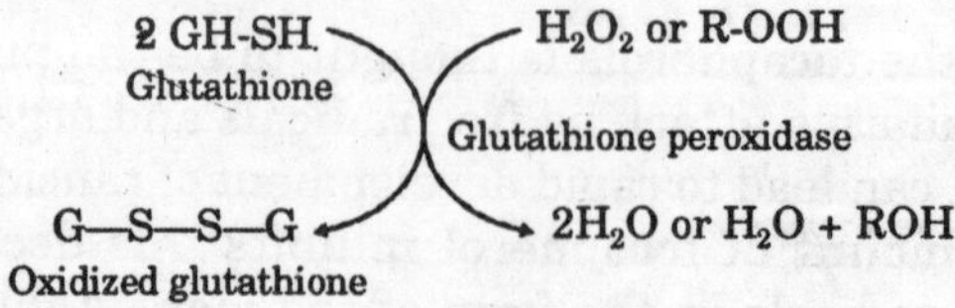

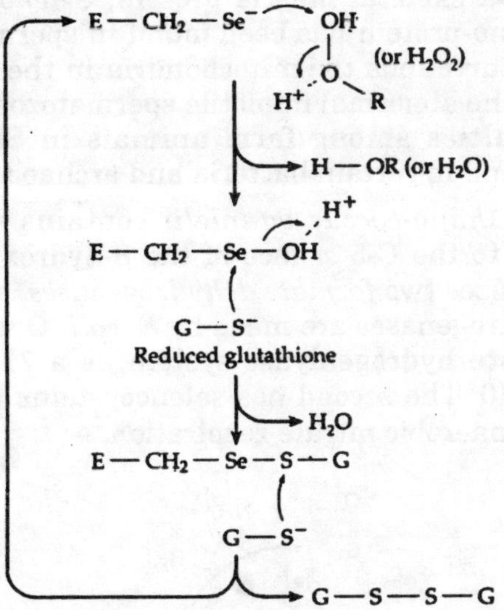

E — CH₂ — Se⁻ OH (or H_2O_2) O H⁺ R

→ H — OR (or H_2O)

H⁺

E — CH₂ — Se — OH

G — S⁻
Reduced glutathione

→ H_2O
E — CH₂ — Se — S — G

G — S⁻

→ G — S — S — G

Three types of *iodothyronine deiodinase* remove iodine atoms from thyroxine to form the active thyroid hormone triiodothyronine and also to inactivate the hormone by removing additional iodine. In this case the $-CH_2-Se^-$ may attach the iodine atom, removing it as I^+ to form $-CH_2-Se-I$. The process could be assisted by the phenolic –OH group if it were first tautomerized.

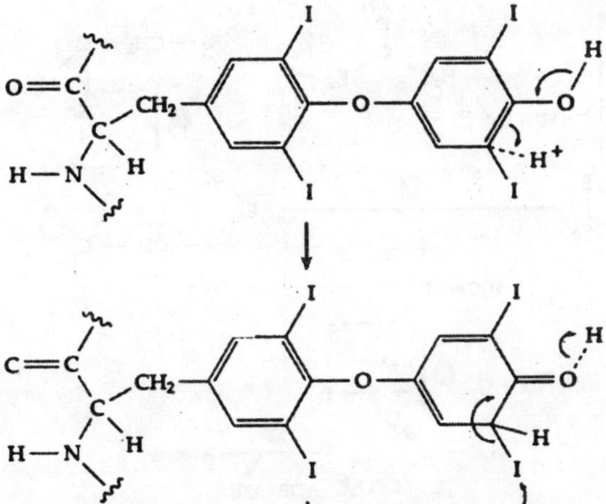

Triiodothyronine

E — CH₂ — Se⁻

E — CH₂ — Se — I

A recently discovered human Selenoprotein is a *thioredoxiit reductase* which is present in the T cells of the immune system as well as in placenta and other tissues. The 55-kDa protein has one selenocysteine as the penultimate C-terminal residue. Another mammalian Selenoprotein, of uncertain function, is the 57-kDa *Selenoprotein P*. It contains over 60% of the selenium in rat plasma and is also present in the human body. Selenoprotein P contains ten selenocysteine residues. Some of these may be replaced by serine in a fraction of the molecules. A smaller 9.6-kDa skeletal muscle protein, *Selenoprotein W*, contains a single selenocysteine. Another seleno-protein has been found in sperm cells, both in the tail and in a keratin-rich capsule that surrounds the mitochondria in the sperm midpiece. Lack of this protein may be the cause of the abnormal immotile spermatozoa observed in Se-deficient rats and of reproductive difficulties among farm animals in Se-deficient regions. Several selenoproteins have been found in certain bacteria and archaea.

A *hydrogenase* from *Methano-coccus vannielii* contains selenocysteine. This enzyme transfers electrons from H_2 to the C-5 *si* face of the 8-hydroxy-5-deazaflavin cofactor F_{420}. The same bacterium synthesizes two *formate dehydrogenases*, one of which contains Se. Two Se-containing formate dehydrogenases are made by *E. coli.* One of them, which is coupled to a hydrogenase in the formate hydrogenlyase system, is a 715-residue protein containing selenocysteine at position 140. The second has selenocysteine at position 196 and functions with a nitrate reductase in anaerobic nitrate respiration.

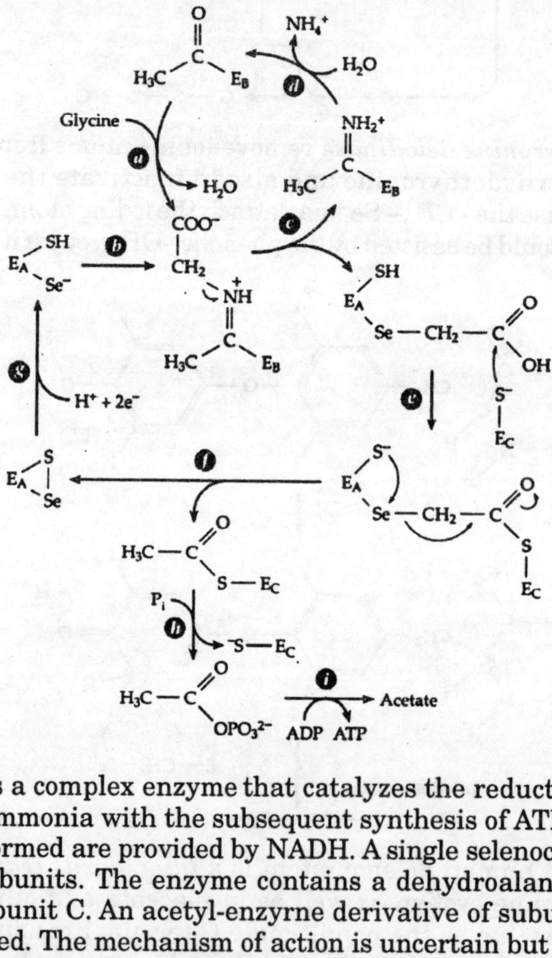

Glycine reductase is a complex enzyme that catalyzes the reductive cleavage of glycine to acetyl phosphate and ammonia with the subsequent synthesis of ATP. Electrons for reduction of the disulfide that is formed are provided by NADH. A single selenocysteine residue is present in the small 12-kDa subunits. The enzyme contains a dehydroalanine residue in subunit B and a thiol group in subunit C. An acetyl-enzyrne derivative of subunit C, perhaps of its -SH group, has been identified. The mechanism of action is uncertain but the have been suggested. The subunits are designated E_A, E_B, and E_c. Particularly hard to understand because formation of a thioester in this manner is not expected to occur spontaneously and must be linked in some way to other steps.

A selenium-containing *xanthine dehydrogenase* is present in purine-fermenting clostridia. Like other xanthine dehydrogenases, it converts xanthine to uric acid and contains nonheme iron, molybdenum, FAD, and an Fe-S center. The selenium is probably present as Se^{2-} bound to Mo as is S^{2-} in xanthine oxidase. A related reaction is catalyzed by the Fe-S protein *nicotinic acid hydroxylase* found in some clostridia. Splitting of the Mo(V) EPR signal when ^{77}Se is present in the enzyme shows that the selenium is present as a ligand of molybdenum. Another member of the family is a purine hydroxylase that converts purine 2-hydroxypurine, or hypoxanthine to xanthine.

Table 10.4. Selenium-Containing Proteins

Enzyme	*Source*	*Mass (kDa)*	*Subunit composition*	*Other cofactors*
Glutathione peroxidases				None
Cellular	Mammals	21 × 4		
Plasma	Humans			
Phospholipid hydroperoxide	Pig, rat	18		
Iodothyronine deiodinases	Vertebrates			
Thioredoxin reductase	Humans	55 × 2		
Selenoprotein *P*	Mammals	57		None
Selenoprotein *W*	Rat	19.6		
Formate dehydrogenase	Bacteria- *E. coli* Archaea	600	$\alpha_2\beta_4\gamma_{2-4}$	Heme *b* Mo; molybdopterin
Hydrogenase	Bacteria		$\alpha_2\beta_4\gamma_2$	FAD; NiFeSe
Archaea				
Glycine reductase	Some clostridia		ABC	
Selenoprotein A			12	
Selenoprotein B			200	
Carbonyl protein C			250	Fe?
Nicotinic acid hydroxylase	Clostridia		300	FAD, FeS, Mo
Purine hydroxylase	Clostridia			FAD, FeS, Mo
Thiolase (contains Selenomethionine)	*Clostridium kluyverii*	39 × 4	None	
Carbon monoxide dehydrogenase	*Oligotropha carboxidovorans*	137 × 2	$\alpha_2\beta_2\gamma_2$	FAD; Mo; molybdopterin

COOH NADP⁺ NADPH COOH
H_2O H^+
N O N H

A *thiolase* from *Clostridium kluyveri* is one of only two known selenoproteins that contain selenomethionine 'However, the selenomethionine is incorporated randomly in place of methionine. This occurs in all proteins of all organisms to some extent and the toxicity of selenium may result in part from excessive incorporation of selenomethionine into various proteins.

Selenium is found to a minor extent wherever sulfur exists in nature. This includes the sulfur-containing modified bases of tRNA molecules. In addition to a small amount of nonspecific incorporation of Se into all S-containing bases there are, at least in bacteria, specific Se-containing tRNAs. In *E. coli* one of these is specific for lysine and one for glutamate. One of the modified bases has been identified as 5-methyl-aminomethyl-2-selenouridine. It is present at the first position of the anticodon, the "wobble" position. Selenium has its own metabolism. Through the use of ^{75}Se as a tracer, normal rat liver has been shown to contain Se^{2-}, SeO_3^{2-}, and selenium in a higher oxidation state.

Glutathione may be involved in reduction of selenite to selenide. The nonenzymatic reduction of selenite by glutathione yields a selenotrisulfide derivative. The latter is spontaneously decomposed to oxidized glutathione and elemental selenium or by the action of glutathione reductase to glutathione and selenium. Selenocysteine can be converted to alanine + elemental selenium. Some bacteria are able to oxidize elemental Se back to selenite. Selenium undergoes biological methylation readily in bacteria, fungi, plants, and animals. This may in some way be related to the reported effect of selenium in protecting animals against the toxicity of mercury. Excess selenium may appear in the urine as trimethylselenonium ions.

$4\,GSH + SeO_3^{2-} + 2\,H^+ \xrightarrow{pH\,7} 3\,H_2O + G{-}S{-}S{-}G + G{-}S{-}Se{-}S{-}G$ (Selenotrisulfide)

Spontaneous: $G{-}S{-}Se{-}S{-}G \rightarrow G{-}S{-}S{-}G + Se^0$

Glutathione reductase: $G{-}S{-}Se{-}S{-}G \rightarrow G{-}S{-}Se{-}H \rightarrow G{-}Se{-}SH$

How is selenium incorporated into selenocysteine-containing proteins? This element does enter amino acids to a limited extent via the standard synthetic pathways for cysteine and methionine. However, the placement of selenocysteine into specific positions in selenoproteins occurs by the use of a minor serine-specific tRNA that acts as a suppressor of chain termination during protein synthesis. The genes for these and presumably for other selenocysteine-containing proteins have the "stop" codon TGA at the selenocysteine positions. However, when present in a suitable "context" the minor tRNA, carrying selenocysteine in place of serine, is utilized to place selenocysteine into the growing peptide chain.

In bacteria, and presumably also in eukaryotes, the selenocysteinyl-tRNA is formed from the corresponding seryl-tRN A by a PLP-catalyzed β-replacement reaction. The selenium donor is not Se^{2-} but selenophosphate $Se\text{-}PO_3^{2-}$ in which the Se-P bond is quite weak. After addition to the aminoacrylate intermediate in the PLP enzyme the Se-P bond may be hydrolytically cleaved to HPO_4^{2-} and selenocysteyl-tRNA.

CHAPTER

11 Optimizing Nutrition for Exercise and Sport

The primary factors that affect exercise performance capacity include an individual's genetic endowment, the quality of training, and the effectiveness of coaching. Beyond these factors, nutrition plays a critical role in optimizing performance capacity. In order for an athlete to perform well, their training and diet must be optimal. If an athlete does not train enough or has an inadequate diet, their performance may be decreased. On the other hand, if an athlete trains too much without a sufficient diet, they may be susceptible to become overtrained. Because optimizing training and dietary practices are critical to peak performance, athletes have searched for various ways to improve exercise performance capacity through the use of ergogenic aids.

An ergogenic acid is any training technique, mechanical device, nutritional practice, pharmacological method, or psychological technique that can improve exercise performance capacity or enhance training adaptations. This includes aids that improve the preparation to performance, the efficiency of physiological and psychological responses to exercise, and/or recovery from exercise. Research has demonstrated that various ergogenic aids can help an athlete optimize performance capacity. This chapter overviews the role that nutrition has on enhancing exercise and sport performance, describes nutritional guidelines that athletes should employ to optimize training adaptations, and evaluates the ergogenic value of various nutrients that have been proposed to improve exercise capacity.

ENERGY DEMANDS FOR ACTIVE INDIVIDUALS

The first nutritional principle to optimize performance of athletes is to make sure that they consume enough calories to offset energy demands so as to maintain energy balance. Daily caloric intake for untrained individuals typically ranges between 1900-3000 kcal/d (*i.e.*, 27 – 43 kcal/kg/d for a 70 kg person) (2 – 5). Obviously, exercise training increases energy expenditure. The longer and more intense an athlete exercises, the greater the energy expenditure. Energy expenditure estimates for athletes have ranged from 3500 kcal/d (50 kcal/kg/d) for individuals training 30 – 60 min/d up to 12,000 kcal/d (*i.e.*, 170 kcal/kg/d) for cyclists competing in the Tour de France bicycle race (cycling 4 – 6 h/d). For most high school and college athletes training 2 – 2.5 h/d, energy expenditure estimates range between 60 – 80 kcal/kg/d. However, athletes often do not consume enough calories to offset energy demands. This may result in a chronic deficit in energy intake and has been implicated as one potential causative factor in overtraining. Athletes particularly susceptible to maintaining negative energy intakes during training include runners, cyclists, swimmers, triathletes, gymnasts, skaters, dancers, wrestlers, and boxers (1 – 3). Additionally, female athletes have been reported to have a high incidence of eating disorders. Consequently, the sports medicine professional should ensure that athletes are well fed and consume enough calories to offset the increased energy demands of training.

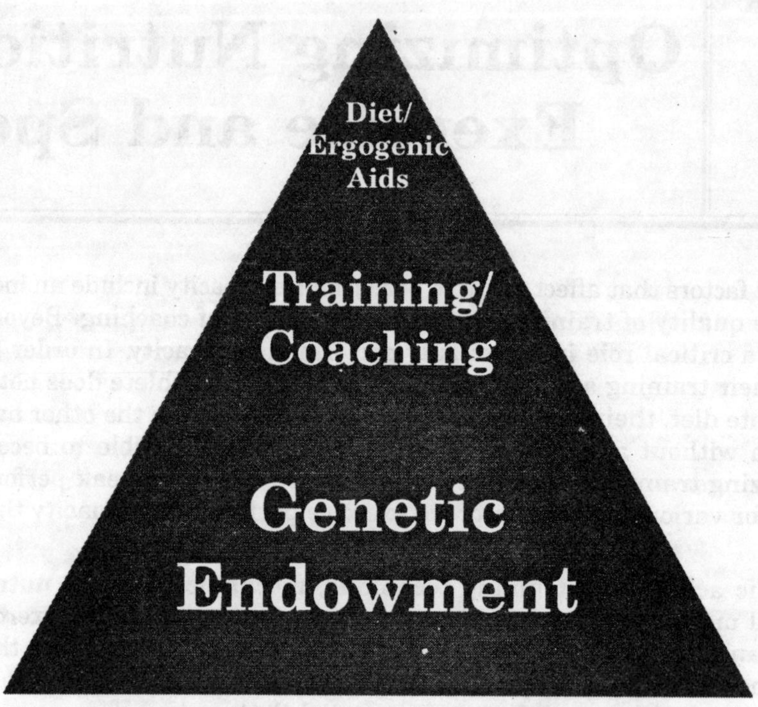

Fig. 11.1. Factors affecting peak performance.

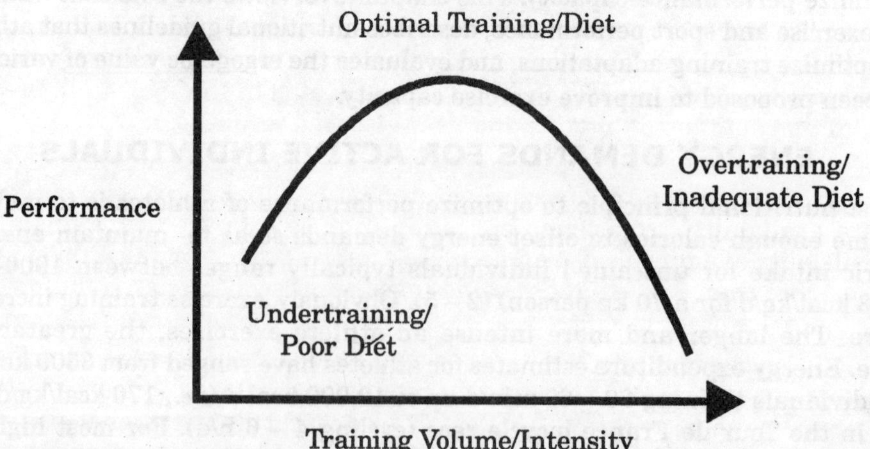

Fig. 11.2. Training stimulus.

GENERAL MACRONUTRIENT GUIDELINES FOR ATHLETES

The second principle in supporting the nutritional needs of athletes is to ensure that athletes consume the proper amounts of carbohydrate, fat, and protein in their diet. Summarizes macronutrient guidelines for athletes. The following discussion provides additional insight on how to structure the diet of athletes in order to optimize performance.

Carbohydrate

Carbohydrate serves as the primary fuel for high intensity intermittent or prolonged exercise. Carbohydrate is stored in the muscle (about 15 g/kg) and liver (about 80 – 100 g). Intense exercise depletes muscle and liver glycogen stores. The stores are replenished from dietary carbohydrate. Unfortunately, when significant amounts of carbohydrate are depleted, it may be difficult to fully replenish carbohydrate levels within one day. Consequently, when athletes train once or twice per day over a period of days, carbohydrate levels may gradually decline leading to fatigue, poor performance, and/or overtraining. Research has indicated that athletes should ingest between 8 – 10 g/kg/d of carbohydrate during intense periods of training in order to help maintain carbohydrate stores.

In order to do so, it is recommended that athletes eat frequently (*e.g.*, 4 – 6 meals/d) and ingest high-calorie carbohydrate foods and/or concentrated carbohydrate drinks. Preferably, the majority of dietary carbohydrate should come from complex carbohydrates with a low to moderate glycemic index (*e.g.*, grains, starches, fruit, etc.). Although this sounds relatively simple, intense training often suppresses appetite and/or alters hunger patterns. Some athletes do not like to exercise within several hours after eating because of sensations of fullness and/ or a predisposition to cause gastrointestinal distress.

Furthermcre, travel and training schedules may limit food availability and/or the access to the types of food athletes are accustomed to eating. This means that care should be taken to coordinate mealtimes with training and make sure athletes have sufficient availability to nutrientdense foods throughout the day for snacking between meals (*e.g.*, drinks, fruit, carbohydrate/protein bars, etc.). Research has also demonstrated that the timing and composition of meals consumed may play a role in maintaining carbohydrate stores during training.

In this regard, it takes about 4 h for carbohydrate to be digested and begin to be stored as muscle and liver glycogen. Consequently, pre-exercise meals should be consumed about 4 – 6 h before exercise. This means that if an athlete trains in the afternoon, breakfast is the most important meal to top off muscle and liver glycogen levels.

Research has also indicated that ingesting a light carbohydrate and protein snack 30 – 60 min prior to exercise (*e.g.*, 50 g of carbohydrate and 5 – 10 g of protein) serves to increase carbohydrate availability toward the end of an intense exercise bout. This also serves to increase availability of amino acids and decrease exercise-induced catabolism of protein. When exercise lasts more than 1 h, athletes should ingest glucose/electrolyte solution (GES) sports drinks in order to maintain blood glucose levels and help prevent dehydration.

Following intense exercise, athletes should consume carbohydrate and protein (*e.g.*, 1 g/ kg of carbohydrate and 0.5 g/kg of protein) within 30 min after exercise, as well as consume a high carbohydrate meal within 2 h following exercise. This nutritional strategy has been found to accelerate glycogen resynthesis, as well as promote a more anabolic hormonal profile that may hasten recovery. Finally, for 2 – 3 d prior to competition, athletes should taper training

by 30 – 50% and consume 200 – 300 g/d of extra carbohydrate in their diet. This carbohydrate-loading technique has been shown to supersaturate carbohydrate stores prior to competition and improve endurance exercise capacity. Thus, the type of meal and timing of eating are important factors in maintaining carbohydrate availability during training.

Table 11.1. Dietary Macronutrient Guidelines for Athletes

Nutrient	*Proposed ergogenic value*	*Summary of research findings*
Carbohydrate	Primary fuel used for anaerobic and high intensity aerobic exercise. Increasing dietary availability proposed to increase glycogen content and increase exercise capacity.	Athletes engaged in heavy training need to consume a diet high in carbohydrate (55-65% of caloric intake). Ingesting 8-10 g/kg/d of carbohydrate during heavy training has been suggested as one strategy to reduce the incidence of overtraining (1, 2, 3, 6).
Protein	Increasing dietary availability of protein has been suggested to be a means of maintaining nitrogen balance and enhancing gains in muscle mass.	Protein blance studies indicate that athletes involved in heavy training need to ingest 1.3-1.7 g/kg/d of protein in order to maintain nitrogen balance (about 1.5 times the RDA for protein). Although most athletes consume enough protein in their diet, some athletes are susceptible to protein malnutrition (dancers, gymnasts, runners, swimmers, etc.). Studies indicate that supplementing the diet with protein above that necessary to maintain protein balance does not increase strength or muscle mass (7,8).
Fat	Primary fuel for moderate- to low-intensity exercise. Some have proposed that increasing fat and/or derivatives of fat may enhance endurance exercise performance. Others suggest that increasing fat content in the diet may help moderate insulin levels leading to a better degree helping to better degree helping to promote fat loss.	It is generally recommended that athletes consume less than 1 g/kg/d of fat in their diet @ <.30% of calories) (3). Studies indicate that individuals who successfully maintain weight loss consume less than 40 g/d of fat (9). Studies generally do not indicate the fat/fat derivative supplements enhance exercise performance enhance exercise performance. However, supplements that increase fat use during exercise may enhance time to exhaustion.

Protein

Historically, there has been considerable debate regarding protein needs of athletes. Initially, it was recommended that athletes do not need to ingest more than the RDA for protein (*i.e.,* 0.8 – 1.0 g/kg/d for children, adolescents, and adults). However, research over the last decade has indicated that athletes engaged in intense training need to ingest about one and a half to two times the RDA of protein (1.3 – 1.7 g/kg/d) in order to maintain protein balance. If an insufficient amount of protein is obtained from the diet, an athlete will maintain a negative nitrogen balance which can increase protein catabolism and slow recovery. Over time, this

may lead to muscle wasting and training intolerance (1, 3, 7). Although most athletes ingest this amount of protein in their normal diet, there are some athletes who are susceptible to protein malnutrition due to greater protein degradation, weight restrictions, and/or an inability to ingest enough calories to offset energy expenditure (*e.g.*, runners, cyclists, swimmers, triathletes, gymnasts, dancers, skaters, wrestlers, boxers, etc.). Therefore, care should be taken to ensure that these types of athletes consume a sufficient amount of protein in order to maintain nitrogen balance (*e.g.*, 1.5 g/kg/d). Research has also indicated that ingesting more protein than necessary to maintain nitrogen balance does not promote greater gains in strength or muscle mass.

Fat

The dietary fat intake recommendations for athletes are similar to those for non-athletes in order to promote health. Generally, it is recommended that athletes consume less than 30% of their daily caloric intake as fat. For athletes attempting to decrease body fat, it is also recommended that they consume 0.5 – 1 g/kg/d of fat. This is because weight loss studies indicate that people who are most successful in losing weight and maintaining the weight loss are those who ingest less than 40 g/d of fat in their diet. Strategies to help athletes manage dietary fat intake include teaching them which foods contain fat so that they can make better food choices and learn to how to count fat grams.

PROPOSED NUTRITIONAL ERGOGENIC AIDS

Nutritional ergogenic aids are the most common type of ergogenic aids used by athletes. They include alterations in the composition of the diet, timing of eating, and/or supplementation of various macro and micronutrients that may enhance performance. Nutritional strategies that improve the preparation for exercise, the efficiency of exercise, performance capacity, and/or enhance the recovery from exercise may be viewed as ergogenic. Consequently, the final nutritional strategy for enhancing training and/or performance capacity in athletes is the appropriate use of effective and safe nutritional ergogenic aids. Although research has demonstrated that some nutritional strategies and nutrients may affect exercise training and/or performance capacity, the majority of nutritional ergogenic aids marketed to athletes do not affect performance. The following section reviews macro and micronutrients that have been proposed to improve exercise capacity.

Carbohydrate and Carbohydrate By-Products

As stated previously, dietary carbohydrate availability can significantly affect muscle and liver carbohydrate stores and performance capacity. For this reason, in addition to the dietary guidelines described, a significant amount of research has been conducted to determine ways to optimize carbohydrate availability during exercise and/or spare muscle glycogen use during exercise. Generally, increasing availability of any form of carbohydrate has the potential to improve exercise capacity by serving as an exogenous fuel source. Describes the proposed ergogenic value and summary of research findings for several forms of carbohydrate and carbohydrate by products that have been proposed to enhance exercise performance. Of the products reviewed, glucose electrolyte solution (GES) sport drinks possess the greatest potential to improve exercise capacity.

Although some clinical and/or exercise benefits have been reported from dihydroxyacetone phosphate (DHAP), fructose 1, 6-diphosphate (FDP), polylactate, pyruvate, and ribose

Table 11.2. Proposed Nutritional Ergogenic Aids for Athletes—Carbohydrate and Carbohydrate By-Products

Nutrient	*Proposed ergogenic value*	*Summary of research of research findings*
Glucose/electrolyte solution (GES) sport drinks	Ingesting sport drinks during exercise (*e.g.*, 6-8 oz of 6-8% carbohydrate solution every 5-15 min) during prolonged exercise has been proposed to help maintain blood glucose availability to the muscle and extened time to exhaustion in moderately intense exercise bouts (*i.e.*, 70% of VO_2 max) lasting 3-4 h. drinks (>15%) may slow gastric emptying.	Numerous studies indicate that ingesting GES drinks during exercise maintains blood glucose levels, helps promote fluid retention, and decreases dehydration. Recommended for exercise bouts lasting more than 60 min, particularly if exercising in hot/humid environments. Ingesting high concentrated
Fructose 1, 6-Diphosphate (FDP)	FDP serves as an intermediate step in glycolysis after the energy requiring steps of converting glucose to glucose-6-phosphate. Theorized to increase blood ATP and 2,3-diphosphoglycerate levels, enhance dissociation of oxygen from hemoglobin, and serve as an efficient source of carbohydrate to enhance exercise capacity. advantage of over other forms of carbohydrate.	Studies indicate that FDP supplementation (0.25 g/kg) can serve as an effective fuel source during exercise. Some studies indicate that exercise capacity may be improved in patient's with peripheral vascular disease. However, recent studies in healthy subjects Indicate that FDP supplementation has no
Dihydroxyacetone Phosphate (DHAP)	Supplementation with DHAP has been suggested to enhance glycolytic and oxidative metabolism. Additionally, DHAP and pyruvate supplementation have been suggested to promote fat loss and extend endurance exercise capacity by serving as a fuel during exercise.	Few well-controlled studies have evaluated the ergogenic value of DHAP. Some studies reported that DHAP supplementation (16-75 g/d) improved maximal VO_2 and promoted fat loss in obese individuals on hypocaloric diets. However, more Research is needed before definitive conclusions can be made.

Nutrient	*Proposed ergogenic value*	*Summary of research of research findings*
Polylactate (PL)	PL is a semisoluble amino acid/lactate salt that Has been theorized to be easily converted to Pyruvate for entrance into the tricarboxylic Acid cycle (TCA) and thereby enhance Carbohydrate availability during endurance exercise.	Few studies have evaluated the ergogenic value of PL. One study reported that in comparison to a placebo and maltodextrin, PL supplementation during exercise (7% GES solution) promoted higher pH and bicarbonate levels with no Differences in performance. Conversely, another study reported that addition of PL to a glucose polymer drink did not affect physiological responses to exercise or performance.
Pyruvate	Supplementation of pyruvate with DHAP has been suggested to promote fat loss and extend endurance exercise capacity by serving as a fuel during exercise.	Studies indicate that calcium pyruvate (6-25 g/d) with or without DHAP (16-75 g/d) supplementation promoted significantly more Fat loss in obese individuals on hypocaloric Diets. However, there is little data to Support that the dosages currently marketed to Promote fat loss (0.592/d) affects) body Composition or exercise responses.
Ribose	Ribose is a naturally occurring five-carbon sugar (pentose) that is primarily found in the body as constituents of riboflavin (vitamin B_2). nucleic acids, nucleotides, and nucleosides. supplementation has been theorized to increase ATP availability and recovery.	Some medical studies indicate that ribose supplementation (10-60 g/d) can increase ATP Availability in certain patient populations, blunt ischemia threshold in heart patients, and enhance predictive value of thallium exercise Tests. Whether ribose supplementation unknown.

Table 11.3. Proposed Nutritional Ergogenic Aids—Lipids and Lipid By-Products

Nutrient	*Proposed ergogenic value*	*Summary of research findings*
Conjugated linoleic acids (CLA)	CLA are essential fatty acids found primarily in fat from whole dairy products. Animal studies indicate that adding CLA to dietary feed decreases body fat, increases bone mass, has anticarcinogenic properties, enhances immunity, and inhibits athlerosclerotic progression. consequently, CLA supplementation in humans has been suggested to help manage body composition, delay loss of bone, and provide health benefits.	Animal studies provide a strong theoretical rationale as to the potential ergogenic and health benefit of CLA supplementation. However, few clinical trials have been conducted. Of the studies available, CLA supplementation does not appear to promote fat loss in humans. However, there is some evidence that CLA may affect bone mass and immune status. Additional research is necessary in humans.
Glycerol	Glycerol has been reported to promote fluid retention by decreasing urine formation. glycerol added to water may be beneficial to hyperhydrate athletes prior to exercise in an effort to prevent dehydration during exercise.	Studies indicate that glycerol feedings (1 g/kg with water) can promote fluid retention. This should help athletes susceptible to dehydration perform better during exercise in the heat. However, the ergogenic effect of glycerol hyperhydration is unclear.
L-Carnitine	Carnitine serves as a transporter of fatty acids from the cytosol into the mitochondria and helps modulate the metabolism of coenzymeA (CoA). studies indicate that fatty acid oxidation is regulated in part by the ability to shuttle fatty acids into the mitochondria for entrance into the TCA cycle. Consequently, L-carnitine supplementation has been theorized as a means of enhancing fat oxidation and sparing muscle glycogen during exercise as well as promoting fat loss.	Numerous studies have evaluated the potential ergogenic value of L-carnitine supplementation In patient and athletic populations. although some well-controlled studies indicate that L-carnitine supplementation (0.5-2 g/d) may increase fat oxidation and improve cardiovascular efficiency during exèrcise, most studies indicate that L-carnitine does not affect energy metabolism, exercise capacity, or body composition.

Nutrient	*Proposed ergogenic value*	*Summary of research findings*
Medium-chain triglycerides	Medium-chain triglycerides (MTC) diffused directly into the mitochondria for entrance into beta-oxidation. Theoretically, MCT feedings should serve as an efficient fuel source for exercise possibly serving to enhance endurance capacity.	Most studies indicate that MCT supplementation does not affect low to moderate-intensity endurance exercise. However, recent research suggests that MCT supplementation with carbohydrate may spare muscle glyogen utilization and increase time to exhaustion during intense cycling time trial performance.
Omega-3 fatty acids antioxidants,	Omega 3FA have been reported to serve as enhance immunity, and to decrease risk of cardiovascular disease. Some have suggested that omega 3FA supplementation in athletes would decrease muscle damage and help maintain immune function.	Several studies have evaluated the potential ergogenic value of omega 3FA supplementation. Although there is evidence that omega 3FA supplementation (1-3 g/d) may affect lipid oxidation and immune responses, most studies indicate no ergogenic benefit on aerobic or anaerobic power.

Table 11.4. Proposed Nutritional Ergogenic Aids—Protein and Amino Acids

Nutrient	*Proposed ergogenic value*	*Summary of research findings*
Arginine, ornithine, lysine	Clinical studies indicate that supplementation of these amino acids may stimulate growth hormone release serving to preserve muscle mass during bed rest. Additionally, some studies indicate that arginine supplementation improves immune status. consequently, some have suggested that supplementation of these amino acids during training may increase muscle mass and strength gains.	Recent studies indicate that supplementation with arginine, ornithine, or lysine (10-25 g/d), either separately or in combination, does not enhance the effect of exercise stimulation on either hGH or various measures of muscular strength or power in experienced weightlifters.
Aspartate, asparagines	These amino acids serve as precursors to oxaloacetate in the TCA cycle. Supplementation has been theorized to spare muscle glycogen use and enhance endurance performance capacity.	Some well-controlled studies support the ergogenic value of aspartate and arginine supplementation on sparring muscle glycogen use and improving exercise capacity. However, other studies have reported limited effects. additional research is necessary.
Branched-chain amino acids (leucine, isoleucine, valine)	Exercise-induced decreases in BCAA levels has been suggested to contribute to central fatigue as well as muscle catabolism. Supplementation of BCAA with sports drinks may increase BCAA availability and decrease the ratio of free tryptophan/BCAA. theoretically, this may minimize serotonin production in the brain and delay central fatigue.	A number of studies have reported that BCAA supplementation can affect physiologica and psychological responses to exercise. However, it is unclear the degree to which these potentially beneficial effects may affect performance. Initial research promising but more studies are needed, particularly during training.
Creatine	The availability of phosphocreatine (PC) stores in the muscle significantly affects the amount of energy generated during brief periods of high intensity exercise. Creatine supplementation has been hypothesized to increase muscle creatine content, help maintain ATP levels during exercise, and accelerate the rate of resynthesis of ATP during and following high intensity, short duration exercises.	The majority of research studies indicate that short-term creatine supplementation (20 g/d for 5 d) significantly increases muscle creatine and PC content, enchances energy availability during exercise, and improves high intensity, repetitive exercise performance. Long-term creatine supplementation has been reported to improve strength and muscle mass gains during training.

Nutrient	*Proposed ergogenic value*	*Summary of research findings*
Glutamine	Glutamine has been reported to affect protein synthesis possibly by increasing cell volume and osmotic pressure. Glutamine availability also directly affects lymphocytic function. Collectively, glutamine supplementation may promote muscle growth and prevent immuno suppression during training.	Exercise decreases serum glutamine levels. This reduction has been suggested to contribute to exercise-induced immunosuppression. Glutamine supplementation has been reported to increase serum levels and there is some evidence that glutamine supplementation can improve postexercise immune profiles. However, although there is strong scientific rationale, no long-term studies have evaluated the effects of glutamine supplementation on training adaptations or body composition.
Beta-hydroxy-beta-methylbutyrate (HMB)	Leucine and metabolites of leucine such as α-ketoisocaproate (KIC) have been reported to inhibit protein degradation. The anticatabolic effects have been suggested to be regulated by the leucine metabolite β-HMB. Adding β-HMB to dietary feed improved carcass quality in sows and steers. It has been hypothesized that supplementing the diet with leucine and/or β-HMB may inhibit protein degradation during resistance-training.	Leucine infusion has been reported to decrease protein degradation in humans suggesting that leucine may serve as a regulator of protein metabolism. Supplementing the diet with 1.5-3 g/d of calcium β-HMB supplementation during resistance-training in well trained athletes are less clear. Greater benefits appear to occur during heavy training.
Tryptophan	Blood levels of the amino acid tryptophan increase during prolonged exercise as fatty acids are mobilized for fat oxidation. Increases in brain concentrations of tryptophan have been reported to contribute to fatigue as well as increase endogenous opioid production. Tryptophan supplementation has been theorized to help athletes tolerate pain and enhance endurance exercise capacity.	Most studies indicate that increases in tryptophan in the blood and bran contributes to central fatigue. Although an initial study suggested that L-tryptophan supplementation improved endurance exercise performance while exercising at 80% of maximal exercise capacity, other studies indicated that L-tryptophan had no effect or promoted an ergolytic effect on performance.

supplementation, it is our view that additional research is necessary to determine the efficacy of these nutrients before they are recommended for athletes.

Lipids and Lipid By-Products

Because fat can serve as a primary fuel source during low to moderate intensity exercise and most people have a considerable amount of fat stored as potential energy, researchers have also investigated the effects of lipids and lipid byproducts on exercise capacity and training. The basic rationale is that if fat oxidation can be increased during exercise, then carbohydrate stores can be spared, and exercise capacity can be improved. Additionally, if a greater amount of fat can be burned during exercise, this may help in weight management. Presents the proposed ergogenic value and summary of research findings for selected lipids and lipid by-products that have been proposed to affect exercise performance. Of the nutrients presented, glycerol supplementation used as a means to hyperhydrate athletes susceptible to dehydration appears to possess the most ergogenic potential.

Although research to date is promising, more studies are needed in humans to determine whether conjugated linoleic acids (CLA) supplementation affects body composition during training and/or may possess a health benefit. Based on current data, there appears to be a limited ergogenic value of L-carnitine and medium-chain triglyceride (MCT) supplementation. Finally, although there may be some health benefits from diets high in omega 3-fatty acids, there is no evidence that omega 3-fatty acid supplementation with these lipids affects exercise performance.

Protein and Amino Acids

Amino acids are the foundation of protein in the body and are essential for the synthesis of tissue, specific proteins, hormones, enzymes, and neurotransmitters. Amino acids are also involved in the synthesis of energy through gluconeogenesis and regulation of numerous metabolic pathways. Consequently, it has been suggested that athletes may require additional protein in their diet in order to enhance muscle and tissue growth, the synthesis of hormones and enzymes necessary for energy metabolism, or serve as a potential energy substrate during exercise. Describes the potential ergogenic value of amino acids that have been purported to affect exercise capacity and/or promote training adaptations.

Of the amino acids reviewed, creatine appears to be one of most effective and safe nutritional supplements to enhance anaerobic exercise capacity, strength, and gains in muscle mass during training. Studies have indicated thataspartate, branched-chain amino acids (leucine, isoleucine, and valine), glutamine, and β-hydroxy β-methylbutyrate (HMB) may affect exercise capacity, enhance recovery, and/or promote greater training adaptations. However, not all studies report ergogenic value and additional research is needed. Although there may be some clinical applications, there appears to be little ergogenic value of arginine, ornithine, lysine, and tryptophan supplementation for athletes.

Vitamins

Vitamins are essential organic compounds that serve to regulate metabolic processes, energy synthesis, neurological processes, and prevent destruction of cells. There are two types of vitamins: fat soluble and water soluble. The fat-soluble vitamins include vitamins A, D, E, and K. The body stores fat-soluble vitamins and therefore excessive intake may result in

Table 11.5. Proposed Nutritional Ergogenic Aids – Vitamins

Nutrient	*Proposed ergogentic value*	*Summary of research findings*
Vitamin A	Constituent of rhodopsin (visual pigment) and is involved in night vision. Some suggest that vitamin A supplementation may improve sport vision.	No studies have shown that vitamin A supplementation improves exercise performance.
Thiamin (B_1)	Part of the coenzyme (thiamin pyrophosphate), which is needed to convert pyruvate to acetyl CoA for entrance into the Krebs cycle. Supplementation is theorized to improve anaerobic threshold and CO_2 transport. Deficiencies may decrease efficiency of energy systems.	Dietary availability of thiamin does not appear to affect exercise capacity when athletes have a normal intake.
Riboflavin (B_2)	Constituent of flavin nucleotide coenzymes involved in energy metabolism. Theorized to enhance energy availability during oxidative metabolism.	Dietary availability of riboflavin does not appear to affect exercise capacity when athletes have a normal intake.
Niacin (B_3)	Constituent of coenzymes involved in energy metabolism. Theorized to blunt increases in fatty acids during exercise, reduce cholesterol, enhance thermoregulation, and improve energy availability during oxidative metabolism.	Studies indicate that niacin supplementation can help decrease blood lipid levels in patients with elevated cholesterol. Niacin supplementation during exercise has been reported to decrease exercise capacity by blunting the mobilization of fatty acids.
Pyridoxine (B_6)	Pyridoxine has been marketed as a supplement that will improve muscle mass, strength and aerobic power in the lactic acid and oxygen systems. It also may have a calming effect that has been linked to an improved mental strength.	In well-nourished athletes, pyridoxine failed to improve aerobic capacity, or improve aerobic capacity, or lactic acids accumulation. However, when combined with vitamins B_1 and B_{12} it may increase serotonin levels and be beneficial in sports like pistol shooting and archery.
Cyanocobalamin (B_{12})	Cyanocobalamin is a coenzyme involved in the production of DNA and serotonin. DNA is important in protein and red blood cell synthesis. Theoretically it would increase muscle mass, the oxygen-carrying capacity of blood and decrease anxiety.	In well-nourished athletes, no ergogenic effect has been reported. However, when combined with vitamins B_1 and B_6, cyanocobalamin has been shown to improve performance in pistol shooting. This may be due to increased levels of serotonin, a neurotransmitter in the brain, which may reduce anxiety.

Nutrient	*Proposed ergogentic value*	*Summary of research findings*
Folacin	Folic acid functions as a coenzyme in the formation of DNA and red blood cells. An increase in red blood cells could improve oxygen delivery to the muscles during exercise.	In well-nourished and folate deficient athletes, folic acid did not improve exercise performance.
Pantothenic acids	Pantothenic acid acts as a coenzyme for acetyl coenzyme A (acetyl CoA). This may benefit aerobic or oxygen energy systems.	Research has reported no improvements in aerobic performance with acetyl CoA supplementation. However, one study reported a decrease in lactic acid accumulation, without an improvement in performance.
β-Carotene	Serves as an antioxidant. Theorized to help minimize exercise-induce lipid perioxidation and muscle damage.	Research indicates that β-carotene supplementation with or without other antioxidants can help decrease exercise-induced perioxidation. Over time, this may help athletes tolerate training. However, it is nuclear whether antioxidant supplementation affects exercise performance.
Vitamin C	Vitamin C is used in a number of different metabolic processes in the body. It is involved in the synthesis of epinephrine, iron absorption, and is an antioxidant. Theoretically, it could benefit exercise performance by improving metabolism during exercise. There is also evidence that vitamin C may enhance immunity.	In well-nourished athletes, vitamin C supplementation does not appear to improve physical performance. However, there is some evidence that vitamin C supplementation following intense exercise may decrease the incidence of upper respiratory tract infections.
Vitamin E	As an antioxidant, vitamin E could help prevent the formation of free radicals during intense exercise and prevent the destruction of red blood cells, improving or maintaining oxygen delivery to the muscles during exercise. Some evidence suggests that it may reduce risk to heart disease or decrease incidence of recurring heart attack.	Current research has reported no ergogenic effect at sea level. However, at high altitudes, vitamin E may improve exercise performance. Additional research is necessary to determine whether long-term supplementation may help athletes tolerate training to a better degree.
Vitamin K	Important in blood clotting. There is also some evidence that vitamin K may affect bone metabolism in post-menopausal women.	Vitamin K supplementation (10 mg/d) in elite female athletes has been reported to increase calcium-binding capacity of osteocalcin and promoted a 15-20% increase in bone formation markers and a 20-25% decrease in bone resorption markers suggesting an improved balance between bone formation and resorption.

Table 11.6. Proposed Nutritional Ergogenic Aids—Minerals

Nutrient	*Proposed ergogenic value*	*Summary of research findings*
Boron	Boron has been marketed to athletes as a dietary supplement that may promote muscle growth during resistance training. The rationale was primarily based on an initial report that boron supplementation (3 mg/d) significantly increased β-estradiol and testosterone levels in post-menopausal women consuming a diet low in boron.	Studies which have investigated the effects of 7 wk of boron supplementation (2.5 mg/d) during resistance training on testosterone levels, body composition, and strength have reported no ergogenic value. There is no evidence at this time that boron supplementation during resistance training promotes muscle growth.
Calcium	Involved in bone and tooth formation, blood clotting, and nerve transmission. Diet should contain sufficient amounts especially in growing children/adolescents, female athletes, and post-menopausal women. Vitamin D needed to assist absorption.	Calcium supplementation may be beneficial in populations susceptible to osteoperosis. Calcium supplementation provides no ergogenic effect on exercise performance.
Chromium	Chromium, commonly sold as chromium picolinate has been marketed with claims that the supplement will increase lean body mass and decrease body fat levels.	Animal research indicates that chromium supplementation increases lean body mass and reduces body fat. Early research on humans reported similar results. However, more recent, well-controlled studies reported that chromium supplementation does not improve lean body mass or reduce body.
Iron	Iron supplements are used to increase aerobic performance in sports that use the oxygen system. Iron is a component of hemoglobin in the red blood cell, which is a carrier of oxygen.	Iron supplements do not appear to improve aerobic performance, unless there is iron-deficiency anemia.
Phosphate	Phosphate has been studied for its ability to improve all three energy systems, primarily the oxygen system or aerobic capacity.	Recent well-controlled research studies reported that phosphate supplementation (4 g/d for 3 d) improved the oxygen energy system in endurance tasks. More research is needed to determine the mechanism for improvement.
Magnesium	Activates enzymes involved in protein synthesis. Involved in ATP reactions. Serum levels decrease with exercise. Some suggest that magnesium supplementation may improve energy metabolism/ATP availability.	Well-controlled research indicates that magnesium supplementation does not effect endurance exercise performance in athletes.
Selenium	Selenium has been marketed as a supplement to increase aerobic exercise performance. Working closely with vitamin E and glutathione peroxidase (an antioxidant), selenium may destroy destructive free radical production of lipids during aerobic exercise.	Although selenium may reduce lipid peroxdiation during aerobic exercise, improvements in aerobic capacity have not been demonstrated.
Vanadium	Vanadium may be involved in reactions in the body that produce insulin-like effects on protein and glucose metabolism. Owing to the anabolic nature of insulin, this has brought attention to vanadium as a supplement to increase muscle mass, enhance strength and power.	Limited research has shown that noninsulin-dependent diabetics may improve their glucose control, however there is no scientific proof that vanadyl sulfate has any effect on muscle mass, strength or power.

toxicity. B vitamins and vitamin C are water soluble. Excessive intake of these is eliminated in the urine. Describes recommended daily allowance (RDA), proposed ergogenic benefit, and summary of research findings for fat and water soluble vitamins. Although research has demonstrated that specific vitamin supplements may pose some health benefit (*e.g.*, vitamin E, niacin, folate, vitamin C, etc.), few have been reported to directly provide ergogenic value for athletes. However, some vitamins may help athletes tolerate training to a better degree by reducing oxidative damage (vitamins E and C) and/or help to maintain a healthy immune system during heavy training (vitamin C).

Theoretically, this may help athletes tolerate heavy training leading to improved performance. The remaining vitamins reviewed appear to have little ergogenic value for athletes who consume a normal, nutrientdense diet. Since analyses of athlete's diets have found deficiencies in caloric and vitamin intake, some sport nutritionists recommend that athletes consume a low-dose one-a-day multivitamin and/or a vitamin-enriched postworkout carbohydrate/protein supplement during periods of heavy training.

Minerals

Minerals are essential inorganic elements necessary for a host of metabolic processes. Minerals serve as structure for tissue, important components of enzymes and hormones, and regulators of metabolic and neural control. Some minerals have been found to be deficient in athletes or become deficient in response to training and/or prolonged exercise. When mineral status is inadequate, exercise capacity may be reduced. Dietary supplementation of minerals in deficient athletes has generally been found to improve exercise capacity. Additionally, supplementation of specific minerals in nondeficient athletes has also been reported to affect exercise capacity. Describes minerals that have been purported to affect exercise capacity in athletes.

Of the minerals reviewed, several appear to possess health and/or ergogenic value for athletes under certain conditions. For example, calcium supplementation in athletes susceptible to premature osteoporosis may help maintain bone mass. Iron supplementation in athletes prone to iron deficiency and/or anemia has been reported to improve exercise capacity. Sodium phosphate loading has been reported to increase maximal oxygen uptake, anaerobic threshold, and improve endurance exercise capacity by 8 – 10%. Increasing dietary availability of salt (sodium chloride) during the initial days of exercise training in the heat has been reported to help maintain fluid balance and prevent dehydration. Finally, zinc supplementation during training has been reported to decrease exercise-induced changes in immune function.

Consequently, somewhat in contrast to vitamins, there appear to be several minerals that may enhance exercise capacity and/or training adaptations for athletes under certain conditions. However, although ergogenic value has been purported for the remaining minerals, there is little evidence that boron, chromium, magnesium, or vanadium affect exercise capacity or training adaptations in healthy individuals eating a normal diet.

Water

The most important nutritional ergogenic aid for athletes is water. Exercise performance can be significantly impaired when 2% or more of body weight is lost through sweat. For example, when a 70-kg athlete loses more than 1.4 kg of body weight during exercise (2%), performance capacity is often significantly decreased. Further, weight loss of more than 4%

Table 11.7. Propsed Nutritional Ergogenic Aids—Others

Nutrient	*Proposed ergogenic value*	*Summary of research findings*
Alcohol	Alcohol has been studied for its use as a psychological stress reducer in precision sports such riflery, archery and dart throwing. It also has been investigated as an energy source.	Some limited research does support an ergogenic effect in precision sports like riflery when about one drink of alcohol is consumed (a blood alcohol of 0.02), 30-60 min prior to competition. However, its use is illegal in these sports and not recommended. Using alcohol in other high anxiety sports such as figure skating, fencing, and gymnastics may result in sanctions and is unethical.
Alkaline salts (bicarbonate)	Sodium bicarbonate has been researched for its effect on improving power in sports that are of a short duration and use the lactic acid energy system.	An average dose of about 3000 mg/kg of body weight taken 1-2 h before exercise has delayed fatigue. However, many athletes report gastrointestinal distress, nausea, bloating, and stomach cramps.
Caffeine	Caffeine has been studied for its ability to stimulate power in the energy systems and mental strength. Caffeine stimulates the central nervous system, increasing arousal. It also stimulates the release of epinephrine which may improve cardiovascular function. Lastly, it may increase free fatty acid use during aerobic exercise and facilitate calcium release from the sarcoplasmic reticulum for stronger muscle contractions in anaerobic events.	Many research studies have concluded that caffeine may improve performance in many of the energy systems. However, does greater that about 5 cups of coffee may exceed legal limits.
Choline	Choline has been studied for its ability to improve oxygen utilization in aerobic events by maintaining or improving acetylcholine levels for enhanced function of the neuromuscular system.	The research has reported conclusively that choline supplementations increase blood choline levels, however, and improvement in performance remains equivocal and therefore more research is needed. Daily doses average 1.5-2.0 g.
Coenzyme Q10	Coenzyme Q10 is found in the mitochondria and is involved in oxygen transport and ATP production. It is also an antioxidant that may help destroy free radicals during intense aerobic exercise.	Coenzyme Q10 has been found to improve heart function, aerobic capacity and exercise performance in patients with heart disease, but not in healthy athletes.
Ephedring, ephedra (Ma Huang)	Ephedrine is a sympathomimetic that may enhanc muscle contractility, improve blood flow out of the heart, open bronchial airways and increase blood sugar. It could improve both aerobic and anaerobic types of exercise.	Current research does not seem to support any improvement in exercise capacity. Side effects such as nervousness, headaches, stomach upset, irregular heart beats and seizures may result.
Inosine	Ergogenic claims for incosine range from, improving ATP production, increasing aerobic capacity, reducing lactic acid and improving the use of blood sugar during exercise.	Most studies investigating inosine have been well controlled and designed. The results concluded that inosine had no egrogenic value and may even impair performance.

Table 11.8. Proposed Nutritional Ergogenic Adis—Plant Extracts

Nutrient	*Proposed ergogenic value*	*Summary of research findings*
Bee pollen	The multiple nutrients in bee pollen have been promoted for their ability as a good energy source for recovery following intense training.	Well-controlled research studies have reported no ergogenic effect on any of the three energy systems, (*i.e.*, ATP-PC, lactic acid, and oxygen systems).
Echinacea	An herb that is purported to enhance immune function and decrease the severity, duration, and incidence of colds and infections. Proposed to help athletes decrease the risk of upper respiratory tract infections during heavy training.	Medical studies generally support the premise that Echinacea may reduce the incidence, severity, and duration of colds and infections. No studies have evaluated whether using Echinacea during heavy training would decrease colds and infections.
Gamma oryzanol (ferulic acid)	Phytosterols theorized to enhance anabolic hormonal responses to training.	Research data are limited. One study reported that 9 wk of supplementation (0.5 g/d) did not affect strength, body composition, or anabolic hormones.
Ginkgo biloba (GB)	A plant extract purported to enhance memory and improve mental concentration and performance. Theorized to help athletes mental alertness and concentration during competition.	Supplementation of GB has been shown to improve symptoms associated with cognitive deficits in patient populations. There is also some evidence that exercise capacity was improved in peripheral vascular diseased patients administered 120 mg/d for 24 wk. No studies have evaluated whether GB affects performance capacity in healthy athletes.
Ginseng	By activating the hypothalamic-pituitary-adrenal cortex axis, ginseng may improve the three energy systems, namely ATP-PC, lactic acid and oxygen systems. Improved nitrogen balance and increased stamina also have been proposed.	Well-controlled research does not support any ergogenic effect for ginseng.
Octacosanol	A long-chain alcohol purported to lower blood lipids and serve as a fuel substrate to enhance exercise performance.	Octacosanol has been reported to lower blood lipid profiles. However, no studies support the purported ergogenic value in athletes.

Nutrient	*Proposed ergogenic value*	*Summary of research findings*
Smilax officianalis (SO)	A compound that contains steroidal saponins purported to enhance immunity as well as provide an androgenic effect on muscle growth.	Some data supports the potential immune enhancing effects of SO. There are no data to support the claimed androgenic effect.
St. Johns Wort (SJW)	Herbal product purported to serve as a naturally occurring alternative to antidepressants. SJW has been marketed to athletes as a supplement to promote a calming/relaxation effect.	No studies have been conducted to evaluate the potential ergogenic value of SJW supplementation in athletes.
Wheat germ oil (WGO)	WGO contains linoleic fatty acids, vitamin E, and octacosanol. Consequently, WGO has been theorized to improve endurance, stamina and vigor, specifically enhanced glycogen metabolism and increased oxygen uptake.	A review of approx 35 studies does not support the use of wheat germ oil as an ergogenic aid 981).
Yohimbine (yohimbe)	Yohimbe has been studied for its ability to increase testosterone levels and improve muscle mass and strength. It also may lead to increased levels of norepinephrine a stimulate used for weight loss.	Well-controlled research studies have failed to report increases in muscle mass, testosterone levels or reductions in body fat when ingesting 15-20 mg/d 82).

of body weight during exercise may lead to heat illness, heat exhaustion, heat stroke, and possibly death. For this reason, it is critical that athletes consume a sufficient amount of water and/or GES sports drinks during exercise. The normal sweat rate of athletes ranges from 0.5 – 2.0 L/h depending on temperature, humidity, exercise intensity, and their sweat response to exercise. Therefore to maintain fluid balance and prevent dehydration, athletes need to consume 0.5 – 2 L/h of fluid in order to offset weight loss. This requires frequent ingestion of 6 – 8 oz of cold water or a GES sports drink every 5 – 15 min during exercise. Athletes should not depend on thirst to prompt them to drink because people do not typically get thirsty until they have lost a significant amount of fluid through sweat.

Additionally, athletes should weigh themselves prior to and following exercise training to ensure that they maintain proper hydration. Every 1 kg of weight lost during exercise is equivalent to 1 L of fluid (about five cups) the athlete should have consumed during exercise. Athletes should train themselves to tolerate drinking greater amounts of water during training and make sure that they consume more fluid in hotter/humid environments. Preventing dehydration during exercise is one of the most effective ways to maintain exercise capacity. Finally, inappropriate and excessive weight loss techniques (*e.g.,* cutting weight in saunas, wearing rubber suits, severe dieting, vomiting, using diuretics, and so on) are extremely dangerous and should be prohibited.

Other Nutritional Ergogenic Aids

Present other nutrition-related compounds that have been purported to possess ergogenic value for athletes. Of the nutrients described, sodium bicarbonate, caffeine, and echinacea appear to have the greatest potential to affect exercise performance and/or training adaptations. Sodium bicarbonate loading (0.3 g/kg of baking soda) prior to exercise has been consistently reported to enhance high-intensity exercise lasting 1 – 3 min in duration (*e.g.,* a 400 – 800-m run). Although some athletes may experience gastrointestinal distress, bicarbonate loading appears to be a highly effective ergogenic aid for athletes as long as they can tolerate the supplementation protocol. Caffeine is a naturally occurring stimulant found in many foods consumed in the normal diet (*e.g.,* coffee, tea, chocolate).

Caffeine ingestion (6 – 9 mg/kg) prior to exercise has been reported to increase fat oxidation, spare muscle glycogen use, and enhance endurance exercise performance. The ergogenic effects of caffeine appear to be more pronounced in nonhabitual caffeine users and habitual users who abstain from consuming caffeine for about a week prior to competition. Although some athletic governing bodies have banned excessive intake of caffeine as an ergogenic aid, studies show that even when taken within the limits allowed by athletic governing bodies, caffeine may provide ergogenic benefit. One word of caution, however, is that caffeine serves as a mild diuretic. Therefore, caffeine intake prior to exercise may hasten dehydration. Echinacea is an herb that has been reported to enhance immune function and decrease the severity, duration, and incidence of colds and upper respiratory tract infections.

Because intense training may compromise immune function in athletes, some have suggested that echinacea supplementation during heavy training may decrease the incidence of colds and infections. Although there are data to support its immunoenhancing effects, we are aware of no studies that have determined whether use of echinacea specifically in athletes would help maintain immune function during intense training. Herbal effects are further discussed. Likewise, little data support the potential ergogenic value of alcohol, choline, coenzyme Q10, ephedrine/ma huang, inosine, bee pollen, gamma oryzanol, ginkgo biloba, ginseng, octacosanol, smilax officinalis, St. John's Wort, wheat germ oil, or yohimbine.

CONCLUSING REMARKS

Dietary and nutritional practices of athletes can significantly affect exercise performance capacity. In order to optimize performance, athletes should (1) eat enough calories to offset energy expenditure (typically 60-80 kcal/kg/d); (2) consume the proper amount of carbohydrate (8-10 g/kg/d), protein (1.5 g/kg/d) and fat (0.5-1 g/kg/d); (3) ingest meals and snacks at appropriate time intervals prior to, during, and/or following exercise in order to provide energy as well as to promote recovery following exercise; and (4) only consider using nutritional supplements that have been found to be an effective and safe means for improving performance capacity. For strength and power athletes, research has indicated that creatine and sodium bicarbonate supplementation possess the greatest ergogenic value.

For endurance athletes, research suggests that carbohydrate loading, GES sports drinks, sodium phosphate loading, and caffeine are among the most advantageous ergogenic aids. In addition, several other nutrients have been reported to affect exercise metabolism, improve exercise performance, enhance recovery, and/or help maintain health status under specific conditions (*e.g.*, glycerol, branced-chain amino acids, glutamine, homatropine methylbromide, vitamin E, vitamin C, iron, zinc, and echinacea). However, additional research is needed to determine the potential ergogenic value of these nutrients in various athletic populations.

Index

□□□